Blockchain Technologies for Sustainable Development in Smart Cities

P. Swarnalatha
Vellore Institute of Technology, India

S. Prabu
Vellore Institute of Technology, India

IGI Global
PUBLISHER of TIMELY KNOWLEDGE

A volume in the Practice, Progress, and Proficiency in Sustainability (PPPS) Book Series

Published in the United States of America by
IGI Global
Engineering Science Reference (an imprint of IGI Global)
701 E. Chocolate Avenue
Hershey PA, USA 17033
Tel: 717-533-8845
Fax: 717-533-8661
E-mail: cust@igi-global.com
Web site: http://www.igi-global.com

Library of Congress Cataloging-in-Publication Data

Names: Swarnalatha, P. (Purushotham), 1977- editor. | Sevugan, Prabu, 1981- editor.
Title: Blockchain technologies for sustainable development in smart cities / P. Swarnalatha and Prabu Sevugan, editors.
Description: Hershey, PA : Engineering Science Reference, an imprint of IGI Global, 2021. | Includes bibliographical references and index. | Summary: "This book offers relevant theoretical frameworks and empirical research findings to nurture investigations in blockchain technology innovations meant as a basis for understanding novel technological, organizational and societal settings involving blockchain technology"-- Provided by publisher.
Identifiers: LCCN 2021043023 (print) | LCCN 2021043024 (ebook) | ISBN 9781799892748 (h/c) | ISBN 9781799892755 (s/c) | ISBN 9781799892762 (ebook)
Subjects: LCSH: Smart cities. | Sustainable development--Data processing. | Blockchains (Databases)
Classification: LCC TD159.4 .B56 2021 (print) | LCC TD159.4 (ebook) | DDC 307.760285--dc23
LC record available at https://lccn.loc.gov/2021043023
LC ebook record available at https://lccn.loc.gov/2021043024

This book is published in the IGI Global book series Practice, Progress, and Proficiency in Sustainability (PPPS) (ISSN: 2330-3271; eISSN: 2330-328X)

British Cataloguing in Publication Data
A Cataloguing in Publication record for this book is available from the British Library.

Practice, Progress, and Proficiency in Sustainability (PPPS) Book Series

Ayman Batisha
International Sustainability Institute, Egypt

ISSN:2330-3271
EISSN:2330-328X

Mission

In a world where traditional business practices are reconsidered and economic activity is performed in a global context, new areas of economic developments are recognized as the key enablers of wealth and income production. This knowledge of information technologies provides infrastructures, systems, and services towards sustainable development.

The **Practices, Progress, and Proficiency in Sustainability (PPPS) Book Series** focuses on the local and global challenges, business opportunities, and societal needs surrounding international collaboration and sustainable development of technology. This series brings together academics, researchers, entrepreneurs, policy makers and government officers aiming to contribute to the progress and proficiency in sustainability.

Coverage

- Green Technology
- Global Business
- Eco-Innovation
- Intellectual Capital
- Knowledge clusters
- ICT and knowledge for development
- Sustainable Development
- Strategic Management of IT
- Technological learning
- Innovation Networks

IGI Global is currently accepting manuscripts for publication within this series. To submit a proposal for a volume in this series, please contact our Acquisition Editors at Acquisitions@igi-global.com or visit: http://www.igi-global.com/publish/.

The Practice, Progress, and Proficiency in Sustainability (PPPS) Book Series (ISSN 2330-3271) is published by IGI Global, 701 E. Chocolate Avenue, Hershey, PA 17033-1240, USA, www.igi-global.com. This series is composed of titles available for purchase individually; each title is edited to be contextually exclusive from any other title within the series. For pricing and ordering information please visit http://www.igi-global.com/book-series/practice-progress-proficiency-sustainability/73810. Postmaster: Send all address changes to above address.

Titles in this Series

For a list of additional titles in this series, please visit: http://www.igi-global.com/book-series/practice-progress-proficiency-sustainability/73810

Innovative Strategic Planning and International Collaboration for the Mitigation of Global Crises
Gabriela Antošová (University College of Business in Prague, Czech Republic)
Information Science Reference • © 2022 • 327pp • H/C (ISBN: 9781799883395) • US $195.00

Handbook of Research on Changing Dynamics in Responsible and Sustainable Business in the Post-COVID-19 Era
Cristina Raluca Gh. Popescu (University of Bucharest, Romania & The Bucharest University of Economic Studies, Romania & The National Institute for Research and Development in Environmental Protection (INCDPM), Romania & National Research and Development Institute for Gas Turbines (COMOTI), Romania)
Business Science Reference • © 2022 • 512pp • H/C (ISBN: 9781668425237) • US $315.00

Achieving Sustainability Using Creativity, Innovation, and Education A Multidisciplinary Approach
Ziska Fields (University of Johannesburg, South Africa)
Information Science Reference • © 2022 • 260pp • H/C (ISBN: 9781799879633) • US $215.00

Handbook of Research on Sustainable Development Goals, Climate Change, and Digitalization
Rui Alexandre Castanho (WSB University, Poland)
Engineering Science Reference • © 2022 • 684pp • H/C (ISBN: 9781799884828) • US $315.00

Sociological Perspectives on Sustainable Development and Poverty Reduction in Rural Populations
Ladislaus M. Semali (Pennsylvania State University, USA)
Information Science Reference • © 2022 • 303pp • H/C (ISBN: 9781799846468) • US $195.00

Handbook of Research on Novel Practices and Current Successes in Achieving the Sustainable Development Goals
Cristina Raluca Gh. Popescu (University of Bucharest, Romania & The Bucharest University of Economic Studies, Romania & University of Craiova, Romania & The National Institute for Research and Development in Environmental Protection (I.N.C.D.P.M.), Romania & National Research and Development Institute for Gas Turbines COMOTI, Romania)
Engineering Science Reference • © 2021 • 461pp • H/C (ISBN: 9781799884262) • US $295.00

Advancements in Sustainable Architecture and Energy Efficiency
Roberto A. González-Lezcano (Universidad CEU San Pablo, Spain)
Engineering Science Reference • © 2021 • 337pp • H/C (ISBN: 9781799870234) • US $195.00

IGI Global
PUBLISHER of TIMELY KNOWLEDGE

701 East Chocolate Avenue, Hershey, PA 17033, USA
Tel: 717-533-8845 x100 • Fax: 717-533-8661
E-Mail: cust@igi-global.com • www.igi-global.com

Table of Contents

Detailed Table of Contents

designs. Stream cipher has been a prominent research area in cryptography, and it is appropriate for use in circumstances of huge data transmission or resource limits. This work provides a quick overview of conventional encryption algorithms based on stream ciphers and cryptographic prerequisites. At present, several attacks faced by stream cipher techniques are also discussed.

Transportation is important for all and sundry. But it's bedeviled with some problems for the environment such as pollution. Intelligent transportation systems (ITS), which emerged to solve this problem, focus on important issues such as increasing safety in traffic. ITS across the world has come to the fore in recent years. In this study, ITS was researched including how it is used in Konya. It serves the public in almost all types of transportation used with the developed IPTS (intelligent public transportation system). The user can instantly access the station and vehicle information of the transportation system to be used through the menu. The system, which is integrated with road, rail, and air transportation, serves the user to avoid any problems from the beginning to the end. All information about online card filling, lost property status, and transportation fees can be accessed through the system. Undoubtedly, Konya has introduced the IPTS to provide a more practical public transportation service.

The development of blockchain has led to the emergence and widespread use of decentralized cryptocurrencies around the globe. As of 2021, the global market capitalization of cryptocurrencies has crossed two trillion dollars. With increasing popularity and adoption, investors have begun to see cryptocurrencies as an alternative to conventional financial assets. However, the volatility associated with cryptocurrencies makes them a highly risky investment. This gives rise to the need for accurate and efficient price prediction models which can help reduce risks associated with cryptocurrency investments. The model aims at predicting the price of two popular cryptocurrencies: Bitcoin and Ethereum. Tensor processing unit (TPU) is used for providing a distributed environment for the proposed model. The results show that the distributed TPU-trained model performed significantly better than the conventional CPU-trained model in terms of training time while maintaining a high degree of accuracy.

and/or firms in China's fintech industry. A variety of challenges and opportunities is analyzed to draw a blueprint for Chinese blockchain development. Finally, this chapter also provides some recommendations to the government.

Although electronic voting has been used for many decades, electoral organizations are still adopting it. Despite technologies that preserve the secrecy, authenticity, and privacy of data, vote-rigging and privacy breaches remain a risk in e-voting systems. Blockchain is a distributed technology that underpins bitcoin. It appears to be a fit for digital election systems, as it provides a chance to keep data secret and prohibit data manipulation. But blockchain alone is not enough to address other concerns that come up during elections such as how to make sure that the person voting has the right to do so while keeping their identity disconnected from their vote. The user can safely vote from the three-levels of authentication: OTP, QR codeshare, and blockchain. Here, the system contains a secured timestamp and the hashcode which is used to identify attacks on the distributed server. These protected services provide secure vote casting and remedies for security issues. The legal limitations are also analyzed in the blockchain distributed ledger.

Blockchain is one of the new modern technologies of the tech world. Bitcoin was first introduced with this technology and is being extremely tested for several cryptocurrencies continuously. The research scholars and businesses have been concentrating on its pertinence in furthering many zones such as education, business, healthcare, logistics, and some more areas. This chapter shows the benefits of blockchain technology in the educational area. The advantage of blockchain technology in the educational sector is support in improving the convenience and value of educational areas. There are some zones in education that are essential: continuous observations of students and student correlated data and transcripts, decrease in physical and administration, and more. Further, the monitoring and tracing capability of the blockchain and some solutions to the problems have been proposed.

Visual blockchain takes the idea of how to output accurate intelligent surveillance data by using computational methods in a smart city. Intelligent surveillance provides location-based visualization from different cameras which are deployed in the smart city. It may result in limited tamper-resistance of the gathered data from cameras. The decentralized nature of blockchain is to integrate video frames as data output by surveillance systems over the smart city by the way of selecting the most appropriate cryptographical algorithms using blockchain prototype to create sustainable computation method of visual blockchain and make reliable solutions for the existing solutions. In this chapter, the contributions are to enrich the blockchain security shield by using the most appropriate cryptographic algorithm to achieve each unique task and create a web-based image prototype through the visual blockchain to minimize the gaps in intelligent surveillance.

An amazing utilization of internet of things (IoT)-based smart technologies has been observed in recent years. Internet of things (IoT) became smarter and autonomous, leading the way by interconnecting and interacting with the environment. Existing technologies such as big data, AI, and cloud computing have been supported by IoT but with some limitations. Blockchain is the better solution to access and share data with the decentralized system. IoT spreads its applications in healthcare, industrial internet of things, supply chain management, traffic management, etc., out of which smart cities take special attention in recent days by integrating with IoT to create a sustainable ecosystem by using information and communication technology (ICT). Smart cities can offer various smart applications for boosting the life prominence of society. These characteristics of blockchain are beneficial to the creation of smart cities. In this regard, the authors have conducted a systematic review of articles. This study highlights the integration of blockchain with IoT.

Foreword

I am happy to write a foreword for this book titled *Blockchain Technologies for Sustainable Development in Smart Cities*. I felt delighted to note the different tools, technologies, key initiatives and challenges in blockchain technologies for sustainable development in smart cities. This book is a substantial compilation of 13 chapters encompassing an overview of blockchain's key theories and concepts. It examines important smart city domains like finance, insurance, healthcare, and supply chain management. It investigates the payment methods, fog/edge computing, green computing, and algorithms, cryptocurrencies and consensus mechanisms for smart city implementation using case studies for each area. It examines Ethereum, Hyperledger, IBM Blockchain, Corda, and smart contracts, as well as policies, regulatory standards, applications, solutions, and approaches. The research difficulties and opportunities for academics, researchers, and companies in adopting blockchain applications while exploring future blockchain ecosystems for smart and sustainable city living are focussed.

I'm delighted to greet the editors and authors on their accomplishments, and to inform readers that they are about to read a significant contribution to the development of smart cities based on blockchain technology with computational intelligence. I'm aware with your research interests and knowledge of blockchain technology. This publication would benefit greatly from the addition of new computational models for sustainable development of smart cities using blockchain technology. This book is an important step forward in the development of this discipline, and it will serve to challenge the academic, research, and scientific communities in a variety of ways.

S. Arunkumar Sangaiah
Vellore Institute of Technology, India

Preface

Cities are indeed being planned in a way that supports a clean environment, folds up smart infrastructure, and allows accessibility for vulnerable groups in order to achieve the goals of the smart city mission. It is critical to investigate the possibility of implementing dynamic and efficient technology to make our cities smarter. This book gives readers a wide grasp of the contemporary urban difficulties that smart cities are addressing. It examines the transformative potential of blockchain, one of the most recent technological advances, in making our cities smarter. It also examines several applications of blockchain technology in the smart city sector, as well as the requirements for adopting a blockchain-based solution and the path forward.

The objective of this book is to analyse and answer the issues and obstacles that comes with using blockchain technology in a variety of applications. Chapters on various aspects of establishing smart cities and blockchain-related subjects have been sent to the editors. The book also looked at the influence of these techniques on the applications that this advanced technology is being used in. This thorough and timely work aspires to be a valuable reference tool. This book provides the available literature to scholars working on the blockchain concept, allowing for further research options in this dynamic sector. It is intended that this content will give developers and managers with the resources they need to embrace and implement these methodologies, platforms, and approaches in order to produce effective solutions.

NEED FOR A BOOK ON THE PROPOSED TOPICS

Blockchain is a decentralised, open ledger that permanently records transactions between two parties without the requirement for third-party authentication. When this technology is combined with IoT and real estate, we might expect to see significant changes in a city's development.

Countries around the world are getting more urbanised, and big cities are becoming magnets for top talent and enterprises, resulting in these cities becoming a significant driver of economic growth. However, this carries the potential of increased traffic and environmental difficulties. Higher demand for resources, greater pressure on urban infrastructures, increased demand for public services, and more pressure on the natural environment are all concerns that arise when cities' populations grow.

Smart Cities, aided by connected technologies, are becoming increasingly important. The COVID-19 pandemic-induced economic crisis demonstrated the need for more efficient and reliable urban management. Advanced technologies, such as blockchain, can help to address these societal concerns while also improving urban administration efficiency. It is critical for a city to guarantee that effective data exchanges are made between many of the different metropolitan stakeholders in order to supply efficient urban services. Without relying on a centralised administrator, blockchain allows network participants to communicate data with privacy, a significant level of dependability and transparency.

The goal of this book is to explain one of the most recent technological advancements, blockchain, and to investigate how it could be used for creating cities smarter and sustainable. It investigates many aspects of a smart city and identifies places where blockchain technology might be applied to improve our cities' living standards and economic progress. This book is aimed at academicians, researchers, professionals, scientists and executives looking for modern exploration on the use of blockchain technologies and their applications. Feature coverage contains a wide scope of topics, including distributed computing, fog/edge computing, transaction processing, the Internet of Things (IoT) and disaster management.

ORGANIZATION OF THE BOOK

Chapter 1: The city of the future is smart, it is entirely interconnected and will regulate traffic, save energy, fight crime, and it will be assisted by Big Data and the Internet of things (IoT). Cities compose of many systems providing services for us. Systems such as Transport, Energy, Health, and others depend on each other. The connections between these systems seem to have been broken and need to be reconnected to provide efficient services for the citizens. To create a Smart City, there is the need to combine Internet Communication Network (5G), Application Software, Database & Blockchain Application, AI, Cloud Computing, GPS, Maps. For the Smart City concept to be instrumental and beneficial to the citizens, there should be a centralized large-scale database center that will tie all these pieces of data generated in the various smart sectors together for it to function as an intelligently coordinated, geographically distributed system. On this score that we propose nine main dimensions to the Smart City concept.

Chapter 2: Blockchain plays vital role in enabling users to securely communicate with reliability and transparency. Cryptographic hashing is the underlying structure of blockchain empowering cryptocurrencies. With the rapid advancement of communication network, cipher algorithms are becoming increasingly important in information security. Multimedia data are widely used in various applications, including health sectors, military, entertainment, education, banking, and communication with recent improvements in internet technologies. Meanwhile, the application background for cipher algorithms is becoming gradually more complicated, posing problems to existing systems and necessitating the development of new appropriate designs. Stream cipher has been a prominent research area in cryptography, and it is appropriate for use in circumstances of huge data transmission or resource limits. This work provides a quick overview of conventional encryption algorithms based on stream ciphers, and cryptographic prerequisites. At present, several attacks faced by stream cipher techniques are also discussed.

Chapter 3: Transportation is important for all and sundry. But it's be deviled with some problems for the environment such as pollution. Intelligent Transportation Systems (ITS), which emerged to solve this problem, focuses on important issues such as increasing safety in traffic. ITS across the world come to the fore in recent years. In this study, ITS was researched on how it is used in Konya. It serves the public in almost all types of transportation used with the developed IPTS (Intelligent Public Transportation System). The user can instantly access the station and vehicle information of the transportation system to be used through the menu. The system, which is integrated with road, rail, and air transportation, serves the user to avoid any problems from the beginning to the end. All information about online card filling lost property status and transportation fees can be accessed through the system. Undoubtedly, Konya has introduced the IPTS to provide a more practical public transportation service.

Chapter 4: The development of blockchain has led to the emergence and widespread use of decentralized cryptocurrencies around the globe. As of 2021, the global market capitalization of cryptocurrencies has crossed two trillion dollars. With increasing popularity and adoption, investors have begun to see cryptocurrencies as an alternative to conventional financial assets. However, the volatility associated with cryptocurrencies makes them a highly risky investment. This gives rise to the need for accurate and efficient price prediction models which can help reduce risks associated with cryptocurrency investments. The model aims at predicting the price of two popular cryptocurrencies; Bitcoin and Ethereum. Tensor Processing Unit (TPU) is used for providing a distributed environment for the proposed model. The results show that the distributed TPU trained model performed significantly better than the conventional CPU trained model in terms of training time while maintaining a high degree of accuracy.

Chapter 5: Blockchain, a relatively recent technological breakthrough has been spreading like wildfire with reports of adoption extending across industries. Although most popularly known in the context of cryptocurrencies, this technology's true strength lies in its ability to disrupt existing, traditional industries. It can be used to run any form of exchange, agreement (trust), or tracking processes. The capability to negotiate decentralized trust is key to its application in SCM. Holding the capacity to transform SC functionality, blockchain technology creates possibilities for security and redundancy mechanisms that are foundational for a revolutionary new approach to SCM. While aiming to explore the incumbent status, potential applications, and future directions of blockchain technology, this study iterates that its advent opens up new and exciting ways to implement a variety of different modern technological systems. The writers offer an overview of what an efficient SCM system should entail, with the aim to overcome any existent limitations via the application of blockchain technology.

Chapter 6: Blockchain technology represents an attractive factor for sectors that are interested in managing and keeping their transactions on a database that records all the movements that take place on each transaction. Therefore, to keep pace with the digital transformation taking place in this field of the real estate sector, this paper aims to submit a proposal in the Iraqi real estate sector to implement this technology. In addition, the paper clarified the current mechanism adopted in Iraq in completing real estate rental transactions and the problems that accompany this method. Then, we explained the advantages provided by the application of blockchain technology through the smart contract, in addition to eliminating the problems of tax evasion in this sector. Despite the negative points that this technology suffers from, it still represents a safe environment for preserving, managing and exchanging information.

Chapter 7: Blockchain is one of the technologies that is gaining significant attention within the healthcare industry. In order for Blockchain to reach its full potential, it's essential that governments and public institutions map the blockchain ecosystem, and establish a blockchain framework for coordinating early adopters. We have proposed a web-based application using blockchain with security algorithms to store patients' and other users' details who sign-up on our website. Blockchain is highly secure as any tamper with data in any block is not possible. Blockchain attributes ensure smooth interaction between different applications and systems along the availability chain. We will also reward the patient with our own implemented bitcoin on their transaction. Ethereum library Web3 is used to make a complete decentralized peer-to-peer chat service without any involvement of a third party. Our application will be able to send encrypted messages both securely and anonymously. Blockchain technology has been seeing widespread interest as a means to ensure the CIA triad in fewer environments.

Chapter 8: There are villages and rural areas in modern India which are still to get electricity connection. And there are those places where there is a connection but low voltage and frequent cut-offs are still a major problem. Electricity companies face a lot of issues regarding the efficient power supply to all their customers as well as the customers have issues regarding their needs of power. Wastage of power during transmission is another big issue. This paper aims to develop and implement algorithms to improve the above-mentioned situation. Improve the consumption of electricity from users' point of view and implement better renewable resources for the generation of electricity at the household level. To make the pricing model more robust, Blockchain is used as it ensures secure transactions. Blockchain can give users more efficiency and authority over their energy sources.

Chapter 9: In recent years, blockchain, with its features of decentralization, security, reliability, transparency of information, and low maintenance cost, has been applied in the fintech industry in more and more cases in China. The maturity of blockchain-enabled technology has improved. This is expected to rebuild the underlying Internet technology or will trigger a new round of development and innovation in the Chinese fintech industry. This chapter aims to investigate the current application of blockchain technology in China's fintech industry by analyzing case studies. Through the comparison and analysis of the cases in the fintech industry, this chapter attempts to examine how and whether digital transformation and other innovative technologies can drive new business models and create valuable experiences for consumers and/or firms in China's fintech industry. A variety of challenges and opportunities is analyzed to draw a blueprint for Chinese blockchain development. Finally, this chapter also provides some recommendations to the government.

Chapter 10: Although the electronic vote is being used around for many decades, electoral organizations are still progressively adopting it. Despite technologies that preserve the secrecy, authenticity, and privacy of data, vote-rigging and privacy breaches remain a risk in e-voting systems. Blockchain is a distributed technology that underpins Bitcoin, appears to be a fit for digital election systems, as it provides a chance to keep data secret and prohibit data manipulation. But blockchain alone is not enough to address other concerns that come up during elections such as: how to make sure that the person voting has the right to do so while keeping their identity disconnected from their vote. The user can safely vote from the three-level of authentication system's OTP, QR codeshare, and Blockchain. Here the system contains a secured timestamp and the hashcode which is used to identify attacks on the distributed server. These protected services provide secure vote casting and remedies for security issues. The legal limitations are also analyzed in the blockchain distributed ledger.

Chapter 11: Blockchain is one of the new modern technologies of the tech world. Bitcoin was the first introduced with this technology and is being extremely tested for several cryptocurrencies continuously subsequently. The research Scholars and businesses have been concentrating on its pertinence in further many zones such as Education, business, healthcare, logistics and some more areas. This chapter educates the Benefits of blockchain technology in the educational area. The advantage of block chain technology in the educational sector is support in improving the convenience and value of educational areas. There are some zones in education that are essential as continuous observations of students and student correlated data and transcripts, decrease in physical and administration and more. In their first explanation of the previous problems in the education area. Further, the monitoring and tracing capability of the blockchain, some solutions to the problems have been proposed.

Chapter 12: Visual blockchain takes the idea of how to output accurate intelligent surveillance data by using computational methods in a smart city. Intelligent surveillance provides location-based visualization from different cameras which are deployed in the smart city. It may result in limited tamper-resistant of the gathered data from cameras. The decentralized nature of blockchain is to integrate video frames as data output by surveillance systems over the smart city. By the way of selecting the most appropriate cryptographical algorithms using blockchain prototype to create sustainable computation method of visual blockchain and make reliable solutions for the existing solutions. In this book chapter, our contributions are to enrich the blockchain security shield by using the most appropriate cryptographic algorithm to achieve each unique task and create a web-based image prototype through the visual blockchain to minimize the gaps in intelligent surveillance.

Chapter 13: An amazing utilization of Internet of Things (IoT)-based smart technologies has been observed in recent years. Internet of Things (IoT) became smarter and autonomous, leading the way for the world day to day life by interconnecting and interacting with the environment. Existing technologies such as Big Data, AI, and Cloud Computing have been supported by IoT but with some limitations. Blockchain is the better solution to access and share data with the decentralized system. IoT spreads its applications in healthcare, industrial internet of things, supply chain management, traffic management, etc. Out of which smart cities takes special attention in recent days by integrating with IoT to create a sustainable ecosystem by using information and communication technology (ICT). Smart cities can offer various smart applications for boosting the life prominence of society. These characteristics of blockchain are beneficial to the creation of smart cities. In this regard, we have conducted a systematic review of articles. This study highlights the integration of blockchain with IoT.

Acknowledgment

It is evident that the preparation of a book of this scope necessitates the participation of a significant number of people. We owe a debt of gratitude to (Ms) Jan Travers, Director of Intellectual Property & Contracts, Gianna Walker, Angelina Olivas, Katie McLoughlin, Assistant Development Editor – Book Development, IGI Global and the IGI Global editorial staff, whose encouragement and support made the book publication initiative possible and contributed to its success. We would like to express our gratitude to the Vellore Institute of Technology Vellore administration for their invaluable help. Peer reviews make the most significant contribution to the development of a book like this. We thank the reviewers for spending many hours reading the work and offering us with insightful comments and suggestions.

We would like to offer our heartfelt gratitude to all the contributors who have submitted high-quality chapters, as well as the professionals who have helped us with timely review comments and suggestions.

Dr. Gopinath Ganapathy, *Professor and Chair, Bharathidasan University, Tamilnadu, India*

Dr. V. Susheela Devi, *Principal Research Scientist, Department of Computer Science and Automation, Indian Institute of Science, Bangalore, Karnataka, India*

Chapter 1
Smart City Technology and Its Implementation: 5G as a Catalyst

Gyasi Kwabena Emmanuel
Vellore Institute of Technology, India

***Purushotham Swarnalatha**
https://orcid.org/0000-0001-6801-9013
Vellore Institute of Technology, India

ABSTRACT

The city of the future is smart. It is entirely interconnected and will regulate traffic, save energy, fight crime, and it will be assisted by big data and the internet of things (IoT). Cities compose of many systems providing services for us. Systems such as transport, energy, health, and others depend on each other. The connections between these systems seem to have been broken and need to be reconnected to provide efficient services for the citizens. To create a smart city, there is the need to combine internet communication network (5G), application software, database and blockchain application, AI, cloud computing, GPS, maps. For the smart city concept to be instrumental and beneficial to the citizens, there should be a centralized large-scale database center that will tie all these pieces of data generated in the various smart sectors together for it to function as an intelligently coordinated, geographically distributed system. On this score, the authors propose nine main dimensions to the smart city concept.

INTRODUCTION

As per the United Nations projections, 66% of the global populace will be living in urban areas by 2050. Nigeria, China, and India are predicted to drive the increase in urbanization. The greater part of the global populace lives in urban communities, which is consistently rising, bringing about huge difficulties like more energy consumption, more traffic, more contamination, more water use, excessive waste generation, and the likes. Urban areas should adapt to these issues, hence the need for Smarter Cities, livable Cities

DOI: 10.4018/978-1-7998-9274-8.ch001

that can think for themselves. The City of the future is Smart, it is completely interconnected and will control traffic, conserve energy, battle wrongdoing, and it will be helped by Big Data and the Internet of things (IoT). 5G which is a novel mobile communication standard, is the pivot for the Internet of things since it can connect Infrastructure on a grand scale. The smartness of a City is measured by the extent to which advancement (technologies) such as the Internet and the Internet of things (IoT) is employed.

RELATED WORKS

Although "Smart City" has been widely and variously defined by many researchers, there is no generally acknowledged definition of a Smart City. Usually, the definition of Smart City is equipped towards the focal point of the researcher. Whiles some definition focuses on social or business aspects others too look at it from the technological viewpoint. All the more comparatively, (Hollands 2008; Caragliu et al. 2011) term Smart City as the incorporation of social, physical, and IT framework (structures) to work on the nature of city services. According to (Giffinger et al. 2007), Smart City has six main dimensions namely; Smart Government, Smart Economy, Smart People, Smart Environment, Smart Mobility, and Smart Living. This definition has been extensively espoused by many researchers including (Papa et al., 2013; Hernández-Muñoz et al., 2011), and the dimension is used as a yardstick for ranking the Smartest City. (Hall et al. 2000; and Washburn et al. 2009) in their research work on Smart Cities underscore the need to integrate software services and applications to enhance city services and the lives of city dwellers. In the same vein, (Kanter and Litow 2009) also share the same idea that for creating a Smart City environment, each independent software for each city domain has to be integrated. In their research work, they maintained that city sub-systems such as energy, transport, water, and education have to be networked and synchronized. (K. E. Skouby et al., 2014) stress the need for Advanced Artificial Intelligence (AAI) systems for handling complex IoT patterns. (Saha et al., 2017) advocate the use of Remote Sensing to gather information and subsequently use it to produce a reference map for a City.

The concept of a smart city has been embraced not only by governments, individuals, or groups but more also by companies and organizations. For example, IBM's Smart City strategy focuses on applying next-generation information technology to all walks of life by deploying sensors and devices in buildings, roads, railways, tunnels, hospitals, dams, bridges, water systems, oil and gas pipelines, energy networks, and many more objects around the globe that make up the Internet of things through the Internet. International Organization for Standardization (ISO) (ISO/IEC2015) stake in Smart City focuses on city outcomes to handle challenges such as political and economic growth, rapid population growth, and climate. The International Telecommunication Union (ITU) (ITU-T 2014) supports the notion of using Information Communication Technology to improve the well-being of individuals, the efficiency of urban operation and services as well as the competitiveness. These two organizations underscore their concept of Smart City to meet present and future generations' needs. (Caragliu et al. 2011) emphasize the important benefits of sustainability and management of natural resources.

Almost all the available pieces of literature seem to focus on just one or two elements that make up a Smart City whiles few of them still depend on the six dimensions of the Smart City proposed by Giffinger in 2007.

MAIN FOCUS OF THE CHAPTER

In this chapter we propose nine main dimensions to the Smart City instead of the six main dimensions proposed by (Giffinger et al., 2007) and then explore the effective ways of integrating these nine dimensions. The chapter likewise outlines what is required of a Smart City and its Infrastructure, discusses some of the various technologies which are the bedrock of Smart Cities, the application of these technologies in different sectors, and how they can be coordinated.

Smart City Technologies

Smart Cities only make sense when technology meets the citizen's needs. The focus of Smart City development is towards applying next-generation information technology to all walks of life and all sectors of the economy. For this chapter's purpose, the information technology aspect has been divided into four sections, namely;

i) The Internet ii) Application Software iii) Big Data, Database & Blockchain, and Cloud Computing, iv) Auxiliary Systems.

The Internet

One of the first technologies which are essential to the advancement of a Smart City is the availability of highly reliable internet facilities. As the name suggests, a City is said to be smart if it uses various kinds of electronic Internet of things (IoT) sensors to gather data and subsequently manipulates the data to control resources, assets, and services effectively. Internet technologies such as Power Line Communication (PLC), WiFi, 2G, 3G, and 4G are used to interface with gadgets that can perceive, distinguish objects, assemble data, and exchange data with different devices. Although these internet technologies can connect devices, the overwhelming demand by the Internet of Things (IoT) devices has exceeded the generation capacity of these networks. The internet latency became a setback for the Smart City development concept. With the birth of the fifth generation of wireless/mobile broadband (5G WMB), the hope of Smart City realization has been rekindled. The International Telecommunication Union (ITU) classifies 5G mobile network services into three classifications, in particular: Enhanced Mobile Broadband (eMBB), Ultra-reliable, and Low-Latency Communications (uRLLC), and Massive Machine Type Communication (mMTC) **(Huawei Technologies, 2016).** Figure 1 shows the physical Infrastructure of 5G.

Figure 1. 5G physical infrastructure

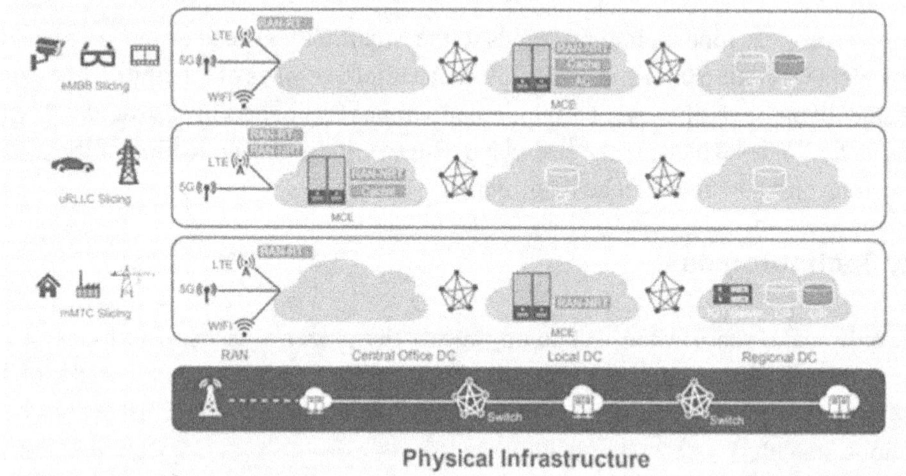

The extent of Enhanced Mobile BroadBand is to satisfy individuals' need for an inexorably computerized way of life and centers around administrations that require high data transmissions (high bandwidth), like High Definition (HD) Video, Augmented Reality (AR), and virtual reality (VR) or computer-generated reality. Ultra-reliable, low-latency communications are designed to meet the demanding expectations of the digital industry and target latency-sensitive services such as automated and assisted driving and remote management. Massive Machine Type Communication takes care of the demands for a further developed digital society and aims at services such as high requirements for connection density, like Smart City and Smart Agriculture **(Sarah Tourville, 2018).**

The Application Software

In the technological setup, software applications play a significant role in the smooth running of devices. Without software applications, all devices are just ordinary objects. The software application helps the end-users to perform single or multiple tasks. Both frontend Application Programming Interface (APIs) and Backend Application software are required to integrate IoT devices. For Smart City devices to be operational, there is a need for a highly effective software platform that will enable interoperability between a city's multiple systems, managing a sizeable amount of data, deal with a large variety of sensors, support the required scalability, and more importantly ensure the privacy of citizens. Table 1 shows the factors that need to be considered in developing Smart Cities software applications.

Table 1. Factors to consider in smart city software application

Area	Description	Tools
Privacy	Protecting Data collected from citizens, cities, and enterprises	Cryptography, Anonymization, and Access Tokens
Data Management	Managing all the Data collected in the platform	NoSQL, Relational Databases and Processing tools
Heterogeneity	Ensuring the interoperability of devices and applications	Standards, Ontology, and a City Unified Model
Energy Management	Managing the Electricity used by devices deployed in the city	Dashboards, Controllers, and control loops
Communication	Enabling communication among heterogeneous devices	M2M Techniques
Scalability	Allowing the growth of devices and users connected to the platform	Distributed tools and Algorithms and P2P applications
Security	Protecting the city data, services, and Infrastructure	Cryptography, Access Token, and Devices
City Models	Defining a model, describing the city	Semantic Web and Ontologies
Platform Maintenance	Maintaining the city systems and Infrastructure	Monitoring and alert tools

(Cheng et al., 2015) proposed a Smart City application platform design called City-Data and Analytics Platform (CiDAP), which centers on IoT networks and information gathering. It is a Big Data stage that spotlights the utilization of city information gathered from the sensor-appended gadgets in the City. It guarantees setting mindfulness and insight in applications and services. CiDAP platform measures huge volumes of datasets accumulated from an Internet of Things Middleware. Figure 2 shows the CiDAP platform design. The architecture has five head parts, to be specific; IoT – Agents, IoT – Brokers, Big Data Repository, and Big Data Processing.

The IoT–Agents associate with the Internet of Things Middleware and fill in as an entryway to the gadgets at the disposal of the stage as every information source links to an IoT – Agent.

IoT- Brokers serve as a bound together interface to the IoT – Agents, guaranteeing admittance to the middleware data. It communicates with the Big Data Repository and sends data for storage, and with city model server to send information, to be utilized straight by applications.

The Big Data Repository is liable for putting away crude information gathered from the devices in the city just as handled information from the Big Data processing components. CiDAP stage utilizes CouchDB, which is a NoSQL database, as it stores files such as JSON. It also has an internal tool that simplifies processing, making new organized perspectives and tables to index data or transform data into new formats.

Big Data Processing employs Apache Spark for processing intricate and concentrated data stored in the Big Data Repository (for Example, data mining or data aggregation). The Big Data Processing component likewise measures accurate information through batch processing or real-time data using data streams.

The City Model Server is the stage's interface to outer applications. City Model API permits applications to execute both basic and intricate queries and prefer explicit bits of information from the stage. Complex queries demand aggregated historical data, while simple queries demand the latest data from devices. Subscription is the mechanism for applications to receive data from the devices periodically **(Santana et al., 2017)**.

Figure 2. Shows the city-data and analytics platform (CiDAP) (Cheng et al., 2015)

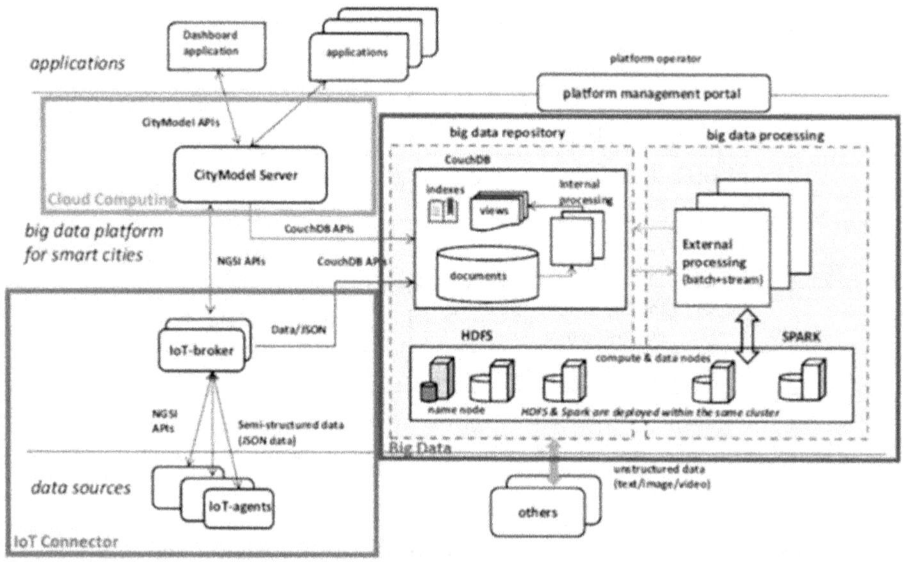

Big Data, Database, and Cloud Computing

The delivery of computing services like servers, databases, storage, software, networking, intelligence, and analytics over the Internet ('the Cloud') to provide faster advancement, flexible assets, and economies of scale is referred to as Cloud Computing **(Janbask Training, 2021)**, Cloud Computing provides elasticity to the services and data storage. It offers the Hadoop Eco-system, a platform for data storage and processing. Sensors of IoT devices measure and record various sorts of data, namely, Structured, Semi-Structured, and Unstructured data. Data of these kinds come under what we called Big Data. Big Data refers to data that is so huge, fast, or complicated that it is strenuous or impossible to store and process using the traditional Relational Database Management System (RDMS) methods. Big Data is portrayed by Volume, Velocity, Variety, Value, Veracity, as well as Validity. The volume alludes to the sheer size of the data. Velocity is the speed at which the data is being generated. Variety refers to the different kinds of data generated from various sources **(Cat McClintock, 2016).** The Value deals with the mechanism to bring the current meaning out of the data, while the Validity refers to the precision and rightness of the data for the intended purposes. Since the data generated from the City are so massive, fast, and comes in real-time and in different formats, they are stored in NoSQL databases. Big Data technologies such as Hadoop distributed file systems (HDFs) and some Hadoop components are required to help the storage and processing of these forms of data gathered from Smart Cities real-time data sensors used in the Internet of Things environment.

Auxiliary Systems

For this paper, auxiliary systems here refer to all other technologies that are implemented in the improvement of Smart Cities—Systems and technologies such as Remote Sensing Technology, GIS, Maps, GPRS.

Remote Sensing: Is the science (or better still, art) of obtaining data about the Earth's surface without really being in touch with it. Remote Sensing is executed by recording reflected or emitted energy and processing, analyzing, and applying it **(NRCan.gc, 2015)** In Smart City development, remote-Sensing is one of the processes for gathering data about the Earth for decision making. **(Saha et al., 2017)** explain the use of Remote Sensing to collect information about the rooftop and subsequently use it to produce a reference map for a City in India.

Maps (Google Map): provides a mapping service or a planning administration that offers satellite imagery (satellite symbols), aerial photography (airborne photography), street maps, 360° intelligent all-encompassing perspectives on roads or streets, real-time traffic conditions (constant traffic conditions), and route planning for traveling either by foot, car, bicycle, or air. Google Maps plays a very crucial role in Smart City development due to the services mentioned above that it provides.

Application of Smart Technologies in Smart Cities

Cities are composed of many sectors providing services for people. Sectors such as Transport, Energy, Health, and many more depend on each other. Connections between these sectors seem to have been broken and hence, the need for them to be reconnected to provide sustainable and efficient services for the citizens. Smart Technologies promises to be the solution to this interconnection of the various sectors. This paper looks at how Smart Technologies can be applied in sectors such as Transport, Health, Energy, Education and Tourism, Water and Environment, Housing and Engineering, Government and Security, Trade and Finance, Agriculture, to have Smart sectors which when linked will result in smart-city. Figure 3 shows a picture of the various Smart City dimensions.

Figure 3. Smart city dimensions

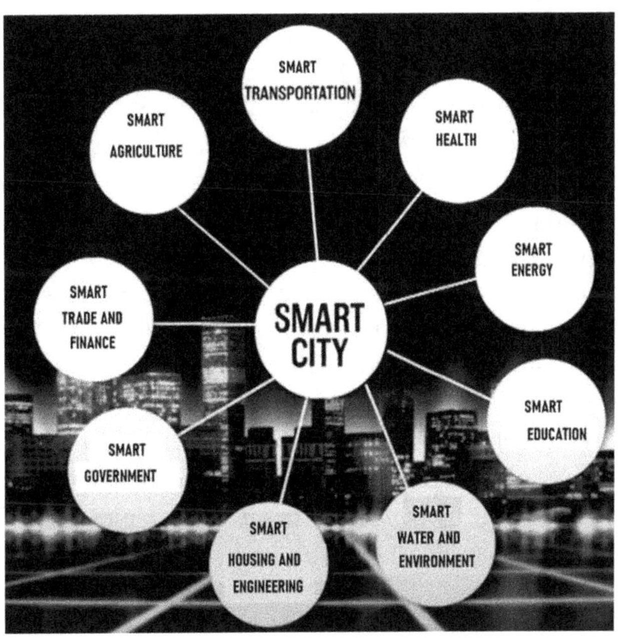

Smart Health Sector

The health sector is not immune to technological innovations. Technologies like IoT, Big Data, Advanced Analytics, Cloud Computing, and many others have changed the old system of healthcare into Smart Healthcare. Smart Healthcare is the use of mobile and electronic technology to make diagnoses of the disease, improve treatment of patients, and share information between patients and health service providers to enhance the quality of life. Internet of Things (IoT) in healthcare technologies is termed the Internet of Medical Things (IoMT). Through this technology, patients can be identified, monitored, and also inform caregivers about the patient's vitals **(Anandarajan et al., 2018)** IoMT provides the needed essentials and critical data of the patient to the health care providers to inform swift responses to any identified issues. IoMT devices help patients keep track of their vitals such as heart rate, glucose levels, sleep patterns, and, more importantly, remind them of medications and activities like routine blood and urine tests in the case of the aged population. There are several other IoT devices at the health centers that are contributing to Smart Healthcare delivery. All these devices produce data in enormous quantities and at a rapid speed.

Big Data technology in healthcare helps foster smooth Machine to Machine (M2M) and Machine to Human (M2H) communication due to the vast volumes of data generated by the IoMT devices. The study of Machine Learning in healthcare has become possible due to the extreme size of available healthcare data.

Cloud Computing technology is making healthcare much smarter as edge computing is employed to detect, monitor, and notify the occurrence of epilepsy episodes in a patient to its caregiver.

Mobile software applications like Telemedicine allow medical services experts' (health care professionals) to assess, diagnose, and treat patients remotely. It also helps health professionals to extend helping hands to their colleagues in other health institutions when their expertise in a particular area is requested.

Smart Education Sector

Various technologies are employed in the education sector to enhance teaching and learning. Cloud Computing, Internet of Things (IoT) devices, Databases, and many other technologies have transformed the traditional way of teaching and learning in a modern way such as; Smart School, Smart Campus, Smart Classroom, Smart Teaching, and Learning, E-Leaning, and E-Library.

Smart School: is a technology-based teaching-and-learning institution that makes good use of several IoT devices and multimedia tools such as movies, audio, photos, slides, and the likes to make the teaching and learning processes more effective and efficient, especially for children. The concept of Smart School at the upper and tertiary level aims at a seamless contact-free institution where students, parents, guardians, and anyone who wishes to access the institution's services can do so with little or no human intervention. All transactions with the school can be done remotely without any physical interaction. Activities such as Admissions, Payment of fees, Selection of Courses, deferring of Admission and courses, request and authentication of Transcripts and certificates, and many more are made possible by the use of IoT devices.

Smart Campus: on the other hand, is a place where IoT gadgets and applications make new encounters or services and work with functional productivity. It, therefore, comprises Smart Teaching and learning Infrastructure, Smart Security, and many other activities in Smart Cities since Universities on their own are small towns or communities themself.

Smart Teaching and learning: refer to the organization of teaching and learning activities with the help of software tools and IoT devices such as Smartboard, Smart screen, Smart projector, Smart computers, and peripherals. These IoT devices, plus speakers and microphones, put together make a Smart classroom.

E-Learning: Electronic Learning, which is popularly known as e-learning, is the utilization of electronic advances to get to instructive educational curriculum outside of a traditional classroom **(prime-alliance.org)**. On the other hand, it is the acquisition of knowledge, training, learning, or Education delivered online through a computer or any IoT device. This type of learning makes good use of technologies such as Databases, IoT devices, Big Data, Cloud Computing, and the likes. Much data is generated through this method since teaching, and learning materials such as videos, audios, pictures, texts (notes), assignments, and certificates are deposited in a cloud database to make it accessible to students and all interest groups.

E-Library: which is also known as Electronic Library or Digital Library alludes to an actual webpage or site that gives admittance to audio (digitized sound), video, and composed materials such as books, journals, articles, and many more at any point in time. Through Cloud Computing, Database, and IoT devices, these materials can be uploaded, stored, downloaded, retrieved, or accessed online around the clock.

Smart Agriculture Sector

Smart agriculture consists of agricultural practices using the Internet of things (IoT), sensors, and other innovations to maximize the productivity of the farm. It is also known as precision agriculture. Smart agriculture encompasses smart farming, Smart harvesting, Smart storage, and Smart processing.

Smart farming is the use of IoT devices in all farming operations. There are several types of farming, namely, 1) Food Crops farming, 2) Fish farming, and 3) Animal Husbandry (Livestock farming).

Food Crops Farming*:* The mass exodus of people moving to the cities has compelled both government and citizens to compete for land and space to put up mansions, schools, hospitals, industries, and construction of roads. This has made it virtually impossible to engage in large-scale mechanized crop farming in the cities and hence, call for a Smart way of farming in the Cities. A farming method such as vertical farming is an alternative solution to this scarcity of land for farming in the Cities. Vertical farming is the act of developing or growing plants in upward or vertically stacked layers, regularly including controlled environmental farming or agriculture aimed at improving plant development and soilless cultivation methods like aqua-farming, hydroponics, and aeroponics. **(wikipedia.org).** These upward cultivating strategies use structures or buildings, shipping containers or steel trailers, burrows or tunnels, and abandoned or deserted mines shafts to cultivate crops. Since they are executed under controlled conditions, technologies such as IoT gadgets are utilized to work with the checking of temperature, air, and monitoring of water quantity of the crops, in the environment. Figure 4. Shows lettuce filled in an indoor vertical cultivating framework.

Figure 4. Lettuce grown in an indoor vertical farming system.

1. *Fish Farming*: IoT advancements are utilized in fish cultivating processes through sensors networks to measure values like pH, temperature, and more parameters **(iot-marketplace.com).** Through these IoT devices, fish farming management can be quickly and remotely monitored from anywhere outside the location of the fish pond.
2. *Animal Husbandry*: This is also known as Precision livestock farming. In this cultivating technique, IoT and sensors can be utilized to oversee domesticated animals' wellbeing (livestock health), track animals as assets to minimize loss, and assists domesticated animals with creating posterity all the more proficiently. For instance, a radio frequency ID tag (RFID) can be utilized to follow steers IN and OUT of the dairy cattle range. Nanotron's swarm bee is another IoT technology that can be used to monitor the health of animals to provide continuous area following (real-time location tracking). This is used along with analytic software to track the location and movement of animals.

It keeps farm managers informed of the health of their animals since a sick animal will not move as frequently as it used to do when healthy.

Smart Harvesting, Smart processing, and Smart storage are all Smart Agricultural processes that use technology. In general, smart agriculture uses cutting-edge innovations like IoT, GPS, Big Data, and other connected devices to collect data from the field and then analyze the data to make accurate decisions to increase crop yield.

Smart Transport Sector

The transport sector has seen tremendous changes with the help of technologies such as IoT, Big Data, GPS, Navigational Maps, and the likes. The Internet of Things (IoT) devices, which are made up of sensors and actuators are changing the transport sector by integrating technical and business mobility patterns, automation, and data analytics. Areas, where technology has significantly impacted the transport-sector, include connected cars, traffic management, vehicle tracking system, public transport management, and many more.

1. *Connected Cars*: Modern cars are fitted with more than 50 microcomputers which are interconnected to monitor engine and brakes, exhaust gas consumption, and tire pressure. Technologies like Vehicle-to-Vehicle (V2V) and Vehicle-to-Infrastructure (V2I) permit clients to make parking spot reservations, forestall a collision, and advance smooth traffic stream.
2. *Traffic Management*: IoT devices with sensors are used to manage traffic, and this is known as Smart traffic management. Smart traffic signals have now replaced traditional video surveillance to overcome traffic congestion.
3. *Vehicle Tracking System*: Fleet operators use a vehicle tracking system for tasks such as routing, security, dispatching, fleet management, and onboard information. IoT, GPS, Google Maps have helped improve their exhibition, including the capacity to catch driver conduct to limit sitting.
4. *Public Transport Management:* Public transit management systems use IoT technology as it offers benefits like constant vehicle following (real-time vehicle tracking), calculating for surprising occasions (awful weather, road closure), and give customized travel information.

Self-driving vehicles use Technological advancements like Internet of Things (IoT) gadgets, Artificial Intelligence (AI), Big Data, GPS, and Google Maps to make driving possible.

Smart Financial Sector

Smart Banking and Smart payments are very vital in the financial sector. For these two key areas to be active and efficient, there should be dedicated and reliable internet connectivity.

Smart Banking: Optimizes client commitment models to drive client retention with high-quality services such as E-banking, ATM transaction, Mobile banking, and many more. The success or failure of these activities rests on robust internet availability since, without functional speed internet connectivity, transactions in any of these methods are bound to fail. An ATM requires functional internet connectivity to be operational. In the same way, E-banking hinges on functional internet connectivity and a Database system. Technology such as 5G Mobile Network promises to be competent in addressing all the shortfalls in these methods of transactions. Smart City citizens need not waste their time in a long queue to withdraw money or transact any banking business.

Smart Payment: The upsurge of electronic banking and shopping has witnessed the advancement in various online business installment frameworks, and technology has been made to increase, improve, and provide secured e-payment transactions. Electronic installment techniques, for example, the utilization of Credit Cards, Debit Card, E-Wallet, Net-banking, Mobile Payments, Amazon Pay, are completely worked with by the Internet.

E-Commerce: This is also known as online business involves buying and selling jobs and items using the Internet, and the exchange of money and information to execute these transactions (shopify.com). Online business is similarly called Electronic Commerce or Internet Commerce. With the flood of individuals in urban communities making clog, merchants and potential purchasers have brought about a Smart method of selling and shopping, which is an internet-based transaction.

Smart Water and Environment

Smart Water Management: This is a machine intended to gather significant and noteworthy information on the flow, pressure, and appropriation of the City's water. The management involves the use of technology to detect the water network's leakage, Smart meter technology, pressure management, Smart water quality monitoring, Smart network operations, and maintenance **(Alshattnawi et al., 2017).**

Smart Waste Management: Internet of Things devices with sensors are set in squander containers (waste receptacles) to measure fill levels and to inform city assortment administrations when receptacles are fit to be emptied **(Catania et al., 2014)**. The implementation of this technology offer solutions to tons of thousands of waste generated by city dwellers at a reduced cost and in a more efficient many. Quick emptying of filled bins helps reduce sicknesses and disease outbreaks, such as Cholera.

Smart Weather Forecasting: Meteorologists use sensors to measure weather parameters such as Temperature, Pressure, Relative Humidity (RH), Wind speed and direction, Rainfall, Sunshine, Clouds, Visibility, and many more to make a weather forecast. Automatic Weather stations are equipped with Weather sensing devices that automatically measure, record, and transmit data to the forecasting office for analysis. The weather forecast has dramatically improved through these self-recording instruments due to the accurate measurement of these parameters by the various IoT devices. Timely and accurate weather warning helps citizens to make well-informed choices. For instance, Farmers depend on the onset of rain to start planting their crops.

Smart Housing and Engineering

Smart Home: This refers to a home that utilizes Internet of things gadgets and sensors to empower the remote monitoring and management of appliances and systems, like smart TV, Smart Lighting, Microwave, Heating, Air-conditioning, Surveillance cameras (CCTV), Refrigerator, Washing Machine, Smart Door Locks, and many more **(techtarget.com).**

Smart Governance

Smart governance refers to the use of technology by government officials to facilitate and supports better planning and decision-making. In order words, it is the creative use of technologies such as the

Internet, Intelligent Analytics, Cloud Computing, Big Data, Maps, GPS, IoT and sensor devices (e.g., smartphones, tablets, drones, cameras), and many other intelligent technologies to transform the way services are delivered to the people. Smart governance aims to use technology to improve democratic processes and transform the ways public services are carried out. Smart Security and Smart Communication are essential to Smart governance.

Primarily, Smart governance is effective management and coordination of all the Smart sectors for the benefit of citizens.

Smart Energy Sector

Smart Grid and smart meters are the two most critical areas in the Energy sector. Apart from these, one other equally important area is a smart street lighting system.

Smart Grid: Technologies such as IoT devices are used to monitor power generation to improve fault detection and allow self-healing of the transmission network. A Smart transformer regulates voltage and keeps in touch with the smart grid regularly **(Adekunle et al., 2015).** This is to permit remote administration (when necessary) and to deliver information and feedback regarding the power supply and the transformer. It gives the specific measure of power required and reacts to power fluctuation immediately inside the power grid to guarantee uninterrupted optimized voltage.

Smart Meter: It is an IoT gadget that records the consumption of electric energy and transmits the data (information) to the power supplier for checking and charging (monitoring and billing). It allows energy consumers to monitor and regulate power usage.

Smart Street Lighting: This is a sensor-driven street light that responds or adapts to the movement of pedestrians, cyclists, and cars. These street lights dim when no activity is detected and brightens when it recognizes an activity. It is also known as an intelligent street light. Through this mechanism, streets are well lit only when necessary, and this helps save electric power.

OBSERVATIONS

Because cities are globalized and provide good education and better services such as Healthcare, Roads, lucrative businesses, and others people find cities very attractive and as a result, nearly 200,000 individuals troupe to the urban communities consistently. In the next few years, 70% of the World's populace will be living in urban communities, and this dramatic development in urbanization will make urban communities battle to offer fundamental types of assistance to residents because there are restricted normal assets. Table 2 shows the Top 10 Smart Cities in the World, as indicated by the fifth release of the IESE Cities in Motion Index (CIMI), 2019. New York City secured the 1st position with a total population of the City estimated to be 8,336,817, which represents 2.54% of the total population of the entire USA. Smart Bins for waste management, Smart Water Metering, Smart Mobility, Smart Street Lighting, and Provision of free WiFi for all residences, are among many other technologies being implemented there. It is fascinating to observe that a renowned city like Dubai could not feature in the top 10 positions, but instead, a small Icelandic city called Reykjavik managed to secure the 5th position. This attests to the fact that the smartness of a City is the interconnectivity of its various sectors to the Internet and the effective use of technology. Table 2 shows the top 10 Smartest Cities in the world.

Table 2. Top 10 smartest cities in the world

NO	NAME OF SMART CITY	POPULATION OF THE CITY (Wikipedia)	COUNTRY OF THE CITY	POPULATION OF THE COUNTRY (Wikipedia)	(%) IN CITY	TECHNOLOGY IMPLEMENTED
1	New York	8,336,817 *(2019 Est.)*	USA	328,239,523 *(2019 Est.)*	2.54	1. Smart Bins (waste management) 2. Smart water Metering (Water) 3. Smart Mobility (Transportation) 4. Smart street Lighting (Energy) 5. Free WiFi to all residence
2	London	8,908,081 *(2018 Est.)*	United Kingdom	67,886,004 *(2020 Est.)*	13.12	1. Open and Secure City Data share 2. Smart Mobility (Transportation) 3. Smart street 5G WiFi connectivity 5. Air quality monitoring with sensors
3	Paris	2,148,271 *(2020 Est.)*	France	67,076,000 *(2020 Est.)*	3.20	1. Smart Mobility (Transportation) 2. Air quality monitoring with sensors 3. Smart housing (Energy generation) 4. Smart Benches (Data Collection)
4	Tokyo	13,929,280 *(2019 Est.)*	Japan	126, 150,000 *(2019 Est.)*	11.04	1. Smart Security (Face Recognition-AI) 2. Self-Driving Cars (Smart Transportation) 3. Smart Manufacturing 4. Smart Mobility
5	Reykjavik	131,136 *(2020 Est.)*	Iceland	364,134 *(2020 Est.)*	36	1. Smart Environment 2. Smart Mobility (Transportation) 3. Smart WiFi connectivity
6	Singapore	5,703,600 *(2019 Est.)*	Singapore	5,703,600 *(2019 Est.)*	100	1. Smart Traffic cameras 2. Smart Payment of Parking fee 3. Free WiFi to all residence 4. Smart Mobility (Transportation) 5. Smart GPS Surveillance cameras
7	Seoul	9,733,509 *(2020 Est.)*	South Korea	51,709,098 *(2019 Est.)*	18.82	1. Smart Data Collection with sensors 2. Smart Parking lot (Transportation) 3. Smart Surveillance cameras (sensors) 4. Smart Weather monitoring (Env.)

Continued on following page

Table 2. Continued

NO	NAME OF SMART CITY	POPULATION OF THE CITY (Wikipedia)	COUNTRY OF THE CITY	POPULATION OF THE COUNTRY (Wikipedia)	(%) IN CITY	TECHNOLOGY IMPLEMENTED
8	Toronto	2,731,571 *(2018 Est.)*	Canada	38,014,184 *(2020 Est.)*	7.19	1 Smart Mobility (Transportation) 2 Smart Waterfront 3. Smart street Lighting (Energy) 4. Smart Housing
9	Hong Kong	7,500,700 *(2019 Est.)*	China	1,427,647,786 *(2018 Est.)*	0.53	1. Smart Mobility (Transportation) 3. Smart Farming (Agriculture) 3. Smart Industry (Engineering) 4. Smart Security (Face Recognition-AI) 5. Smart street 5G WiFi connectivity
10	Amsterdam	1,308,873 *(2018 Est.)* Urban Area	Netherlands	17,424,978 *(2019 Est.)*	7.51	1. Smart Grid (Energy) 2. Smart Housing 3. Smart Data sharing 4. Smart Mobility (Transportation)

It is challenging and nearly impossible to serve the population that will be living in the cities, and besides, governments have to discover Smart ways of dealing with the restricted regular assets and how they can all the more likely help individuals to sustain our growing cities. Cities compose of numerous frameworks offering types of assistance for their dwellers. Systems such as Transport, Energy, Health, and others depend on each other. The connections between these systems seem to have been broken and need to be reconnected to provide efficient services for the citizens. Congestion creates lots of costs for cities in the sense that every driver spends around five months of their life in traffic, and traffic creates carbon emission and emissions create pollution. Every year, 7 million people die because of complications caused by pollution. The idea behind smart cities is to have Sustainable, Efficient, and Liveable cities. To achieve these ideas, we have to consider three elements, namely, Systems, Information, and People. Almost every City now has all the sub-systems that are needed by living organisms. To create a Smart City, we need to combine Internet communication network (5G), Software and Database Application, Internet of Things (IoT) devices, Sensors, Actuators, Cloud Computing, GPS, Maps, Artificial Intelligent (AI), Big Data, Blockchain Technology, and Remote Sensing. For the Smart City concept to be instrumental and beneficial to the citizens, there is a need to have a centralized large-scale database center that will tie all these pieces of data generated in the various Smart sectors together for it to function as an intelligently coordinated, geographically distributed system. Blockchain technology should be implemented in securing the database. The Database should be managed by those with expertise under the supervision of the Central government. It should be made accessible to both institutions and the citizens upon request. The Database has to be continuously updated to reflect the changing time. Figure 5 shows the connectivity of the Smart City sectors through the 5G Mobile Broadband and how they can be managed effectively in a centralized database system.

Figure 5. Connectivity of smart sectors the data center (Demestichas et al., 2015)

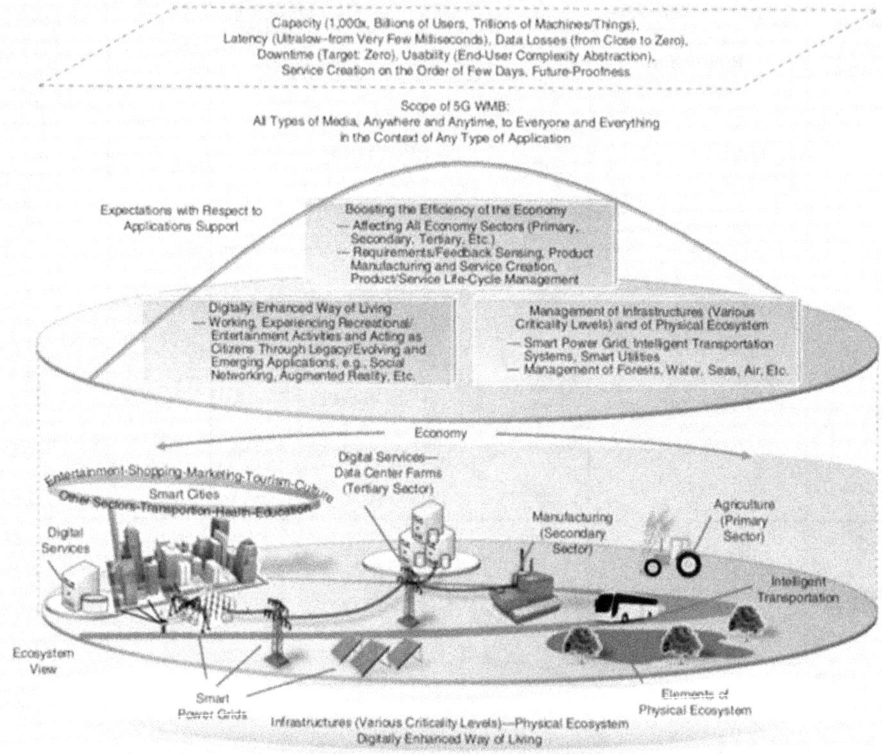

Implementation of Blockchain Technology at the Backend of the Integrated Database Storage and Management System (IDSMS)

Since the largest record of each individual in the country will be stored in this Integrated Database, it is likely to be the toast of hackers. Securing this database is non-negotiable and implementing Blockchain technology is one of the surest means by which the government could mitigate the risk of potential hacking. With Blockchain technology, public ledgers and databases can be secured on a network. Its robust security can prevent unauthorized access and minimize the risk of a data breach. It can also streamline various operations and reduce the burden of manual work.

CONCLUSION

Initiative for developing Smart City has begun in almost all our cities as virtually, every sector in our City today is at least implementing one smart city concept such as; Smart payment (Finance Sector), digitization of patient health records (Health Sector), smart metering (Energy Sector), e-learning (Education Sector). However, most of these sectors are functioning in isolation. Integrating these sectors into a complete and centralized database has been a significant challenge. Another difficulty faced by these sectors that are implementing the Smart City concept is unreliable internet connectivity. This makes the coming into force, a 5G mobile network a game-changer. In the 5G paradigm, single network infrastructure can meet various service requirements. The mobile network will offer expanded types of assistance: eMBB, uRLL, and mMTC request various necessities for network control capacities. Smart City software platforms can uphold the turn of events and coordination of Smart City applications. Developing our cities along the nine main dimensions of Smart Cities namely; Smart Transport Sector, Smart Health Sector, Smart Energy Sector, Smart Education Sector, Smart Water and Environment Sector, Smart Housing and Engineering Sector, Smart Government Sector, Smart Trade, and Finance Sector, and Smart Agriculture Sector is the best approach in realizing the Smart city concept. The concept of a Smart City is fully achieved only when digital technologies make a meaningful difference in people's lives.

REFERENCES

Adekunle. (2015). Islamic Education in Nigeria. From Rote Learning to E-Learning. *European Conference on E-Learning, Academic Conferences International Limited, 422.*

Alshattnawi. (2017). Smart Water Distribution Management System Architecture Based on Internet of Things and Cloud Computing. *International Conference on New Trends.* ieeexplore.ieee.org

Anandarajan. (2018). *Protecting the Internet of medical things: A situational crime-prevention approach. In Cogent Medicine.* Taylor & Francis.

Caragliu, A., Del Bo, C., & Nijkamp, P. (2011). Smart cities in europe. *Journal of Urban Technology, 18*(2), 65–82.

Catania. (2014). An approach for monitoring and Smart planning of urban solid waste management using the smart-M3 platform. *Proceedings of 15th Conference of Open.* ieeexplore.ieee.org

Cheng, B. (2015). Building a big data platform for smart cities: Experience and lessons from Santander. In *2015 IEEE International Congress on Big Data (BigData Congress'15).* IEEE.

Demestichas. (2015). Intelligent 5G Networks: Managing 5G Wireless\/Mobile Broadband. *IEEE Vehicular Technology Magazine, 10*(3), 41–50. doi:10.1109/mvt.2015.2446419

E-commerce Definition - What is Ecommerce. (n.d.). https://www.shopify.com/encyclopedia/what-is-ecommerce

Fish Farming - The IoT Marketplace. (n.d.). https://www.the-iot-marketplace.com/solutions/fish-farming

GNetworkArchitecture – A high-level perspective. (2016). Huawei Technologies.

Grid Control | PRIME Alliance. (n.d.). https://www.prime-alliance.org/?page_id=312

Hernández-Muñoz, J. M., Vercher, J. B., Muñoz, L., Galache, J. A., Presser, M., Luis A Hernández, G., & Pettersson, J. (2011). Smart cities at the forefront of the future internet. In *The Future Internet Assembly* (Vol. 6656, pp. 447–462). Springer.

Hollands, R. G. (2008). Will the real smart city please stand up? Intelligent, progressive, or entrepreneurial? *City, 12*(3), 303–320.

ITU-T. (2014). *Smart Sustainable Cities: An Analysis of Definitions.* Tech. Report. FG-SSC-10/2014.

Kanter & Litow. (2009). *Informed and interconnected: A manifesto for smarter cities.* https://papers.ssrn.com/sol3/papers.cfm?abstract_id=1420236

McClintock. (2016). *Big Data Brings Big Changes to Product Design.* https://www.ptc.com/en/cad-software-blog/big-data-brings-big-changes-to-product-design

P. P. T. & S. K. L. (2019). Smart City Services - Challenges and Approach. *2019 International Conference on Machine Learning, Big Data, Cloud and Parallel Computing (COMITCon)*, 553-558. doi: 10.1109/COMITCon.2019.8862243

Papa, R., Gargiulo, C., & Galderisi, A. (2013). Towards an urban planners' perspective on smart city. *TeMA J. Land Use. Mobil. Environ.*, *6*(1), 5–17.

Remote-Sensing. (n.d.). https://www.nrcan.gc.ca/maps-tools-publications/satellite-imagery-air-photos/remote-sensing-tutorials/fundamentals-remote-sensing-introduction/9363

Saha. (2017). A Remote Sensing Approach to Smart City Development in India: Case of Bhopal City, Madhya Pradesh. *Conference: the Special Collection.* DOI: doi:10.1145/3055219.3055232

Santana. (2017). Software Platforms for Smart Cities: Concepts, Requirements, Challenges, and a Unified Reference Architecture. ACM Comput. Surv., 50(6).

Tourville, S. (2018). *Council Post: The Power And Promise Of Immersive.* https://www.forbes.com/sites/forbesagencycouncil/2018/12/07/the-power-and-promise-of-immersive-technology-in-brand-storytelling/

Training, J. (2021). *Cloud Computing Interview Questions & Answers.* https://www.janbasktraining.com/blog/cloud-computing-interview-questions-answers/

Vertical farming. (n.d.). In *Wikipedia.* https://en.wikipedia.org/wiki/Vertical_farming

What is a smart home or building. (n.d.). https://internetofthingsagenda.techtarget.com/definition/smart-home-or-building?

Chapter 2
Analysis on Cryptographic Design Techniques of Stream Ciphers and Attacks

Megala G.
https://orcid.org/0000-0002-8084-8292
Vellore Institute of Technology, India

Prabu S.
Pondicherry University, India

Swarnalatha P.*
Vellore Institute of Technology, India

Venkatesan R.
https://orcid.org/0000-0002-4336-8628
SRM TRP Engineering College, India

ABSTRACT

Blockchain plays a vital role in enabling users to securely communicate with reliability and transparency. Cryptographic hashing is the underlying structure of blockchain empowering cryptocurrencies. With the rapid advancement of communication network, cipher algorithms are becoming increasingly important in information security. Multimedia data are widely used in various applications, including health sectors, military, entertainment, education, banking, and communication with recent improvements in internet technologies. Meanwhile, the application background for cipher algorithms is becoming gradually more complicated, posing problems to existing systems and necessitating the development of new appropriate designs. Stream cipher has been a prominent research area in cryptography, and it is appropriate for use in circumstances of huge data transmission or resource limits. This work provides a quick overview of conventional encryption algorithms based on stream ciphers and cryptographic prerequisites. At present, several attacks faced by stream cipher techniques are also discussed.

DOI: 10.4018/978-1-7998-9274-8.ch002

INTRODUCTION

Advanced technologies such as blockchain based Internet of Things (IoT), cloud computing, interconnected smart cities play a significant role in managing efficient urban development with interoperability. Blockchain play a major role in enabling participants to securely communicate with reliability and transparency. When exchange of data is huge in volume, there is a need of lightweight efficient cryptographic techniques. The stream cipher is a subset of symmetric cryptosystems that benefits from the speed and scale of hardware design. It has always been a popular and essential research area in cryptography, and it is appropriate for use in circumstances of huge data transmission or resource limits. Cipher algorithms are becoming increasingly important in the security of information with the rapid advancement of network and communication technology. Instantaneously, the application background for cipher algorithms is becoming increasingly complicated, posing problems to existing cipher algorithms and necessitating the development of new appropriate designs shown by (Dey & Sarkar, 2021). Cloud computing is becoming more popular for everyday storage and computing requirements. Its applications range from keeping confidential files to preserving government statistics, enterprise business data, information from Internet of things, remote sensing data, and relevant information from cloud-based social media apps, among others.

Cloud users, on the other hand, become sufferers of the security breaches exposed by cloud storage due to a lack of understanding and sufficient security measures. To prevent security risks, both the Cloud Storage service Provider (CSP) and the users should utilize suitable approaches and maintain a high standard. Multiple users transfer their information to store it to the cloud, which is spread across a number of devices. Cloud storage companies' insiders have insight to the information's contents. Users of cloud data have little direct control over their data once it has been uploaded, therefore maintaining confidentiality, privacy, and integrity are the most difficult challenges. Cryptography techniques can help Cloud Storage accomplish its data secrecy, data integrity, and data and privacy protection goals. To do this, user data is encrypted with a key before being outsourced, allowing for regulated, allowed access. Unauthorized users cannot access the data by employing powerful encryption procedures like stream ciphering and block ciphering. The security issues in cloud storage using stream ciphering algorithms are examined in this work in order to improve its security. The security measures created by various organizations are discussed further. This survey allows researchers to get a larger overview of online storage security vulnerabilities and identify outstanding issues that may be investigated further. Due to a flaw in the encryption process, an attacker might brute force his way into the network and launch a meet-in-the-middle (MITM) assault. Spoofing on anything is commonly involved in MITM attacks. Authentication mechanisms based on public-private key pairs, such as RSA and ECDH, can be employed at various stages of the stack to assist assure authorized communication.

This chapter provides a comprehensive review on stream ciphers, categorizes and encapsulates the design guidelines of classic stream ciphers in clusters, and provides a brief the strengths and drawbacks of diverse stream ciphers in relation to security and incorporation to facilitate stringent requirements and also provide a valid research framework for successful designs. Furthermore, it attempts to predict how stream ciphers will be designed in the future. Multimedia data including health information is frequently huge and expensive. Conventional block encryption methods including such Advanced Encryption Standard (AES), Data Encryption Standard (DES), RSA, BLOWFISH, and IDEA have been proven to be ineffective for multimedia applications in real-time applications due to their slowness in handling large amounts of data.

SPECIFICATION OF PRIMITIVES

Stream Ciphering Technique

Stream ciphers are fast, ease encryption procedures that encrypts the plaintext by handling bit by bit or byte at a time applying time-varying function. These stream ciphers use an initial internal state vectors on initializing constants, varying counters and number used once (nonce) values. Related to hardware, the stream ciphers are being frequently speedier when compared to block ciphers and demand reduced sophisticated equipment. Stream ciphers are ideal for quick implementations that use little resources. These two aspects assist the defensive in implementing resistance techniques on devices that lack the resources to install a block cipher. Though these stream ciphers has less error propagation which is negligible, which is noted as beneficial during the occurrence of problem while in transmission. Fixed permutations are applied to the whole n-bit block units in block cipher mode. Some of the block cipher modes of operations are actually a stream cipher. In modern days, stream ciphering are becoming more and more block oriented. Wireless signals that were more ideally predisposed to a streaming paradigm than data transmitted in bigger chunks of fixed size can also benefit from stream ciphers. Stream ciphers are basically classified into three different types such as synchronous stream cipher, Self Synchronized Stream Cipher (SSSC) and One Time Pad (OTP) ciphers. The generated stream of keys is independent of plain text as well as cipher text stream of bits in synchronized ciphers. For successful decryption of those cipher, the same identical key must be used at same condition. Decryption of cipher stream remain same without any effect when a single bit is alone changed. However the synchronization can be broken if a single bit is deleted or inserted and hence the decryption fails for the remaining ciphers. In asynchronous or SSSC ciphers generates the keystream based on the previous ciphers produced by this cipher and the key value.

Synchronous stream ciphers are said to have no error proliferation, on the other hand SSSC ciphers has some negligible error. The largest amount of improper decryptions would have been restricted to the stated number of prior cipher text characters if a character was updated. When there are only minor faults in the transmission, SSSC or asynchronous ciphers, could exacts the yield bitstream by itself. An efficient automated re-synchronization progression happens through the random key stream generator when a recipient identifies a transmission problem. In low-bit-error-rate situations, SSSC ciphers are often utilized. Since it is only dependent on a set number of preceding numbers, it is feasible to decrypt the data with negligible plaintext loss even though the ciphertext is modified in route. Stream ciphers as well as block ciphers are both examples of symmetric key cryptography, which means that a certain keys are used for en/decryption. Stream ciphers are similar to block ciphers in that they utilize conceptual tools. The major tool is substitution, which involves each of the plaintext bits being XORed with the generated keystream so as to produce the corresponding ciphertext bits. Because stream ciphers can't perform transposition on their own because they can't cache one block of input and transposition the information, effective stream ciphers generate a feedback process in which the current key source is dependent on the system's prior action. Information in transit, data at rest, and information in processing are the three types of data that must all be safeguarded. Data in motion is protected by encrypting data across the network. Data at rest, such as those on file systems, is typically protected by block ciphers. There exists a lot of stream ciphers such as Salsa20 (Ding, 2019), Chacha (Cheng et al., 2020), RC4, Trivium (Potestad et al., 2020), sosemanuk, Quad, ISAAC, fish, seal, snow, panama, rabbit, phelix, py, scream, etc., A few of those streams are listed in table 1 and are discussed with its fundamentals and operations involved in it.

Table 1. Details of stream ciphers

Stream Cipher	Key length bits	Initialization vector (IV) bits	Internal state bits	Possible attacks
Salsa20	256	64 bit nonce & 64 bit stream positing	512	Side channel attack
RC4	2048	No IV	2064	Key derivation
Phelix	256 + 128 bit nonce	12	-	Differential attack
Py	40 to 256	64	8320	Linear cryptanalytic attack
Panama	256	128	1216	Hash collision attack
Chacha	256	96 bit nonce & 32 bit counter	512	Side channel analysis
Trivium	80	80	288	Brute force attack

Salsa20

Salsa20 is a stream cipher constructed on the Add, Rotate and XOR (ARX) procedures, which include 32 bit addition operation, followed by consistent distances rotation processes on a 512-bit internal state and then Exclusive OR (XOR) operation. On a few architectures, it employs a novel round function that improves diffusion and performance. In micro-services architecture involving software implentation, the Add-Rotate-XOR (ARX) procedures prevent timing attacks. A Key value of 256 bits, a nonce value of 64 bits, and then a counter value of 64 bits are all mapped to a of the Salsa key stream consisting of 512 bits like a block by the core salsa function. Salsa20 and ChaCha have the unique feature of being able to quickly search towards any point in the real-time based application key stream value. A 4x4 matrix represents the internal state. The internal state is represented as X consisting 256 bit key value denoted as k1,k2,..k8 (i.e. eight words of key), 128 bit constant values, 64 bit streaming positions denoted as t1,t2 and 64 bit nonce values denoted as N1,N2 which is an additional streaming values.

$$X = \begin{bmatrix} x0 & x1 & x2 & x3 \\ x4 & x5 & x6 & x7 \\ x8 & x9 & x10 & x11 \\ x12 & x13 & x14 & x15 \end{bmatrix} = \begin{bmatrix} c1 & k1 & k2 & k3 \\ k4 & c2 & N1 & N2 \\ t1 & t2 & c3 & k5 \\ k6 & k7 & k8 & c4 \end{bmatrix}$$

Column round operation is performed on the internal states each column by performing circular rotation with 1 step, 2 step, 3 step and 4 step on second column, third column, fourth column and first column respectively. Similarly row round operation is performed on each rows of internal states and thus generates X'.

Main function of salsa is quarter round operation, QR() performing addition modulo and rotations on four word input producing a four word output. QR(w0,w1,w2,w3) is defined as

$$q_1 = w_1 \oplus \left(\left(w_0 + w_3 \right) \lll 7 \right) . q_2 = w_2 \oplus \left(\left(q_1 + w_0 \right) \lll 9 \right)$$

$$q_3 = w_3 \oplus \left(\left(q_2 + q_1 \right) \lll 13 \right)$$

$$q_0 = w_0 \oplus \left(\left(q_3 + q_2 \right) \lll 18 \right)$$

Where w0,w1,w2,w3 are the four words inputted to the QR() which produces four words as q0,q1,q2,q3 which is being replaced on the positions of w0,w1,w2,w3 respectively.

This quarter round operation is first performed on first column four words (x0,x4,x8,x12) of the internal state X. Then quarter round operation is performed on second column words (x5,x9,x13,x1), then on third column words (x10,x14,x2,x6) and on fourth column words (x15,x3,x7,x11). Subsequently, this quarter round operation is performed on first row words (x0,x1,x2,x3), then on second row words (x5,x6,x7,x4), then on third row words(x10,x11,x8,x9) and finally on fourth row words (x15,x12,x13,x14). The mentioned eight quarter round operations are performed in each round of salsa20 and finally generates a modified 512 bit internal state value denoted as X'. Salsa20 perform 20 rounds of operations shown in (Paul, 2020). This generated X' bit stream and the initial internal state value X are added to produce the 512 bit diffused keystream block. As this mixing round operations are invertible, the reverse process is performed to generate key for decryption. Fig. 1 represents the changes in bit positions of salsa20 operation after performing each round operation.

Figure 1. Differences in bit positions of salsa

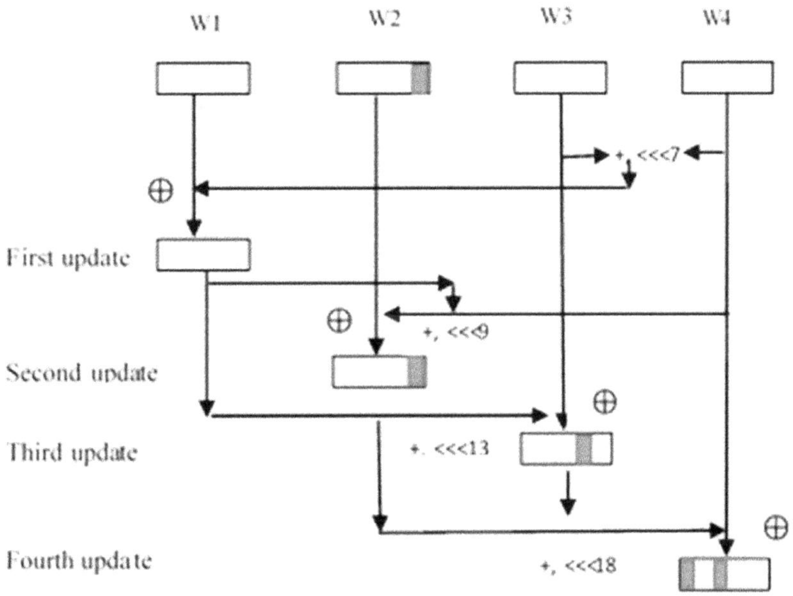

ChaCha

ChaCha is the successor of salsa20 stream cipher, which aim to increase the diffusion per round while achieving the same or slightly better performance. It is similar to salsa20. ChaCha's initial state includes a 128-bit constant, a 256-bit key, a 32-bit counter, and a 96-bit nonce, arranged as a 4×4 matrix of 32-bit words.

$$X = \begin{bmatrix} x0 & x1 & x2 & x3 \\ x4 & x5 & x6 & x7 \\ x8 & x9 & x10 & x11 \\ x12 & x13 & x14 & x15 \end{bmatrix} = \begin{bmatrix} C1 & C2 & C3 & C4 \\ k1 & k2 & k3 & k4 \\ k5 & k6 & k7 & k8 \\ t1 & N1 & N2 & N3 \end{bmatrix}$$

ChaCha replaces the Salsa20 quarter round function, QR(w1, w2, w3, w4) with

$$w1 = w1 + w2; w4 = w4 \oplus w1; w4 = w4 <<< 16;. w3 = w3 + w4; w2 = w2 \oplus w3; w2 = w2 <<< 12$$ .or

$$w1 = w1 + w2; w4 = w4 \oplus w1; w4 = w4 <<< 8;. w3 = w3 + w4; w2 = w2 \oplus w3; w2 = w2 <<< 7$$

The quarter round function in chacha introduces more diffusion quickly when compared to salsa20. It also performs optimization. Instead of rotating columns and rows, circular left rotation is performed on columns and diagonals of the keystream words. Here double rotation round operations are performed as quarter_round operation on first column words $(x0, x4, x8, x12)$. quarter_round operation on second column words $(x1, x5, x9, x13)$. quarter_round operation on third column words $(x2, x6, x10, x14)$. followed by four quarter_round operation on diagonals as $(x0, x5, x10, x15), (x1, x6, x11, x12), (x2, x7, x8, x13)$ *and on* $(x0, x5, x10, x15)$. Hence the bit positions are diffused after each round operations of chacha which is shown in Fig.2.

On platforms in which AES based acceleration is not supported by the CPU, ChaCha20 usually outperforms than more popular Advanced Encryption Standard (AES) algorithm. Most of the mobile devices which are having ARM processors prefers chacha encryption than AES for specific circumstances (Loup, 2017).

Figure 2. Differences in positions after the first round of chacha

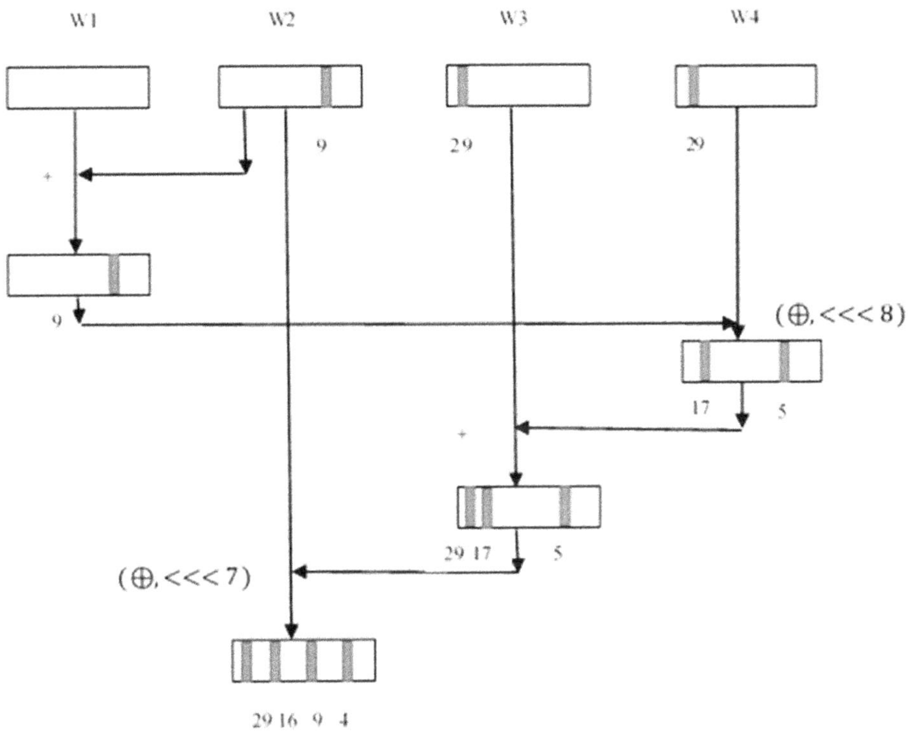

Stream ciphers have the following advantages:

- The encryption process is incredibly quick due to simple XOR operations that make the encryption and decryption process instantaneous.
- Stream ciphers have a low hardware complexity.
- It offers random access to ciphers by decrypting specified parts of ciphers because each plaintext bit corresponds to ciphertext bits in stream ciphers.

RC4 (Rivest Cipher 4)

RC4 is a one-byte stream cipher that encrypts plain text one byte at a time. The key scheduling method and the pseudo random generation algorithm are the two fundamental mathematical notions that it is based on. It creates a keystream, which is used for encrypting and decrypting shown in (Madarro et al., 2021). The keystream generation procedure uses an internal state consisting of two 8-bit index part pointers and a key scheduling algorithm to permute different key lengths up to a maximum of 2048 bits. The key stream is then subjected to pseudo random generation. Permutation is used to initialize the internal state array A for 256 iterations. Here the keylength varies from one to 256 bytes. The bytes of key are mixed by adding two pointing array values A[i] and A[j] with modulo 256. The sum of these two values is used as the index positing pointing key stream byte K value (i.e. A[(A[i]+A[j]) mod 256)]. The values of A[i] and A[j] are swapped. The two aforementioned steps are performed for 256 iterations.

This generated internal state is modified by pseudo random generation to produce key stream byte for each iteration. The steps involved in pseudo random key stream generation are

Step 1) Initialize the pointers i and j to zero.

Step 2) For each iteration:

Step 2.1) Update i value by incrementing and taking modulo 256 on it.

Step 2.2) Update j value by adding it with A[i] and perform modulo 256 (i.e. j=(i+A[i]) mod 256).

Step 2.3) Swap the values of A[i] and A[j].

Step 2.4) Generate Keystream value, K = A[(A[i]+A[j]) mod 256)] and output K value.

Thus, for each cycle, a key stream is formed, which is XORed with the plaintext stream to produce the ciphertext in the encryption process. Although the RC4 stream cipher is incredibly efficient in software, it is not compatible with 32-bit architecture. Several stream ciphers use linear-feedback shift registers (LFSRs), which are effective in hardware but inefficient in software. Because RC4 does not employ LFSRs, it is appropriate for software implementation because it only requires byte transformations. The state array, A[0] until A[255], the key, key[0] through key[k-1], and i, j, and K are all stored in 256 bytes of memory.

A bitwise AND with 255 can be used to perform a modular reduction of a value modulo 256. In RC4, there is no separate nonce value, and a single key stream is used to encrypt several plain streams, making it vulnerable to attacks like bit twisting and associated key attacks. To improve the protocol's security, the hashing of the key with the nonce is used to create a new RC4 key.

AES – CTR Mode Operation

AES-CTR is a type of stream cipher in which the keystream is generated by the encrypted data of a counter. To encrypt and decrypt data in blocks of 128 bits, the AES algorithm required distinct keys of 128/192/256 bits. AES is a cryptographic algorithm that is employed in both software and hardware to protect digital data in various forms such as data, sound, video, and so on from assaults or snooping. The output of the Counter is utilized as the input for the Encryption unit in CTR (counter) mode, and an initialization vector has been used to initialize the counter. The input data is encrypted by XORing it all with the Encryption component's result as shown in Fig.3. The data that results from this procedure is known as ciphertext. Decryption is the reversal of the encryption process. It encrypts data with longer keys, such as 128, 192, and 256 bits. As a result, the AES algorithm becomes more resistant to tampering. It is the most widely used security protocol for a wide range of applications, including wireless transmission, banking transactions, e-commerce, and protected storage systems.

Figure 3. AES encryption in counter mode

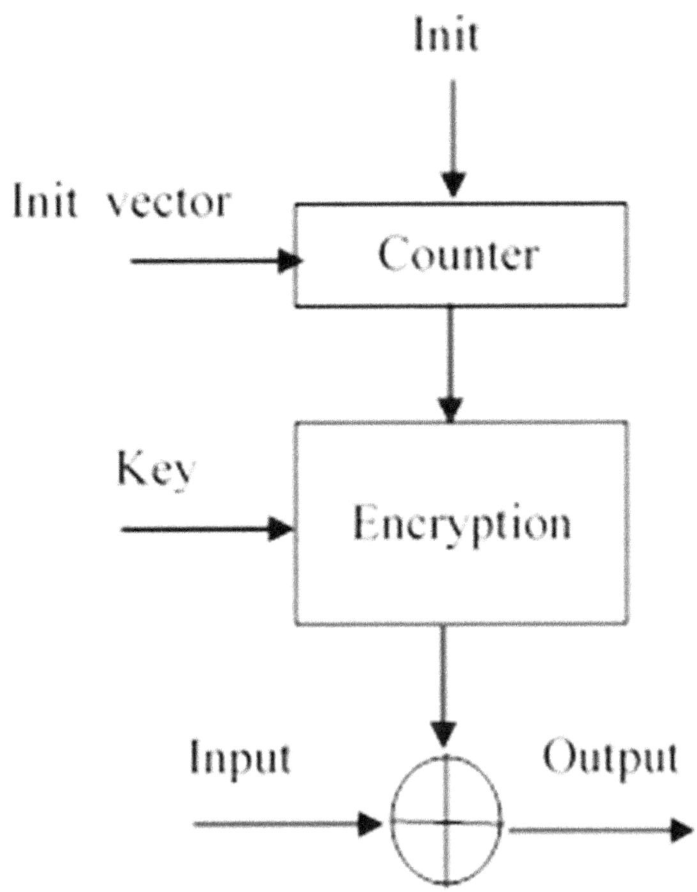

In a stream cipher, the same secret key cannot be used more than once. Changing the key regularly is a serious problem. Modern day stream ciphers use an Initialization vector (IV) which need not be random or unpredictable. Strength of stream cipher does not depend on IV which can be known to adversary. The constraint is that the (key, IV) pair cannot be used twice. Several IVs can be used for the same key which eases the problem of supplying fresh keys.
AES's disadvantages include:

- Its overly simplistic algebraic structure.
- Each block gets encrypted in exactly the same way.
- Software implementation is difficult.
- When it comes to performance as well as security, AES encryption in counter (CTR) mode is challenging one for software implementation.
- Despite hardware support and high memory utilization, AES is sluggish and fails in cache timing attacks.

At comparable speeds, AES demands more hardware. And from the other extreme, unlike many other stream ciphers (such as Trivium and RC4), it permits random access to the keystream and also enabling parallelization (like ChaCha20 does).

ATTACKS ON STREAM CIPHERS

Mostly, XOR operation is operation is performed with the plaintext and key stream to produce cipher text in stream cipher algorithms. If utilized correctly, stream ciphers, which combine plaintext bits with a cipher bit stream using an XOR operations, can be exceedingly secure. These are susceptible to attack, though, if certain safeguards are not taken in (Salam et al., 2021). Effective decryption must never be depended on to show authenticity and also the keys should not ever be utilized repeatedly.

Guessing Attack

The adversary make some guesses on the state values corresponding to different values. Based on the determined states, the resulting key sequence is compared with the executing keystream. Bits are retrieved based mostly on parity operations on the data stream in a guessing attack. It's tough to guess the extraction of bit shown in (Jiao et al., 2020) if computing the quadratic residual modulo is tricky.

Exhaustive Search Attack

Searching all possibilities of keys in all possible states to match the resulting cipher with the keystream in an exhaustive manner. Adversary randomly selects the candidate state and checks whether it produces the cipher output and then the adversary tries to find the initial state. Pseudo random generator is used to generate the keystream generator with the vector space of dimension GF(2).

Algebraic Attack

Algebraic attack is based on the Linear Feedback Shift Register (LFSR). Based on the information gained from LFSR, the adversary tries to find the internal state value. Adversary tries to generate the linear equation system and solve it to find the values of secret key bit stream. The main goal of characterizing the security of cryptographic algorithm is to find solution on solving and simplifying the multivariate non-linear equations. Hence algebraic attack becomes complex on establishing the difficult significant multivariate equations and so it is a challenging task.

Meet in the Middle Attack

The adversary splits the ciphers into two or more sub ciphers and then performs MITM attack on each sub cipher to guess intermediate values. The adversary also performs partial encryption, and then constructs the propagating patterns from the intermediate ciphers so as guess the partial keys. The nonlinearity involved within the keys are guessed in a multi-dimensional way.

Bit Flipping Attack

If some part of the plain text is known to adversary, then he performs meet in the middle attack in an exhaustive manner by XOring that portion with the cipher text so as to obtain the key. This can be prevented by comprising message authentication codes together with cipher to detect the tampering.

Side Channel Scan Attack

In side channel attack, the adversary tries to attack the implementation part rather than attacking the logic of the algorithm. The adversary tries to obtain the correspondence of the bit positions in scan chain with the hardware flip flops so as to gain the knowledge of the internal state of cipher. Once initial state is obtained, the adversary inverts the state to find the key. The adversary tries to collect the information of what the computing hardware performs and when it is performing cryptographic operation and then reverse engineering is performed on the obtained information.

Correlation Attack

Exploiting the existing statistical dependency among the keystream as well as LFSR's output is known as correlation attack. This attack is implemented on nonlinear generators to recover internal state bits.

Distinguishing Attack

The attacker distinguishes the Pseudo random generator output and the random bit strings. Then the attacker tries to recover the keys based on the input seed value given to the pseudo random generator and the output it produces.

Chosen IV Attack

Based on the key and initialization vector (IV), generally stream ciphers produce the pseudo random sequence from PRNG. Mainly, IV is used to generate randomness in the bit stream without modifying the secret key value. It is insecure if the same IV is used more than once. This chosen IV attack exploits the stream cipher weakness in key scheduling approach. In this attack, the adversary tries to extract information from memory and the internal state bits of LFSR.

Cube Attack

Cube attack is the relative attack used to recover the secret key value by exploiting the presence of lower degree polynomial representations of the outputs as a function of key and plain text stream. Adversary computes the subsets of the plain text and perform sum operation among the bits thus deriving the linear equations and solving it to recover key.

DISCUSSIONS

There are typical applications which is mainly used to preserve privacy in blockchain. This technology has the features such as multiple sign transactions, trusted third party, reliability, zero knowledge proofs, anonymity, decentralization, privacy, etc. Table 2 shows the drawbacks of the applications of blockchain based technology.

Table 2. Blockchain technology applications

Application	Features concentrated	Drawbacks
Cryptonote, Confidential transaction process using paillier	Operating encrypted information	Storage cost and computation cost is high.
Monero	Anonymity and exactness	Less scalable
Coin shuffle	Multiple sign	Information leakage
Coin join	Multiple sign	Vulnerable to denial-of-service attack
Mixcoin	Liability	More time consuming
Xim	Mixes coins of two parties with many rounds	Computation cost is high
Blot	Zero knowledge proof	Information leakage of transaction contents
Tumble bit	Protect privacy	Transaction data leakage

Encrypted digital crypto currencies using blockchain are Monero, zerocash and zerocoin discussed by (Wang et al., 2020). These currencies are mainly used to provide the anonymity by hiding complete information about transactions. Considering the confidentiality, privacy and integrity are ensured using better block or stream cipher approaches. The stream cipher is better suited to real-time handling of huge chunky multimedia including speech, video, and images. Chaos-based procedures could be used to create a key stream with the desired qualities for cryptographic operations. They're attracting a lot of attention because of their prospective benefits, including as strong nonlinearity and less cost. Stream ciphers are commonly utilized in situations when both speed and simplicity are required. If a 128-bit block cipher, including certain AES, were used instead of a stream cipher to encrypt 32-bit messages, 96 bits of additional padding bits would be retained.

Encryption techniques either in block cipher mode or stream cipher mode provides high confidentiality for the information shared among the users performing communication. Symmetric ciphers such as AES, a traditional encryption algorithm used in many communications among IOT. Though it is secure, there is a computational overhead in securing large scale multimedia content for Internet of Multimedia Things and real time applications (Buchanan, 2017 and Vasileios, 2018). So many authors prefer stream ciphering for real time IoMT applications for its high bitrate overhead and computational complexity with relative bandwidth utilization. Hence researchers are interested to design light weight cryptographic algorithms to provide security on multimedia contents in an IOT based communications for authorized users. Those users are being authorized in a challenge and response manner using zero knowledge proof and record of transactions of users or their access log are stored in chain of blocks. Cryptographic concepts play a major role in providing secure end to end communications. Substitution, permutation and XOR operation are the three major operations used by many authors to provide light weight secure algorithms. Chaotic based diffusion are also performed by (Lin et al., 2018) in a multi-dimensional way by researchers to prevent the keystream recovery. Chaotic involves many maps such as Arnold cat map, logistic map, bakers map, skew tent map, etc., to map the finite field GF(2) to GF(2) so as to provide high diffusion. Chaos based systems including the non-linear registers are designed by (Ding et al., 2019) to create the novel stream cipher providing good characteristics with less computation complexity.

Applications of blockchain technologies are, a secure electronic voting device developed by (Rathee et al., 2021) involving trusted IoT devices. Performance metrics such as trust, message modification, throughput, latency in authentication and attack analysis are measured. A smart automation system involving several IoT devices interconnected use the cloud services and are being managed by controllers. There exists security challenges in this architecture (Alotaibi, 2019) such as data privacy, anonymity, availability and integrity violations. Blockchain concepts are mainly used in financial services (Wazid et al., 2020), to pay money automatically by the customers in a secured manner. Intelligent transport systems are build comprising vehicles, roadside unit, on board units, fog or cloud servers with several sensors. This system provides travel communications using blockchain technology in decision making with high accuracy. Smart health care systems are developed using blockchain mechanism (Rahman et al., 2020) to provide high confidentiality of patient records, doctor and medicine information. These records are need to be encrypted using cryptographic algorithms and tagged in the blockchain. Hence lightweight encryption protocols using stream cipher approaches are to implemented along with key management and access, authentication and access control policies.

CONCLUSION

Blockchain for IOT trust infrastructure necessitates high computing resources with high throughput. Efficient cryptographic techniques are involved to necessitate lightweight blockchain based solutions. A quick overview of numerous conventional encryption algorithms is provided based on stream ciphers, as well as cryptographic prerequisites for the design of chaotic-based cryptosystems. The basic idea of implementation of stream ciphers are reviewed with its mathematical operations. Attacks on stream is also discussed. Other than brute force attacks, there exists effective side channel attacks, algebraic attack on the stream ciphers. The stream ciphers Salsa and Chacha involving quarter round functions provide more diffusion and thus it is more secure compared to other stream ciphers.

REFERENCES

Alotaibi, B. (2019). Utilizing blockchain to overcome cyber security concerns in the internet of things: A review. *IEEE Sensors Journal*, *19*(23), 10953–10971. doi:10.1109/JSEN.2019.2935035

Buchanan, W. J., Li, S., & Asif, R. (2017). Lightweight cryptography methods. *Journal of Cyber Security Technology*, *1*(3-4), 187–201. doi:10.1080/23742917.2017.1384917

Cheng, S., Wang, L., Ao, N., & Han, Q. (2020). A selective video encryption scheme based on coding characteristics. *Symmetry*, *12*(3), 332. doi:10.3390ym12030332

CryptoLUX. (n.d.). *Lightweight cryptography.* https://www.cryptolux.org/index.php/LightweightCryptography

Dey, S., & Sarkar, S. (2021). A theoretical investigation on the distinguishers of Salsa and ChaCha. *Discrete Applied Mathematics*, *302*, 147–162. doi:10.1016/j.dam.2021.06.017

Ding, L. (2019). Improved related-cipher attack on Salsa20 stream cipher. *IEEE Access: Practical Innovations, Open Solutions*, *7*, 30197–30202. doi:10.1109/ACCESS.2019.2892647

Ding, L., Liu, C., Zhang, Y., & Ding, Q. (2019). A new lightweight stream cipher based on chaos. *Symmetry*, *11*(7), 853. doi:10.3390ym11070853

ECRYPT. (n.d.). *eSTREAM: the ECRYPT stream cipher project.* https://www.ecrypt.eu.org/stream/

Jiao, L., Li, Y., & Hao, Y. (2020). A guess-and-determine attack on SNOW-V stream cipher. *The Computer Journal*, *63*(12), 1789–1812. doi:10.1093/comjnl/bxaa003

Knutson, P. (2020). *Cryptanalytic attack on Salsa20.* https://github.com/catsymptote/Salsacryptanalysis

Lin, Z., Yu, S., Feng, X., & Lü, J. (2018). Cryptanalysis of a chaotic stream cipher and its improved scheme. *International Journal of Bifurcation and Chaos in Applied Sciences and Engineering*, *28*(07), 1850086. doi:10.1142/S0218127418500864

Loup's. (2017). *The design of ChaCha20.* https://loup-vaillant.fr/tutorials/chacha20-design

Madarro-Capó, E. J., Legón-Pérez, C. M., Rojas, O., & Sosa-Gómez, G. (2021). Information Theory Based Evaluation of the RC4 Stream Cipher Outputs. *Entropy (Basel, Switzerland)*, *23*(7), 896. doi:10.3390/e23070896 PMID:34356437

Mavroeidis, V., Vishi, K., D, M., & Jøsang, A. (2018). The impact of quantum computing on present cryptography. *International Journal of Advanced Computer Science and Applications*, *9*(3). Advance online publication. doi:10.14569/IJACSA.2018.090354

Potestad-Ordóñez, F. E., Valencia-Barrero, M., Baena-Oliva, C., Parra-Fernández, P., & Jiménez-Fernández, C. J. (2020). Breaking Trivium Stream Cipher Implemented in ASIC Using Experimental Attacks and DFA. *Sensors (Basel)*, *20*(23), 6909. doi:10.339020236909 PMID:33287234

Rahman, M. A., Hossain, M. S., Islam, M. S., Alrajeh, N. A., & Muhammad, G. (2020). Secure and provenance enhanced Internet of health things framework: A blockchain managed federated learning approach. *IEEE Access: Practical Innovations, Open Solutions*, *8*, 205071–205087. doi:10.1109/ACCESS.2020.3037474 PMID:34192116

Rathee, G., Iqbal, R., Waqar, O., & Bashir, A. K. (2021). On the Design and Implementation of a Blockchain Enabled E-Voting Application Within IoT-Oriented Smart Cities. *IEEE Access: Practical Innovations, Open Solutions*, *9*, 34165–34176. doi:10.1109/ACCESS.2021.3061411

Salam, I., Ooi, T. H., Xue, L., Yau, W. C., Pieprzyk, J., & Phan, R. C. W. (2021). Random Differential Fault Attacks on the Lightweight Authenticated Encryption Stream Cipher Grain-128AEAD. *IEEE Access: Practical Innovations, Open Solutions*, *9*, 72568–72586. doi:10.1109/ACCESS.2021.3078845

Wang, D., Zhao, J., & Wang, Y. (2020). A Survey on Privacy Protection of Blockchain: The Technology and Application. *IEEE Access: Practical Innovations, Open Solutions*, *8*, 108766–108781. doi:10.1109/ACCESS.2020.2994294

Wazid, M., Das, A. K., Shetty, S., & Jo, M. (2020). A tutorial and future research for building a blockchain-based secure communication scheme for internet of intelligent things. *IEEE Access: Practical Innovations, Open Solutions*, *8*, 88700–88716. doi:10.1109/ACCESS.2020.2992467

Chapter 3
The Past, Present, and Future of Intelligent Transportation Systems:
Konya Metropolitan Municipality Applications

Hatice Canan Gungor
https://orcid.org/0000-0002-5563-9552
Necmettin Erbakan University, Turkey

ABSTRACT

Transportation is important for all and sundry. But it's bedeviled with some problems for the environment such as pollution. Intelligent transportation systems (ITS), which emerged to solve this problem, focus on important issues such as increasing safety in traffic. ITS across the world has come to the fore in recent years. In this study, ITS was researched including how it is used in Konya. It serves the public in almost all types of transportation used with the developed IPTS (intelligent public transportation system). The user can instantly access the station and vehicle information of the transportation system to be used through the menu. The system, which is integrated with road, rail, and air transportation, serves the user to avoid any problems from the beginning to the end. All information about online card filling, lost property status, and transportation fees can be accessed through the system. Undoubtedly, Konya has introduced the IPTS to provide a more practical public transportation service.

INTRODUCTION

Recently, the increase in population and industrialization has begun to create both advantages and disadvantages in people's daily life. Transportation, which is one of these disadvantages, caused a decrease in the quality of life of people in their daily lives. Intelligent Transportation Systems (ITS), which emerged to solve this problem, focuses on important issues such as increasing safety in traffic, increasing mobi-

DOI: 10.4018/978-1-7998-9274-8.ch003

lization, protecting the environment by preventing unnecessary energy use and reducing travel times. According to the studies conducted, 25% of the CO2 gas generated as a result of fuel use originates from transportation. It is estimated that this number will increase in the future. ITS across the world come to the fore in recent years Turkey has also entered into important topics.

ITS' infrastructure, pioneered the big cities of Turkey has started to be formed in all of our cities. Istanbul and Konya, which are the leading cities, compete with other world cities in this regard. Using the topographical structure of the city as an advantage, Konya has made and is making serious investments in this regard. Thanks to cameras and sensors located at most points of the city, operations such as intersection management, traffic safety and density control are carried out. These data are instantly shared with national and local channels and mobile applications to raise the awareness of passengers. In this study, ITS was researched how it is used on Konya city. The answers was focused on application was enough or not for the public. ITS in the context of the world-leading Konya Turkey stands out with this innovative attitude. Continuing on its way with the slogan of technological municipalism, the metropolitan municipality guides other world cities. The metropolitan municipality, which has managed to use the topographic advantage of the city very well in intelligent city and Intelligent transportation, attaches importance to its investments and infrastructure works. It serves the public in almost all types of transportation used with the developed IPTS (Intelligent Public Transportation System). The application has a design that users of all ages can easily use. The user can instantly access the station and vehicle information of the transportation system to be used through the menu. It can monitor the location of the vehicle and the traffic density life. The system, which is integrated with the road, rail, and air transportation, serves the user to avoid any problems from the beginning to the end. All information about online card filling lost property status and transportation fees can be accessed through the system. It is clearly that Konya Metropolitan Municipality has introduced the Intelligent Public Transportation System (IPTS)to provide a more practical public transportation service. With IPTS, it is possible to access where the public transportation vehicles are, how many minutes they will arrive at the station, the public transportation lines passing through the station, route and stop information, and many more information about public transportation. IPTS provides great convenience with its variety of access channels and the citizens use IPTS more than 55 million times annually. The Intelligent Traffic Management System, implemented at 86 junctions in the city center of Konya, uses data such as vehicle density and average vehicle speed to manage traffic lights. Thanks to the Intelligent Traffic Management System, the vehicles on the intersection arms are counted instantly by means of cameras, and the green light times at the intersection are automatically adjusted within the determined limits in line with these counts. The system ensures that the directions with high vehicle density are green for a longer period of time, while minimizing the duration of the green light in the directions where the vehicle density is low.

Thanks to the system; The number of vehicles waiting at the intersections, the waiting times of the vehicles at the intersection and energy consumption are reduced and driving comfort is increased.

In line with the strategies of "Promoting Non-Motorized Transportation", "Reducing Automobile Addiction" and "Traffic Demand Management" determined in the Konya Transportation Master Plan, the Konya Cycling Master Plan has been prepared in order to create an uninterrupted network in the city center for the bicycle transportation network for bicycle transportation.

All these studies have also reduced the environmental pollution caused by transportation. Thus, the sustainable environment phenomenon is promising for the province of Konya.

The rapidly increasing world population has brought many problems with it. The disproportionate population growth, which is one of these problems, has caused a serious decrease in the quality of life in city centers. Transportation, which can be described as the common action of every person, started to pose various problems in the daily life of people with the increase in population (Karami & Kashef., 2020). The fact that urban transportation takes a long time, traffic congestion and poor quality of the time spent during the journey have negatively affected people (Li, Liang, Hong, & Zhang., 2021).

Advancing technology and the further increase of the problem have caused the birth and development of Intelligent Transportation Systems. In the field of transportation, the main mission of Intelligent Transportation Systems has been to meet the increasing demand in both passenger and freight transportation services in a more efficient, safe and environmentally friendly manner (Meric, 2018). Studies and developments in this context reveal the importance and future of (ITS).

ITS focuses on important issues such as increasing safety in traffic, increasing mobilization, protecting the environment by preventing unnecessary energy use and reducing travel times. According to studies, 25% of the CO2 gas generated as a result of fuel use is from transportation (Guler, 2015). This ratio is expected to increase further in the future.

USAGE AREAS OF INTELLİGENT TRANSPORTATION SYSTEMS

ITS was originally designed to improve the operation and service quality on highways. Over time, with the effect of different ideas and studies, it acquired side missions and aimed to solve urban problems. Personal vehicle and pedestrian information has been developed over time for public transport, ports and subways (Silva, Khan & Han, 2018).

Figure 1. Traffic detector (https://atus.konya.bel.tr.)
gain and ecological improvement is achieved by using the system.

Playing an active role in the management of highway systems, ITS uses cameras and sensors along the way. By using these equipment, the monitoring of the traffic conditions, the determination of the factors in the traffic and preventing the traffic flow and the emergency management are done quickly by the teams in charge. Used magnetometers, microwave radar, ultrasound and infrared teams provide real-time tracking (Li, Boonaert, Doniec & Lozenguez, 2021). This technology allows to avoid traffic problems that will endanger passenger safety and provides quality transportation. Thanks to the data shared with the drivers using the road on local channels and the internet, the quality of transportation is increased.

Thanks to the cameras used in the arteries in city centers, the traffic density is detected and the system automatically adjust ITS the traffic lights. There will be a decrease in waiting time and a decrease in the number of waiting vehicles, so economic

These cameras are used for regulating and controlling urban traffic signals. The junctions in each artery are controlled by these signal circuits and coordinate the change operation. The peak hour determined over time is planned and applied by various algorithms.

Thanks to the systems installed at the motorway entrances and exits, tolls are collected electronically. This process reduces the need for cash, minimizes waiting time, road usage, etc. provides convenience in collecting data. The process of determining the number of vehicles using the road and determining the location of some vehicles on the road becomes easier. Identifying violations and applying penal sanctions becomes easier with license plate information and images taken from the devices.

There are systems used for the management of the vehicle fleet and tracking the vehicles in public transportation. Thanks to these systems, the position of the vehicle, its speed while in motion and the violations it performs are instantly sent to the management. Thus, passenger safety is provided and the effect of management on the driver increases (Jan, Farman, Khan, Talha, & Din, 2019). It is used to inform the passengers thanks to the data obtained instantly and allows more accurate programs to be made. Thanks to the sensors in the vehicle, the fuel-oil level, routine maintenance status and all parameters related to the vehicle can be monitored.

With the data obtained online, the management of the intersections with a road-rail combination is made. It is very important to prevent accidents at the road-rail junction points, which cause great loss of life and property when neglected. These systems provide safe travel by disconnecting the two roads in a planned way with the help of the signaling circuit.

ITS USAGE IN THE WORLD AND TURKEY

It has been observed that there is an increase in transportation problems at the same rate depending on the population density in the city centers. Traffic jams, delays, accidents, and safe driving problems arise as a result of the dense population. ITS, which is used as a solution to these problems, is becoming widespread and developing (Zadobrischi, Cosovanu, & Dimian, 2020).

Standing out with its high population and narrow residential area, Japan is one of the countries that actively use ITS. With a population density of 12 times that of the USA, Japan constantly improves its ITSin order to prevent transportation problems and traffic congestion (Wu, Levinson, & Sarkar, 2019). Japan, which has made great innovations in public transportation systems, sets an example for other countries in this regard (Clarisse, 2014). Thanks to the different sensors in the vehicle, it provides the driver with services such as predicting the intensity of the route to be used by the driver and making the road choice accordingly, informing the driver of the parking lots available in the vicinity.

With the Dynamic Intelligent Speed Adaptation system pioneered by Belgium, it helps the driver to transmit the changing speed limits on the road route and not to exceed this limit. The driver, who knows different weather conditions, vehicle information, and speed limits that vary according to the region of the road and carry out accordingly, gains time and prevents accidents (Lai & Carsten, 2012).

In-vehicle indicator and route studies, whose ITS were completed between 1973-1979, received full support from the Advanced Information and Communication Council formed by 5 ministries in 1994. The development of ITS, for which 670 million dollars was allocated at that time, has come rapidly until today (YARDIM & AKYILDIZ, 2005). Over one million in-car navigation systems were sold by 25 Japanese companies in 1996 (Meric, 2018). Similar studies continued in America and Europe. Thanks to this technological competition, there have been serious developments in ITS.

The first steps were taken in 1992 in Turkey ITS the electronic toll collection system. Thanks to this system, the pricing of vehicles according to the number of road users and the vehicle class were made automatically (ERSOY & BÖRÜHAN, 2015). However, thanks to the emergency phones placed on the highway, played a major role in quickly eliminating problems such as accidents and breakdowns. The Automatic Pass System(OGS), which was installed on the Fatih Sultan Mehmet Bridge in 1999, was installed in order to reduce the loss of time and provide fast transition and the expected performance was achieved. Following this trial, the system was implemented in 129 lanes at 49 points across the country. In addition to OGS, the Pass System with Card system started to be implemented. Special cards have been designed for drivers to carry in their wallets and to load the amount they want. Thanks to these cards, which have a maximum card reading speed of 0.5 seconds, passages were made without the need for officers (National Smart Transport Systems Strategy Document 2013-123 and Action Plan 2013-2015).

THE CONCEPT OF SUSTAINABILITY IN TRANSPORTATION AND KONYA

It is essential to establish a sustainable relationship between these two main factors in order to maintain a healthy and balanced relationship between society and nature. It is a known fact that one of the important causes of global warming, environmental disasters and many disasters is the damage to this balanced and sustainable relationship (Roh, & Jeon, 2021). The concept of sustainability brings with it an understanding of urbanism. The main goal should be to provide them with an environmentalist approach, taking into account the traditional form, industrialization and development demands of the city. Therefore, it is important to consider the concept of sustainability when making transportation decisions in city planning (Beatley, 2006).

The concept of sustainability and the ITS are intertwined. In our country, especially in the province of Konya, it is necessary in terms of sustainability to construct and encourage the use of bicycle paths and to make them a part of uninterrupted transportation with the design of ITS. Sustainable transportation principles by Kuyumcu in 2017, socially;

- Equality
- Accessibility
- Health
- Safety and Comfort
- Individual responsibility
- Continuous and integrated planning

- Environmentally;
- Preventing air, sound and visual pollution
- Land and resource use
- In terms of economy;
- Cost accounting
- classified as.

Konya metropolitan municipality considers the demand of the people in transportation planning. During this study, 550 people were asked about their suggestions about bicycle paths.

A cross-section of notable recommendations made by survey respondents;

"I love cycling. I recommend that every child between the ages of 8-10 be given a free bicycle. I want an underground bike path."

"There are many bike lanes in our city, but none around our house. I want it done."

"Phosphorus directional signs should be increased on bike paths. Bicycle competition should be held every year and free bikes should be given to the first 100 people."

"Bicycle parking areas should be large in busy locations close to business centers, and there should be safe deposit boxes for a small fee and points where possible to take a shower."

"Vehicles and pedestrians use the bike path, but there is no enforcement. There should be sanctions."

"Because the use of bicycles prevents global warming, bicycles should be used instead of vehicles."

"Most of the writings on the bike paths fade over time."

"Sometimes people drive over the bike path. I almost got hit by a passing car because of an irresponsible person. They also park on the bike path."

"Special parks should be built for bicycles. For example, bicycle parking areas can be built right next to gas stations."

With these suggestions, the city's transportation policies are developing in terms of sustainable transportation.

KONYA METROPOLITAN MUNICIPALITY ITS APPLICATIONS YESTERDAY AND TODAY

ITS in the context of the world-leading Konya Turkey stands out with this innovative attitude. Continuing on its way with the slogan of technological municipalism, the metropolitan municipality guides other world cities. The metropolitan municipality, which has managed to use the topographic advantage of the city very well in Intelligent city and Intelligent transportation, attaches importance to its investments and infrastructure works.

It serves the public in almost all types of transportation used with the developed IPTS (Intelligent Public Transportation System).

Figure 2. IPTS interactive (https://atus.konya.bel.tr.)

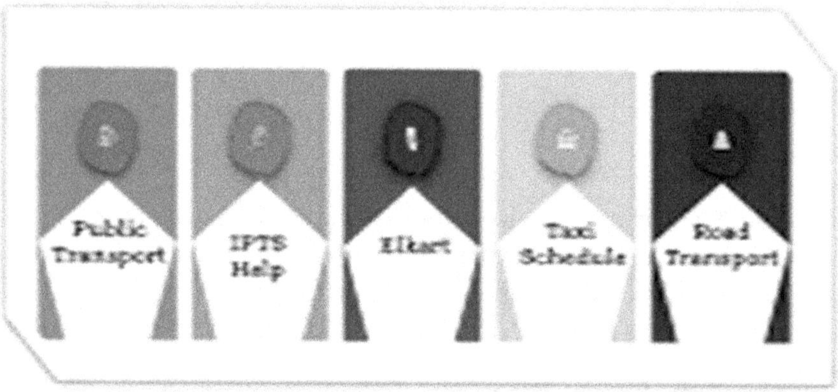

The application has a design that users of all ages can easily use. The user can instantly access the station and vehicle information of the transportation system to be used through the menu. It can monitor the location of the vehicle and the traffic density life. The system, which is integrated with the road, rail, and air transportation, serves the user to avoid any problems from the beginning to the end. All information about online card filling lost property status and transportation fees can be accessed through the system.

With cameras located at different points of the city, the density of the main arteries of the city is automatically determined by the system and the light durations are adjusted. Thanks to this system, the waiting times of the passengers have been reduced and the country has contributed economically and environmental cleaning has been supported.

The structure of the city has a great effect on the preference for bicycles, which are one of the most frequently used transportation vehicles in Konya. It is important that bicycle transportation, which is shown as the cleanest means of transportation, is so widespread that the city is clean and has little damage to the world. The metropolitan municipality, which has an environmental awareness, is working to make its use more widespread (Turkey Cycling Road Master Plan Report, August., 2019 https://ulasim.gov.tr.). Bicycle paths that completely surround the city provide more comfortable and safe usage. Thanks to the investments made in bicycle paths and their surroundings, it increases its usage rate day by day. It has been a popular healthy activity that the public can do as a family. Thanks to the cycling points located in most parts of the city, people can rent a bicycle at low prices even if they do not own a bicycle.

It is planned to increase pedestrian and vehicle safety thanks to the signaling systems in traffic lights and pedestrian crossings. It has been observed that the transmission of traffic light durations to pedestrians and drivers minimizes the negative psychological effects. In addition, traffic lights with audible warnings for visually impaired citizens facilitate the daily life of our disabled citizens.

THE FUTURE OF ITS AND THE CONCLUSION

It is an undeniable fact that ITS, which is spreading and developing rapidly around the world, will affect every aspect of our lives in the future. The further development of the system, whose primary mission is to increase the quality of life and prevent deaths, is desired and expected by the world. As a result of the

Figure 3. Bike stop (https://atus.konya.bel.tr.)

age we are in, artificial intelligence applications and mechanization control have passed into the hands of robots rather than humans. In this system, it will be much easier for ITS to become widespread and to create an integrated system. It is thought that this integrated system will provide a great advantage to humanity to prevent the deterioration of the ecological balance of the world and to avoid natural disasters as a result of age. Connecting autopilot vehicles, which are being used gradually, with a common system and controlling the whole flow by a common center, will greatly prevent accidents and deaths.

According to estimates, it is thought that the world population will exceed 10 billion in 2050. For this reason, it is very important to develop and disseminate ITS for a more livable planet. The fact that we should be aware of the country is that every investment we make for ITS will return us greatly. According to the research, during road transportation;

- The annual fuel cost is 30 billion TL
- Vehicle depreciation cost 22.3 billion TL
- Damage due to accidents is 10.3 billion TL
- 1-year time cost of trips made is 60 billion TL
- The annual cost of emissions from transport is 4.2 billion TL.

In the light of these data, it is thought that with the spread of ITS, approximately 33 billion TL can be gained (2020 Annual Report General Directorate of Highways, https://kgm.gov.tr.). Considering that an annual fund of 2-3 billion TL will be sufficient for the development of ITS, it is seen that it is a very insignificant amount compared to the profit to be made. ITS should be supported and developed to prevent both material damage and loss of life.

REFERENCES

Beatley, T. (2006). Green Urbanism in European Cities. *The Human Metropolis: People and Nature in the 21st-Century City, 297.*

Ersoy, P., & Börühan, G. (2015). Intelligent transportation systems and their applications in road transportation industry in Turkey. *The International Conference on Logistics & Sustainable Transport 2015.*

Guler, M. G. (2015). *2.1.7. Technology Infrastructure Analysis H1*. Yildiz Technical University.

Jan, B., Farman, H., Khan, M., Talha, M., & Din, I. U. (2019). Designing a smart transportation system: An internet of things and big data approach. *IEEE Wireless Communications*, *26*(4), 73–79.

Karami, Z., & Kashef, R. (2020). Smart transportation planning: Data, models, and algorithms. *Transportation Engineering*, *2*, 100013. doi:10.1016/j.treng.2020.100013

Konya intelligent transportation system applications. (2019). Retrieved from https://akillisehir.konya.bel.tr/uygulama/atus-akilli-toplu-ulasim-sistemi https://atus.konya.bel.tr

Kuyumcu, Y. (2017). *Bicycle as a Means of Transportation and Bicycle Path Proposal for Urban Transportation in Çorum* [Master Thesis]. Bahçeşehir University.

Lai, F., & Carsten, O. (2012). What benefit does Intelligent Speed Adaptation deliver: A close examination of its effect on vehicle speeds. *Accident; Analysis and Prevention*, *48*, 4–9. doi:10.1016/j.aap.2010.01.002 PMID:22664662

Li, J., Boonaert, J., Doniec, A., & Lozenguez, G. (2021). Multi-models machine learning methods for traffic flow estimation from Floating Car Data. *Transportation Research Part C, Emerging Technologies*, *132*, 103389. doi:10.1016/j.trc.2021.103389

Li, Z., Liang, C., Hong, Y., & Zhang, Z. (2021). How do on-demand ridesharing services affect traffic congestion? The moderating role of urban compactness. *Production and Operations Management*, poms.13530. doi:10.1111/poms.13530

Meric, E. B. (2018). Intelligent transportation systems (AUS) and development agencies. *Journal of Intelligent Transportation Systems and Applications*, *1*(2), 33–55.

Pham. (2014). Smart cities in Japan. An assessment on the potential for EU-Japan Cooperation and Business Development, Tokyo.

Roh, C. G., & Jeon, H. (2021). Decision-Making Process for Demand Response Public Transportation Service Design—A Case Study in Incheon, Korea. *Sustainability*, *13*(10), 5574. doi:10.3390u13105574

Silva, B. N., Khan, M., & Han, K. (2018). Towards sustainable smart cities: A review of trends, architectures, components, and open challenges in smart cities. *Sustainable Cities and Society*, *38*, 697–713. doi:10.1016/j.scs.2018.01.053

HBSGB Transport. (2013-2015). *National Smart Transport Systems Strategy Document 2013-123 and Action Plan*. Draft. Author.

TurkeyC. R. M. P. R. August 2019., Retrieved from https://ulasim.gov.tr

Turkey Annual Report General Directorate of Highways. (2020). Retrieved from https://kgm.gov.tr

Wu, H., Levinson, D., & Sarkar, S. (2019). How transit scaling shapes cities. *Nature Sustainability*, *2*(12), 1142–1148. doi:10.103841893-019-0427-7

Yardim, M. S., & Akyildiz, G. (2005). Intelligent transportation systems and applications in Turkey. In *6th Transport Congress Proceedings Book*. TMMOB Chamber of Civil Engineers.

Yardım, M. S., & Erel, A. (2003). Data Requirement for Turkish Transport System. *TMMOB Transport Policies Congress Proceedings*, 171-181.

Zadobrischi, E., Cosovanu, L. M., & Dimian, M. (2020). Traffic Flow Density Model and Dynamic Traffic Congestion Model Simulation Based on Practice Case with Vehicle Network and System Traffic Intelligent Communication. *Symmetry*, *12*(7), 1172. doi:10.3390ym12071172

Chapter 4
Cryptocurrency Price Prediction Using TPU-Based Distributed Machine Learning

Yokesh Babu Sundaresan
Vellore Institute of Technology, India

Mohd Hammad Khan
Vellore Institute of Technology, India

Devdutt Sharma
Vellore Institute of Technology, India

Anbarasi Masilamani
Vellore Institution of Technology, India

Sukriti Jaitly
Vellore Institute of Technology, India

Mohak Verma
Vellore Institute of Technology, India

ABSTRACT

The development of blockchain has led to the emergence and widespread use of decentralized cryptocurrencies around the globe. As of 2021, the global market capitalization of cryptocurrencies has crossed two trillion dollars. With increasing popularity and adoption, investors have begun to see cryptocurrencies as an alternative to conventional financial assets. However, the volatility associated with cryptocurrencies makes them a highly risky investment. This gives rise to the need for accurate and efficient price prediction models which can help reduce risks associated with cryptocurrency investments. The model aims at predicting the price of two popular cryptocurrencies: Bitcoin and Ethereum. Tensor processing unit (TPU) is used for providing a distributed environment for the proposed model. The results show that the distributed TPU-trained model performed significantly better than the conventional CPU-trained model in terms of training time while maintaining a high degree of accuracy.

DOI: 10.4018/978-1-7998-9274-8.ch004

INTRODUCTION

The exchange of services and commodities has taken place through a variety of means and mediums throughout the history of human civilization. The majority of these are physical in nature, such as a metal coin or paper currency. However, with the advancement of technology, particularly computer and internet technology, a digital mode of transaction has emerged. When it comes to physical currency transactions, most financial institutions and governments encourage people to use online payment because it is a safe and quick method of doing so. This type of currency usage is an attempt to digitize physical currency through the use of the internet. In recent years, complete and true digital currencies (i.e., those that have no physical form) have also been introduced from time to time in various forms. Most of these digital currencies are completely virtual and intangible, with just a small number of tangible counterparts. Bitcoin and other digital tokens are referred to as cryptocurrency, where the term "crypto" refers to complex cryptography that allows a particular digital token to be generated, stored, and transacted safely as well as, in many cases, anonymously. These digital currencies allow for instantaneous and borderless transactions between their respective holders.

Bitcoin is world's first decentralized cryptocurrency and first practical implementation of blockchain technology. Since its launch in 2009, Bitcoin has revolutionized the technical and financial world. As of 2021, Bitcoin has market capitalization of over 1 trillion USD making it world's largest cryptocurrency by market capitalization (CoinMarketCap, 2021).

Value of Bitcoin has grown tremendously over past decade, going from less than 2500 USD in 2014 to more than 10000 USD in 2020 as depicted in Weighted Price Graph in Figure 1.

Figure 1. Bitcoin weighted price graph

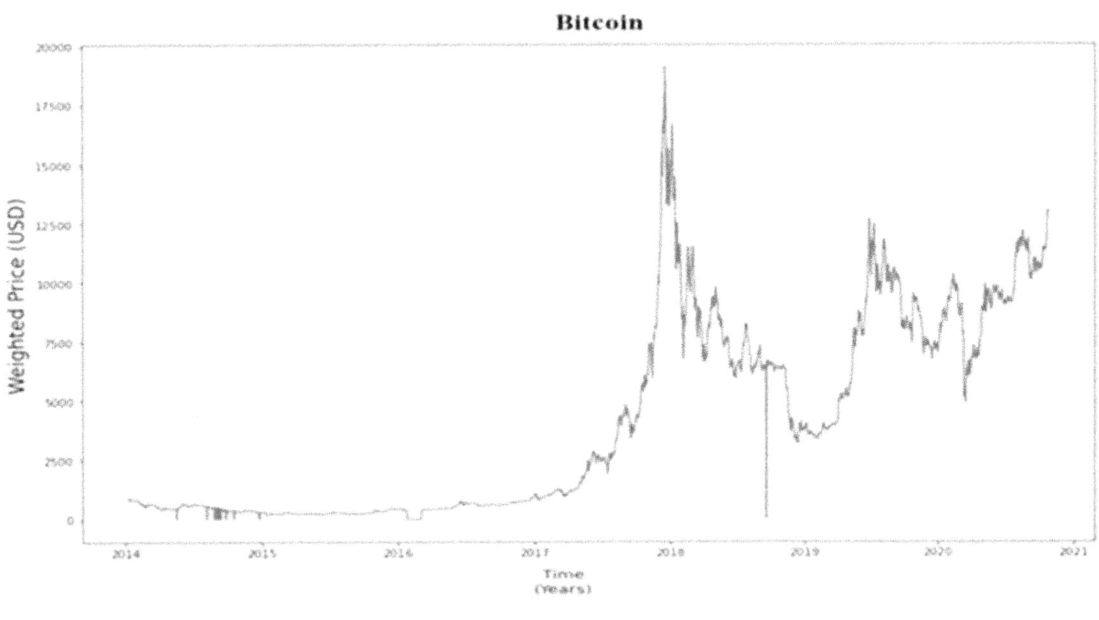

Specialty of Bitcoin is its algorithm that limits its quantity to 21 million and ensures that all the payments are irreversible and anonymous. Apart from immutability and anonymity, another key characteristic of Bitcoin is decentralization. Decentralization implies that there is no central regulatory authority or any other financial institutions in charge of looking after Bitcoin network's administration. Bitcoin is rather managed by a protocol that is maintained in a peer-to-peer system by a community of people, with no requirement for a third party intervention. To ensure consistency and authenticity of transactions, all transactions must first be validated by network nodes before they can be stored in form of blocks on the Bitcoin's blockchain, which is a publicly accessible distributed ledger. This process of validation and addition of new transaction is known as mining. Mining is carried out using a mechanism called Proof-of-Work. Proof-of-Work requires complex computational problem to be solved in order to add new blocks on blockchain. Bitcoins are produced as a prize for participating in Proof-of-Work process, in which users give their system's computational power in exchange for the ability to verify as well as record transactions onto the Bitcoin's blockchain.

Bitcoins are held in a wallet, which contains cryptographic credentials for the user's Bitcoin holdings and allows users to access them when they log into their respective account. Wallet makes use of public-key cryptography, which generates two keys, one public and one private, for each transaction. The public key may be compared to an account number or a name, whereas the private key can be compared to ownership credentials. Once an owner transfers possession of Bitcoin to a new owner, the prior owner uses their private key to publish a record into the system declaring that the ownership has been transferred to the new public key, thus transferring ownership of Bitcoin to the new owner. (Oyamada & Nakadai, 2017). Similar to Bitcoin, Ethereum is a popular cryptocurrency. It is world's second largest cryptocurrency by market capitalization. As of 2021, it has a market capitalization of more than 400 billion USD, second only to Bitcoin (CoinMarketCap, 2021). Ethereum was conceptualized as an open source blockchain by Vitalik Buterin in 2013. The main aim of Ethereum is to provide an efficient platform for development of decentralized applications. One of the key features of Ethereum is Smart Contracts which make it a suitable platform for implementation of various use cases in different domains.

Smart contracts can be defined as self-executing digital contracts on a blockchain. Smart contracts utilize conditional (if-else) logic for execution i.e. they will execute whenever certain predetermined conditions or terms are satisfied. Smart contracts can automate workflows and help participants reach consensus regarding certain actions or outcomes. In order to execute Smart Contracts or conduct transaction on Ethereum blockchain, users need to pay a transaction fees called Gas. Gas helps in allocating resources for a transaction or contract. The price of Gas depends upon the demand-supply dynamics of network miners. Like Bitcoin, Ethereum uses Proof-of-Work mechanism for reaching consensus on state of any data which is recorded on its blockchain. Apart from consensus, Proof-of-Work is also crucial in maintaining the security of Ethereum blockchain by preventing double spending attacks.

Due to its functionality, Ethereum has been instrumental in enabling the development of blockchain based applications in multiple domains. The advantage of Ethereum platform is that it grants decentralized immutability to the applications developed on it. This gives developers and users more control and privacy over their data which would be difficult to achieve while using a private application development platform. As a result of these features, Ethereum has found uses cases in many domains. Decentralized Finance (DeFi) is one of the most popular use cases of Ethereum, DeFi refers to a collection of financial services and products that rely on a blockchain based platform like Ethereum for functioning. Unlike traditional financial institutions, DeFi is not regulated or monitored by any central authority. Using DeFi, users can borrow or lend funds, gain insurance for assets, earn interest on assets etc. The growth of DeFi has been one of the propelling factors behind the widespread adoption of Ethereum in recent years. As popularity of Ethereum has increased, its price has increased manifold. The price of Ethereum has gone from less than 10 USD in 2016 to more than 250 USD in 2020. This can be inferred from Figure 2, which represented the weighted price graph of Ethereum from 2016 to 2020.

Figure 2. Ethereum weighted price graph

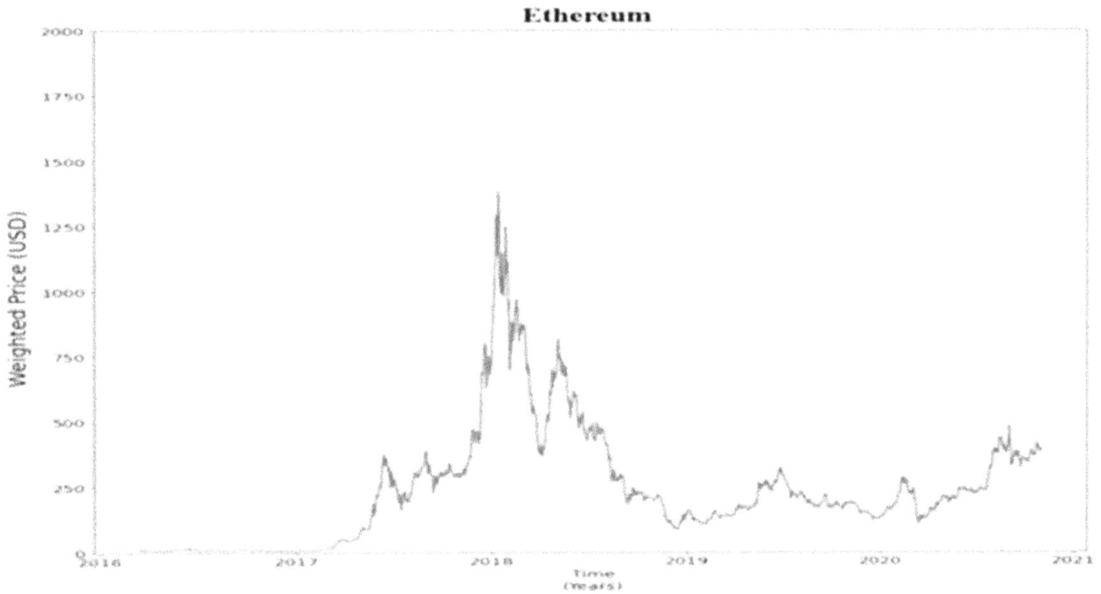

As awareness about blockchain is increasing, numerous organizations and governments are seeking ways to adopt usage of cryptocurrencies. For instance, El Salvador has issued a bill in its parliament to adopt Bitcoin as a legal tender which will enable people to use Bitcoins to buy essentials throughout the country. Furthermore, Ethereum's DeFi services are expected to rival prominent financial corporations in terms of users in near future. Rapid growth of cryptocurrencies has attracted numerous investors. Many cryptocurrency investors except to get huge returns from their investments as usability of cryptocurrencies increases. Due to their decentralized nature, cryptocurrencies are also seen as a hedging alternative for traditional assets and securities.

In spite of tremendous growth and enthusiasm, volatility of cryptocurrencies acts as a major determent for widespread usage and potential investments. The volatility of cryptocurrencies can be attributed to their decentralized nature. As cryptocurrencies are not regulated, their value can be influenced by a variety of economic and social factors. Many of these factors are not easy to foresee or anticipate in advance. Due to this it is difficult to quantify their effect on the price of cryptocurrencies. As a result cryptocurrencies can see sharp rise or fall in their value in a very short timeframe. This can cause panic among investors leading them to make imprudent decisions. Losses incurred from such decisions can make investors reluctant about making future investments in cryptocurrencies. Furthermore, erratic price movements might also disrupt the operations of applications developed using blockchain based platforms dissuading users from using the applications again. In order to predict future movements of cryptocurrency prices, machine learning models can be developed. Machine Learning models make use of a variety of algorithms and techniques to carry out different tasks. With help of relevant datasets, machine learning models can be trained to solve a problem. In field of cryptocurrency price prediction, researchers have had considerable success in predicting future prices by training and testing machine learning models. However developing such machine learning models requires large volumes of data comprising of multiple features and parameters.

When handling such large volume of data, the process of training often requires higher computational power than what conventional computer machineries can provide. This makes it difficult to develop efficient and accurate models in a predefined time frame. In order to meet the computation power requirements, numerous AI (Artificial Intelligence) accelerators have been developed. These accelerators comprise of a large number of computing units that can carry out multiple calculations simultaneously. However training process needs to be further optimized in order to utilize these accelerators to their maximum possible potential. One way of optimizing the training process is distributing the workload among several machines. However distribution of the machine learning process requires efficient parallelization along with synchronization between all systems or computing units. Hence it can be quite difficult to create custom frameworks for distributed machine learning. To counter this problem researchers and developers have come with ready to use frameworks like Spark. However such frameworks have limited functionality. In order to provide more functionality to developers, machine learning platforms such as TensorFlow have been developed.

TensorFlow is a machine learning platform having a library of numerous machine learning algorithms that can be used for development and deployment of a large variety of machine learning models. Apart from conventional machine learning algorithms, distributed machine learning modules are also available on TensorFlow. Using these modules, process of training can be distributed across multiple GPUs (Graphical Processing Unit) or TPUS (Tensor Processing Unit) depending upon the user's requirements. As a result, TensorFlow makes it much easier to build and deploy distributed machine learning models. In this chapter, price prediction model for Bitcoin and Ethereum has been proposed. The proposed model makes use of LSTM and GRU network for training the model. Subsequently the model is distributed (or parallelized) using AI accelerator called TPU. The performance of parallelized TPU trained model is compared with conventional CPU trained model using relevant parameters. This helps in measuring and analyzing the performance of parallelized model in predicting cryptocurrency prices.

BACKGROUND

Overview of TPU

The TPU (Tensor Processing Unit) is an application specific integrated circuit (ASIC) that is designed to handle huge machine learning workloads. It was first announced by Google in 2016 and was made available for third party use in 2018. It is capable of performing a huge number of low precision computations. When training large neural networks, TPUs can reduce the time to accuracy.

TPUs are highly efficient in terms of spatial optimization while also lowering energy consumption levels. The instruction set of TPU is based on Complex Instruction Set Computing. Complex Instruction Set Computing (or CISC for short) is a type of computer architecture that is built for the task of implementing high-level commands. TPUs are able to do more complex jobs as a result of these instructions. Furthermore, TPUs are equipped with a Matrix Processor, which is capable of processing hundreds of thousands of matrix operations in a single clock cycle. Think of it as printing a complete text at once, rather than printing words one at a time or line by line. Moreover, a specialized network that allows for the building of machine learning is included, as well as supercomputers known as TPU-pods are part of the system. Made up of four custom-built application-specific integrated circuits that were especially intended for machine learning using TensorFlow, TPUs provide a genuinely powerful 180 trillion floating point operations per second while using 64GB of high-bandwidth memory. TPUs are thus ideal for both training and inference machine learning models, which makes them ideal for deep learning. In addition to precision and speed, TPUs have a greater energy efficiency than traditional processing chips, using much fewer watts and generating far less heat than conventional processing chips. Because they are built on TensorFlow, the neural networks need relatively little code, making the deployment of TPUs that much simpler.

Working of TPU

When applied to geometric vectors, scalars, and other similar items in an algebraic fashion, a tensor is defined as a geometric object that maps these objects in an algebraic manner to the resulting tensor. In simpler terms, a tensor can be thought of as an extended vector, 3-D matrix, cube of numbers, or even a single number. It can also be thought of as a higher-dimensional structure that is more difficult to visualize in its simplest form. The rank of a tensor indicates how large it is in terms of dimensions. A tensor's dynamics is one of its most important characteristics: it is a mathematical entity that exists within a structure and interacts with other mathematical entities. When other entities in the structure are transformed in a predictable manner, the tensor obeys a relevant transformation rule and its numerical values move about with the other entities. This "dynamical" aspect of a tensor is the distinguishing characteristic that distinguishes it from a simple matrix.

In the field of computer science, Tensor can be defined as a data structure that is equivalent to the NumPy array, which is the primary data structure utilized by machine learning techniques. Tensor is the fundamental unit of operation in TensorFlow, and it practically makes use of NumPy.

TPU has been developed specifically for neural networks as a matrix processor that enables it to carry out high volumes of additions and multiplications at very high speeds. While executing calculations. TPU loads a matrix of adders and multipliers with parameters retrieved from memory. Subsequently data is loaded, and multiplication operations are executed. After each operation result is passed over to next multipliers simultaneously taking summation. Hence the output is total of all multiplication results that occurred between parameters and data. All the calculations and data passing is carried out without accessing the memory.

TPUs address several computational issues with relative ease. TPUs have 65,536 8-bit integer multipliers, which are used in conjunction with the quantization technique, which is the process of approximating an arbitrary value between a predetermined minimum and maximum value using an 8-bit integer multiplier. Overall, this technique reduces floating-point calculations involving 32-bit or even 16-bit quantities to 8-bit integers, which is a significant improvement over previous techniques. Quantization is the first and most powerful tool that TPUs can employ to lower the cost of neural network predictions while maintaining or improving the accuracy of the predictions.

Parallel Processing is a term that refers to the use of many processors at the same time. Generally, simple computations, such as multiplication, are made possible by typical RISC processors, which process a single operation, known as a scalar operation, for each instruction they receive. Unlike RISC processors, TPU has a Matrix Multiplier Unit. As a matrix processor rather than a scalar processor, TPU can process hundreds of thousands of operations (matrix operation) in a single clock cycle. A matrix processor is analogous to printing papers one page at a time rather than character by character or line by line, as is the case with a laser printer. Since TPUs are primarily utilized through cloud computing, the speed of the PCIe bus, data transfer between the Cloud TPU and host memory is somewhat slow when compared to the speed of computation in the cloud. Because of the frequent "ping-ponging" between the host and the Cloud TPU, the latter stays idle for the majority of the time, waiting for data to arrive via the PCIe bus. To counter this programming paradigm for Cloud TPU delegated a large portion of the training to the TPU in order to improve efficiency. The on-chip high bandwidth memory is used to store all of the model parameters in the similar manner. Several training stages are carried out in a loop, which allows for the cost of initiating calculations on Cloud TPU to be amortized. When data is "infed" to the Cloud TPU hardware, the TensorFlow server collects and pre-processes it; cores on a Cloud TPU synchronously run an identical program that is stored in their own unique High-Bandwidth Memory (HBM). Thus reducing data processing time and increasing energy efficiency.

Long Short-Term Memory (LSTM) Network

Long short-term memory (LSTM) is a type of Recurrent Neural Network (RNN) architecture with ability to learn long-term dependencies. A typical LSTM network is made up of structures called cells. LSTM cells contain mechanisms called gates that are responsible for information flow and memory manipulation. These gates help cells retain information for arbitrary time periods. Each LSTM cell contains three gates: input gate, forget gate and output gate. The forget gate is used to remove the information which is no longer useful to store in the cell. The new useful information is added in the cell using the input gate. The output gate is used to release the useful information stored in the cell as output. Use of gates helps LSTMs solve vanishing gradient problem which is persistent in traditional RNNs.

Furthermore each LSTM cell has two states: cell state and hidden state. Cell state is also known as long term memory. It is the memory or information aggregated from previous intervals. Forget gate is used for modifying cell state. The cell state can be adjusted by the output produced by input gate. Hidden state is also known as short term memory. It is the output of the cell and is determined by output gate. Figure 3 depicts the architecture of a typical LSTM cell. The flow of information in a LSTM cell can be easily understood by studying Figure 3. As LSTM is a variant of RNN, it has feedback connections. These connections enable LSTM to process entire sequences of data values. Due to presence of a cell state, LSTMs can retain information in memory for long period of time enabling them to learn long-term dependencies. LSTM is well suited for tasks that require processing a large sequences or series of data points. As a result LSTM is widely used robot control, speech recognition and time series prediction.

Figure 3. LSTM cell architecture

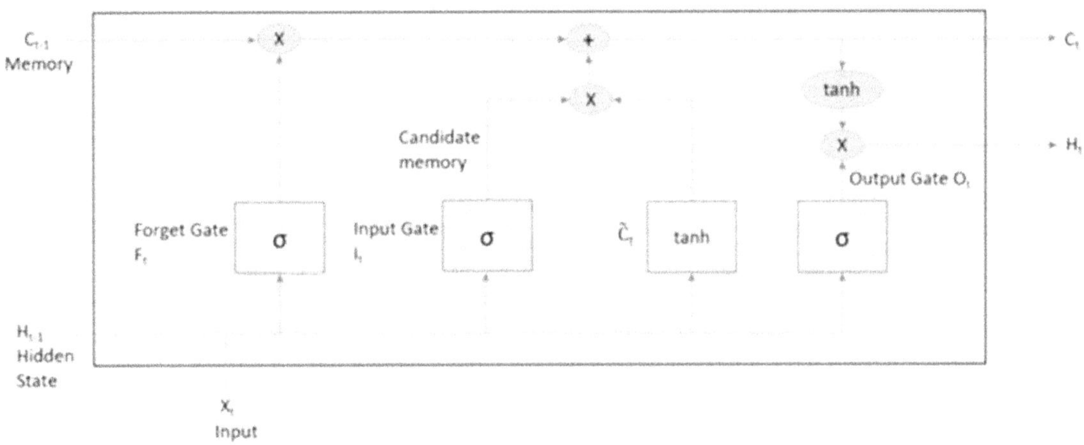

Gated Recurrent Unit (GRU)

Gated Recurrent Unit (GRU) is a variant of RNN architecture with gating mechanism. Similar to LSTM, GRU makes use of gates for managing and controlling the movement of information between different cells of the network. GRU and LSTM have many common properties. Like LSTM, GRU can learn long term dependencies to retain information for longer time intervals and solve vanishing gradient problem. Unlike LSTM, GRU has two gates: update gate and reset gate. The update gate determines the amount of past information (collected from prior steps) that is needed to be passed onto the next state. Reset gate determines the amount of previous information that needs to be discarded. Furthermore GRU lacks a cell state and utilizes hidden state for carrying both short term and long term memory simultaneously. Hidden state for any GRU unit is decided by update gate by combining the output of reset gate and previous hidden state. Update gate decides what portions of previous hidden state and new proposed hidden state (output of reset gate) need to be combined to arrive at final Hidden state of the cell. Subsequently, final Hidden state is given out as output by the GRU unit.

The movement of information and process calculating Hidden state can be understood by studying the architecture of GRU unit depicted in Figure 4. Due to simpler architecture, GRUs are easier to modify than LSTM. In many domains, performance of GRUs are on par with LSTM. Moreover, in few cases GRU has achieved better performance and shorter training time than LSTM when smaller datasets were used (Yang et al., 2020).

Figure 4. GRU cell architecture

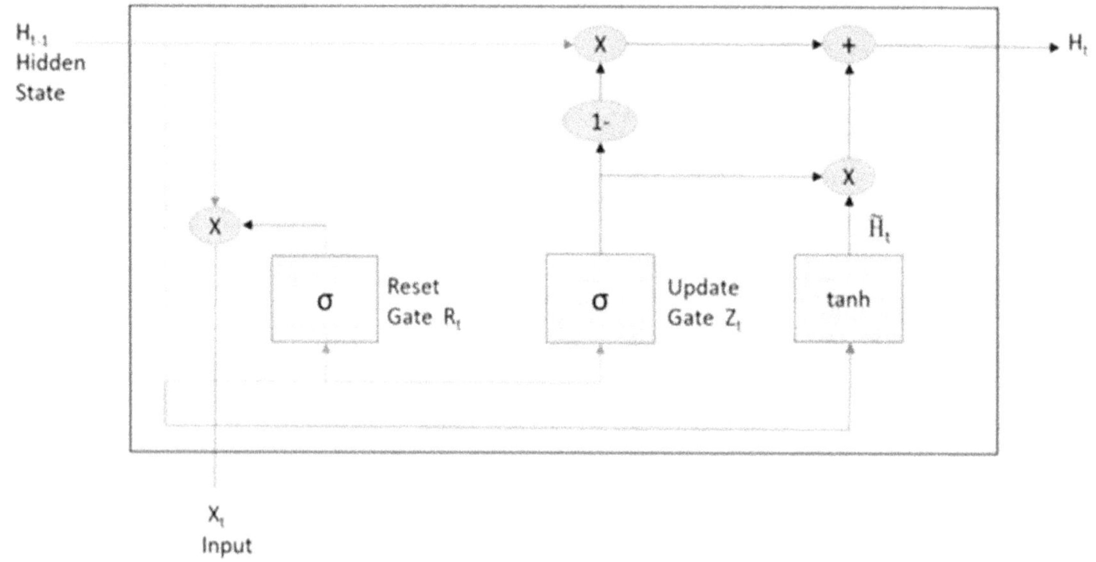

Volatility and fluctuations in prices of cryptocurrencies have been studied keenly by researchers. Studies have shown how various social and economic factors influence cryptocurrency prices. Bitcoin price is positively correlated with Bitcoin's Wikipedia page views and network hash rate (Georgoula et al., 2015). News related to Bitcoin regulation and security breaches have a stronger impact on Bitcoin's volatility than other macroeconomic news (Lyócsa et al., 2020). Additionally leverage effect has a stronger influence on the future Bitcoin volatility than Economic Policy Uncertainty (Yu, 2019).

Stablecoins (cryptocurrencies linked to an external asset) negatively impact the price of Bitcoin and Ethereum (Kristoufek, 2021). Furthermore the significance level of long-run impact of trading volume on Ethereum and Bitcoin price stands at 10% and 1% respectively (Sovbetov, 2018). The daily log returns of Bitcoin have a significant correlation with daily Google trend values of the search term "Bitcoin". However no such correlation can be established between daily log returns of Ethereum and daily Google trend values of search term "Ethereum" (Dempere, 2019). Many of the aforementioned factors and correlations are difficult to model and quantify using conventional statistical methods. Hence in order to develop reliable and accurate price prediction models for cryptocurrencies various machine learning algorithms and techniques have been utilized.

Researchers have used RNN and LSTM network along with ARIMA model to predict the direction of Bitcoin price, and have achieved accuracy of 52.78% using LSTM model (McNally et al., 2018). Another approach employed Convolutional Neural Network (CNN), LSTM and GRU to analyze the factors affecting Bitcoin price. Gold price was used as a parameter to analyze its effect on Bitcoin price (Aggarwal et al., 2019). Additionally, time series based price prediction model using parameters such as volume and stock market capitalization have been developed for Bitcoin and Ethereum. LSTM network has been employed for training the model (Biswas et al., 2021). Moreover, Bi-directional LSTM network which has ability to take input in two directions has been used for creating price forecasting model that can help investors make decisions regarding Bitcoin related investments (P et al., 2021). Ethereum price prediction model developed using conventional machine learning algorithms like Linear Regression and

Support Vector Machine (SVM) has shown high level of accuracy (M. et al., 2020). Another SVM based model helped identify the role of blockchain information in predicting price of Ethereum (Kim et al., 2021). Feedforward neural networks such as Multi-layer perceptron (MLP) have also been utilized for Ethereum price prediction (Kumar & Rath, 2020).

Several cryptocurrency price prediction models have employed ensemble learning. Ensemble learning utilizes combination of multiple machine learning algorithms to achieve a specific task such as price prediction. An Ensemble-based LSTM Bitcoin price prediction model was able to predict the sudden changes in Bitcoin price which could help reduce investment related risks (Shin et al., 2021). Researchers have also utilized recently developed ensemble learning models such as Variational Mode Decomposition (VMD) and Stacking-ensemble learning (STACK) for Bitcoin price forecast (da Silva et al., 2020). In order to utilize subjective information related to cryptocurrencies for price prediction, researchers have used natural language processing techniques such as sentiment analysis. Sentiment analysis is used for processing subjective information in order to obtain social sentiments regarding an object. This can help develop price prediction models using various social and economic factors. Researchers used sentiment analysis to transform news articles into a graph (SentiGraph) which was used to analyze the effect of news articles on Bitcoin price (Yao et al., 2019). In another approach, network sentiments were identified as one of the important factors in predicting price of Bitcoin (Serafini et al., 2020). Similarly, predictive power of tweets with respect to Bitcoin price has also been explored. It was found that fluctuations in Bitcoin price could be predicted by performing sentiment analysis on Bitcoin related tweets (Sattarov et al., 2020). While majority of works focus on improve the accuracy and reliability of price prediction models, only few works have focused on accelerating the training process of the models. An Apache Spark framework-based model utilized Linear Regression for Bitcoin price prediction. The model performed twice as fast as the GPU based model (Purbarani & Jatmiko, 2018). Another approach found that GPU implementation of machine learning based price prediction models was 67.7% faster than CPU implementation in terms of training time (McNally et al., 2018).

TPUs can be utilized for gaining extra computational power for performing efficient calculations and to overcome latency. Distributed Machine Learning can be effectively implemented using TPUs where multiple nodes can work together to speed up the training and testing process of machine learning models. Work analyzing the energy efficiency and performance of different AI (Artificial Intelligence) accelerators have shown that TPUs are one of the fastest AI accelerators available in the world. TPUs achieved near optimal utilization in two computational operators and performed significantly better than other accelerators in end to end training (Wang et al., 2020). Due to their efficiency, TPUs have found applications in multiple domains of artificial intelligence. A high-speed object tracking and prediction model developed on TPU architecture performed significantly better than other architectures and accelerators (Sengupta et al., 2020). Researchers have used TPUs to improve performance of deep convolutional neural networks for facial emotion recognition and have achieved accuracy of 90% (D'Agostino & Schmidt, 2020). In the field of medical science, TPUs were used for reconstruction of magnetic resonance (MR) images using alternating direction method of multipliers (ADMM). TPUs outperformed CPUs and GPUs in terms of computational time (Lu et al., 2020). To detect weeds in organic farming, a real time CNN based weed detection model was developed. The model was accelerated using TPU which resulted in higher energy efficiency (Czymmek et al., 2021)

TPU based distributed evolution strategy for the meta-learning was found to be highly parallel and scalable when working with large tasks with no increment in memory cost (Sheng & He, 2020). Researchers have also proposed TPU based parallel algorithms for discrete Fourier transformations. Use of multiple TPU cores significantly reduced computational time of the proposed algorithms for given problem (Lu, Chen, Hechtman, Wang & Anderson, 2021). The use of TPUs for development of cryptocurrency price prediction models has not been explored till now. Hence in this chapter, TPUs have been used for accelerating the training and testing process of cryptocurrency price prediction model. This will help in evaluating and analyzing the efficiency of TPUs in reducing training time for machine learning based price prediction models.

MAIN FOCUS OF THE CHAPTER

Methodology

The methodology used in the chapter involves five key steps, namely data collection, data preprocessing, model creation and parallelization, model implementation followed by performance comparison and analysis. The sequence of these steps can be interpreted from Figure 5 which represents the architecture flow chart of the proposed model.

Figure 5. Architecture flow chart of proposed model

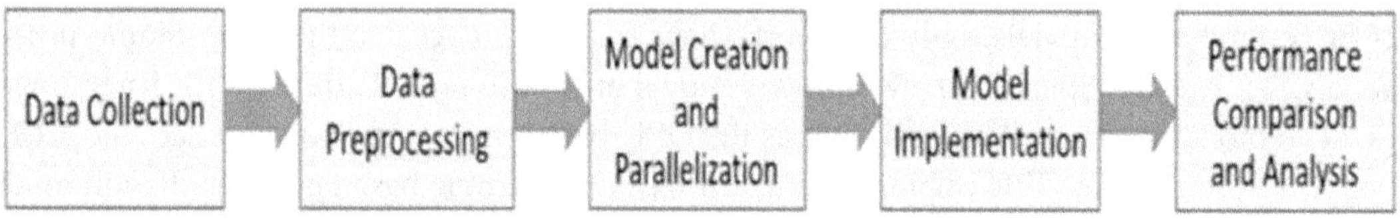

The data collection process involved search and selection of appropriate datasets for development of the model. After data collection, process of data preprocessing is performed. Data preprocessing primarily involves cleaning of data (i.e. removal of duplicate, unformatted and incorrect data points) and normalization (i.e. restructuring and rescaling of data). Following data preprocessing, the task of model creation is performed. Model creation involves selection of appropriate machine learning algorithm and technique which will aid in enhancing the efficiency and accuracy of the model. Subsequently the model is parallelized (or distributed) to speed up the training process. TPUs are used for parallelization of the training process. In the fourth step, created model is implemented using relevant programming language. Machine learning libraries and platforms are used for employing selected machine learning algorithms or techniques for training and testing model. Finally, the performance of parallelized TPU trained model is compared with CPU trained model. Relevant parameters are used for evaluating the efficiency and accuracy of the proposed model. These steps are explained in detail in following sections.

Data Collection

After an extensive search, time series based Bitcoin and Ethereum price data was selected for development of proposed model. Time series based data provides a good indication of the price movements that occur over a long period of time which makes them suitable for developing price prediction models.

The relevant data was acquired from Quandl. Quandl is a popular online data marketplace which provides a large variety of financial data. The acquired data contained sequence of data points (Bitcoin and Ethereum Price) collected at regular time intervals (24 hours in this case). For Bitcoin, data consisted of 2474 days (data points) from 24th January 2014 to 1st November 2020. Similarly, for Ethereum, data consisted of 1704 days (data points) from 4th March 2016 to 1st November 2020. A sample record (of Bitcoin) from the acquired data is depicted in Figure 6.

Figure 6. Sample record of Bitcoin from the acquired data

	Open	High	Low	Close	Volume (BTC)	Volume (Currency)	Weighted Price
Date							
2014-01-07	874.67040	892.06753	810.00000	810.00000	15.622378	13151.472844	841.835522

Data Preprocessing

Data preprocessing customizes the dataset as per the requirements of the model. Model specific customization facilitates the splitting of dataset into training and test sets. After the split, dataset can be easily fed into the model for carrying out testing and training. Through data preprocessing, features required for training of the model are also selected. It is essential to select only relevant and appropriate features for the model to ensure accuracy and reliability.

For the proposed model two features were selected from the dataset; Date and Weighted Price. Collection of records in the Date feature helps in understanding the movement of price over a given time frame. Weighted Price (or Volume Weighted Average Price) is the ratio of total value of a traded financial asset to total volume of asset related transactions for any given trading period. It can provide the average trading price for any traded asset over a particular time period. Investors commonly use weighted price to evaluate the performance of their financial assets. Weighted price is particularly useful for passive investors as it provides clear indications regarding an asset's price movements. Weighted Price can be calculated using the following formula:

$$Weighted\ Price = \frac{\sum(Volume * Price)}{Total\ Volume}$$

Following feature selection, data normalization is carried out. The goal of normalization is to fit the values of data in a common scale without affecting the range of difference. The normalization technique used for the proposed model is Min-max normalization. For any i^{th} data point X_i, the Min-max normalized value Z is obtained by dividing the difference between X_i and smallest data point min(X) by the difference between smallest data point min(X) and largest data point max(X). Through Min-max normalization all data points in the dataset get converted to values ranging from 0 to 1. The smallest data point min(X) is converted to 0 and largest data point max(X) is converted to 1. Following is the formula for getting normalized value Z, for any data point X in dataset.

$$Z_i = \frac{X_i - \min(X)}{\max(X) - \min(X)}$$

Normalization is followed by splitting of preprocessed dataset into training and testing sets. The training set is used for training the proposed model and testing set is used for validating the performance of the model. The training and testing sets contain different data values. This helps in evaluating the performance of the proposed model on unknown data. For the proposed model the preprocessed dataset is split into 70:30 ratio. 70% of the dataset is used for training and 30% is utilized for testing. This ensures that adequate amount of data is used for both training and testing.

Model Creation and Parallelization

Since the dataset used in the proposed model is time series based, univariate time series forecasting has been utilized for making price predictions. Univariate time series forecasting makes use of past values in the dataset to predict future values. Only two factors, namely time and the factor whose values are to be predicted (Bitcoin or Ethereum price in this case) are needed to utilize univariate time series forecasting. As the dataset used in the proposed model is primarily based on two factors (time and price), it is essential to use a forecasting technique that will make accurate and reliable predictions using only two variables which makes univariate time series forecasting adequate for the task. For implementing the selected forecasting technique, LSTM and GRU network are used. LSTM and GRU are well suited for making predictions on time series based datasets due to their ability to remember past information. Both of these networks do not require predetermined context to make predictions. This is due to their ability to learn context necessary to make predictions. Additionally, use of LSTM and GRU networks helps in comparing the performance of proposed model with existing LSTM and GRU based cryptocurrency price prediction models.

Model is created using selected forecasting technique and machine learning architecture. Subsequently model is parallelized. Parallelization of the model is the process of partitioning and training a model across a number of nodes or processors. TPUs are utilized for parallelization of the model. As discussed in the Background section, TPU employ an optimization technique called quantization to scale back execution time. Quantization ensures that precision level similar to that of floating-point calculations is achieved while using integer based calculations. This significantly reduces the computational requirements, accelerating the model training process as a result. Furthermore TPU does not need to access memory multiple times while training the model. Hence TPU has ability to effectively parallelize and speed up the training process of the model.

Model Implementation

The proposed model is implemented using Intel Xeon processor (CPU) with 12.6 GB RAM and TPU v2 with 12.6 GB RAM. For accessing TPU, Google Colab Notebook is utilized as the coding platform. The coding language used is Python. Machine Learning Libraries such as TensorFlow and Keras are used for incorporating LSTM and GRU networks into the model. TPU and Google Compute Engine are initialized using TPU ClusterResolver. TensorFlow's distributed functionality (tf.distribute) makes TPU ready for parallelization and help in distribution of the training process amongst several processing units. Use of tf.distribute ensures that the model is parallelized by making minimal changes in the codebase.

Performance Comparison and Analysis

It is essential to select suitable and relevant parameters for evaluating the performance of the proposed model. Evaluation parameters help in understanding how accurate and efficient the model is in producing desired output. Two parameters are used for evaluating the performance of the proposed model: Training time and Root Mean Square Error (RMSE). Lower values of training time and RMSE are considered better. Training time measures the amount of time taken by the processor to train the model. It is measured in seconds. RMSE is commonly used as an evaluation parameter for measuring performance of price prediction models. RMSE indicate how far the predicted values are from actual values. To find RMSE, the difference between actual and predicted values is squared. Subsequently average of the squared values is calculated. Finally root of the average is taken. For the proposed model, value of training time and RMSE are tabulated for further analysis and comparison.

SOLUTIONS AND RECOMMENDATIONS

The results obtained after the development, training and testing of the proposed model were tabulated. These results are presented in form of tables in this section. Additionally, accuracy graphs were drawn which visualize the obtained results and help in interpreting them. Table 1 depicts the RMSE values obtained after training and testing the proposed model using LSTM and GRU networks. The proposed model was trained on both Bitcoin and Ethereum datasets as can be interpreted from Table 1.

Table 1. RMSE values obtained for proposed model using LSTM and GRU network architectures

	RMSE			
	TPU Trained Model		CPU Trained Model	
	Bitcoin	Ethereum	Bitcoin	Ethereum
LSTM	0.1772	0.1240	0.1574	0.1178
GRU	0.1536	0.1407	0.1315	0.1445

Table 2 depicts the training time taken by proposed model for Bitcoin and Ethereum datasets when LSTM and GRU networks were used. Training time for both TPU and CPU trained model were tabulated for comparison.

Table 2. Training time taken by proposed model

	Training Time Taken (seconds)			
	TPU Trained Model		**CPU Trained Model**	
	Bitcoin	**Ethereum**	**Bitcoin**	**Ethereum**
LSTM	42	34	70	63
GRU	35	40	80	60

As can be inferred from Table 1, RMSE values obtained for the model ranged from 0.18 to 0.11. Lower values of RMSE were obtained for Bitcoin when using GRU network while LSTM produced lower RMSE values for Ethereum. This implies that GRU is more suited for Bitcoin price prediction and LSTM is more appropriate for Ethereum price prediction. While TPU trained model obtained higher RMSE values than CPU trained model in all cases except one (TPU trained GRU model for Ethereum), the main gain was in terms of training time. As depicted in Table 2, the training time taken by TPU trained model was significantly shorter than the training time taken by CPU trained model. This was true for both LSTM and GRU networks. The LSTM based TPU trained model had 40% and 46% decrease in training time for Bitcoin and Ethereum datasets respectively. Similarly, the GRU based TPU trained model had 56.25% and 33.33% decrease in training time for Bitcoin and Ethereum datasets respectively. The highest percentage decrease in training time was achieved for GRU based TPU trained model for Bitcoin (56.25%). The results obtained from proposed model were compared with existing models. Table 3 depicts the comparison of RMSE values obtained by proposed model and some existing works.

Table 3. Comparison of proposed model with existing works

	Model / Technique Used	**Bitcoin (RMSE)**	**Ethereum (RMSE)**
Proposed Model	LSTM (TPU Trained)	0.1772	0.1240
	GRU (TPU Trained)	0.1536	0.1407
(McNally et al., 2018)	LSTM	0.08	-
	ARIMA	0.5374	-
(Kim et al., 2021)	ANN	-	0.068
	SVM	-	1.306

The proposed model produced higher RMSE values than existing LSTM (McNally et al., 2018) and ANN (Kim et al., 2021) models. However the produced RMSE values were lower than existing ARIMA (McNally et al., 2018) and SVM (Kim et al., 2021) models. Furthermore the RMSE values obtained by the proposed model were not too large to severely impact the accuracy of the model. The reduction in accuracy was compensated by the gains made in form of percentage decrease in training time. The proposed model was able to predict the direction of the price movements for majority of data points in the testing datasets. This is depicted in following graphs. Figure 7 depicts the accuracy graphs obtained when GRU network is used for training the proposed model on CPU and TPU for Bitcoin dataset. Through line graphs of Real BTC value and Predicted BTC value, the accuracy of the model can be easily inferred.

Figure 7. Accuracy graph for bitcoin using GRU network (a) CPU model (b) parallelized TPU model

Figure 8 contains the accuracy graphs obtained for GRU network based model for Ethereum dataset. Figure 8 aids in interpreting the accuracy of GRU based model in making price predictions for Ethereum.

Figure 8. Accuracy graph for ethereum using GRU network (a) CPU model (b) Parallelized TPU model

Figure 9 presents the accuracy graphs obtained when LSTM is used for making price predictions on Bitcoin dataset. Accuracy graphs of both CPU and TPU trained models are placed adjacent to each other to enable comparison.

Figure 9. Accuracy graph for bitcoin using LSTM network (a) CPU model (b) parallelized TPU model

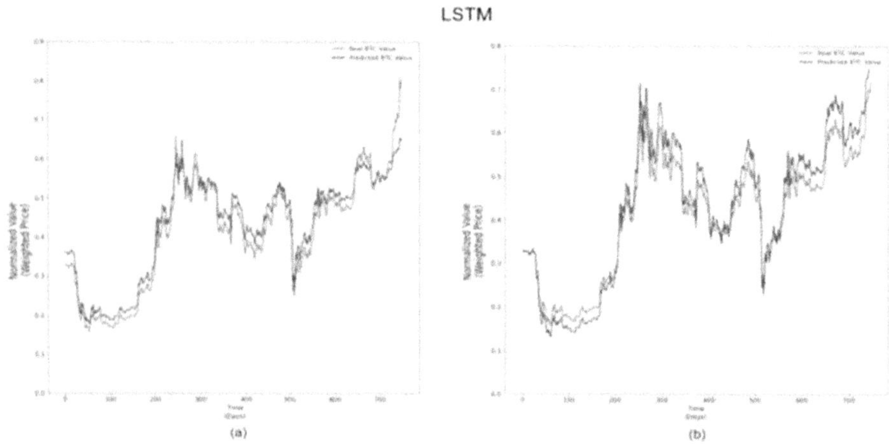

Figure 10 displays the accuracy graphs obtained for Ethereum when LSTM network is utilized for making price predictions. The accuracy of CPU and TPU trained models can be easily inferred from Figure 10.

Figure 10. Accuracy graph for ethereum using LSTM network (a) CPU model (b) parallelized TPU model

To summarize the results, TPU were able to effectively parallelize the proposed model and reduce training time. The parallelized model saw a slight dip in accuracy which was balanced by the gains made in form of shorter training time. This depicts that adequate price prediction models for cryptocurrencies can be developed using TPU enabled distributed machine learning. The results also demonstrate the consistency and efficiency of LSTM and GRU networks in making time series based predictions. The proposed model provides accurate indications regarding upcoming movements in the price of cryptocurrencies and can be trained in short timeframe.

CONCLUSION

In this chapter, it has been demonstrated how TPUs can be utilized to implement distributed machine learning for development of cryptocurrency price prediction models. Both LSTM and GRU networks were consistent with the predictions ensuring that the model was reliable and accurate. Through the use of distributed machine learning, the training time of the model was significantly reduced. Furthermore the speed up achieved in training process did not adversely impact the accuracy of the model. This indicates that TPUs can be effectively used for rapid development of highly accurate price prediction models for different cryptocurrencies. These models will help investors to make less risky and more profitable investments. Furthermore it will help in boosting public's confidence on blockchain technology leading to wider usage of the technology in different domains.

FUTURE RESEARCH DIRECTIONS

Cryptocurrencies are the most popular use case of blockchain technology. Through the use of cryptocurrencies completely decentralized and public financial systems can be developed. In future these decentralized financial systems can become good alternatives to traditional banking systems. Such decentralized systems will be particularly useful in times of calamities such as natural disasters or wars were traditional financial systems and bodies may not be able to function. Hence it is essential to find solutions for drawbacks associated with cryptocurrencies to make them more acceptable and reliable. Machine Learning can help to solve some problems related to cryptocurrencies. Through the use of machine learning reliable price forecasting models can be developed which can help ameliorate the effects of volatility as demonstrated in this chapter. Furthermore AI accelerators such as TPU can help in accelerating the development and training process of the models. While the model proposed in the chapter achieved adequate level of accuracy and significant speed up in training time. It utilized a limited number of features and evaluation parameters. In future more complex price prediction models with multiple features and parameters could be developed. Additionally multivariate time series forecasting technique could be utilized to find effect of multiple parameters on cryptocurrency prices. These complex models could then be parallelized to further investigate the efficiency of TPUs.

REFERENCES

Aggarwal, A., Gupta, I., Garg, N., & Goel, A. (2019). Deep Learning Approach to Determine the Impact of Socio Economic Factors on Bitcoin Price Prediction. *2019 Twelfth International Conference On Contemporary Computing (IC3)*. 10.1109/IC3.2019.8844928

Biswas, S., Pawar, M., Badole, S., Galande, N., & Rathod, S. (2021). Cryptocurrency Price Prediction Using Neural Networks and Deep Learning. *2021 7Th International Conference On Advanced Computing And Communication Systems (ICACCS)*. 10.1109/ICACCS51430.2021.9441872

CoinMarketCap. (2021, October 11). *Cryptocurrency Prices, Charts And Market Capitalizations*. https:// coinmarketcap.com

Czymmek, V., Moller, C., Harders, L., & Hussmann, S. (2021). Deep Learning Approach for high Energy efficient Real-Time Detection of Weeds in Organic Farming. *2021 IEEE International Instrumentation And Measurement Technology Conference (I2MTC)*. 10.1109/I2MTC50364.2021.9459943

D'Agostino, R., & Schmidt, T. (2020). EMOTIONET: A Multi-Convolutional neural Network hierarchical approach to facial and Emotional classification using TPUs. *2020 IEEE International Conference on Engineering, Technology and Innovation (ICE/ITMC)*. 10.1109/ICE/ITMC49519.2020.9198366

da Silva, R. G., Dal Molin Ribeiro, M. H., Fraccanabbia, N., Mariani, V. C., & dos Santos Coelho, L. (2020). Multi-step ahead bitcoin price FORECASTING based on VMD and ensemble learning methods. *2020 International Joint Conference on Neural Networks (IJCNN)*. 10.1109/IJCNN48605.2020.9207152

Dempere, J. (2019). *Factors Affecting the Return and Volatility of Major Cryptocurrencies. In 2019 Sixth HCT Information Technology Trends*. ITT. doi:10.1109/ITT48889.2019.9075117

Georgoula, I., Pournarakis, D., Bilanakos, C., Sotiropoulos, D. N., & Giaglis, G. M. (2015). Using time-series and sentiment analysis to detect the determinants of bitcoin prices. SSRN *Electronic Journal*. doi:10.2139/ssrn.2607167

Kim, H., Bock, G., & Lee, G. (2021). Predicting Ethereum prices with machine learning based on Blockchain information. *Expert Systems with Applications*, *184*, 115480. doi:10.1016/j.eswa.2021.115480

Kristoufek, L. (2021). Tethered, or Untethered? On the interplay between stablecoins and major cryptoassets. *Finance Research Letters*, *101991*. Advance online publication. doi:10.1016/j.frl.2021.101991

Kumar, D., & Rath, S. K. (2020). Predicting the Trends of Price for Ethereum Using Deep Learning Techniques. In S. Dash, C. Lakshmi, S. Das, & B. Panigrahi (Eds.), *Artificial Intelligence and Evolutionary Computations in Engineering Systems. Advances in Intelligent Systems and Computing* (Vol. 1056). Springer. doi:10.1007/978-981-15-0199-9_9

Lu, T., Chen, Y., Hechtman, B., Wang, T., & Anderson, J. (2021). Large-Scale Discrete Fourier Transform on TPUs. *IEEE Access: Practical Innovations, Open Solutions*, *9*, 93422–93432. doi:10.1109/ ACCESS.2021.3092312

Lu, T., Marin, T., Zhuo, Y., Chen, Y., & Ma, C. (2020). Accelerating MRI Reconstruction on TPUs. *2020 IEEE High Performance Extreme Computing Conference (HPEC)*. 10.1109/HPEC43674.2020.9286192

Lyócsa, Š., Molnár, P., Plíhal, T., & Širaňová, M. (2020). Impact of macroeconomic news, regulation and hacking exchange markets on the volatility of bitcoin. *Journal of Economic Dynamics & Control*, *119*, 103980. doi:10.1016/j.jedc.2020.103980

M., P., Sharma, A., V., V., Bhardwaj, V., Sharma, A., Iqbal, R., & Kumar, R. (2020). Prediction of the price of Ethereum blockchain cryptocurrency in an industrial finance system. *Computers & Electrical Engineering, 81*, 106527. doi:10.1016/j.compeleceng.2019.106527

McNally, S., Roche, J., & Caton, S. (2018). Predicting the price of bitcoin using machine learning. *2018 26th Euromicro International Conference on Parallel, Distributed and Network-Based Processing (PDP)*. 10.1109/PDP2018.2018.00060

Oyamada, M., & Nakadai, S. (2017). Relational mixture of experts: Explainable demographics prediction with behavioral data. *2017 IEEE International Conference on Data Mining (ICDM)*. 10.1109/ICDM.2017.45

P, N., Tom, R., Gupta, P., Shanthini, A., John, V., & Sharma, V. (2021). Prediction of Bitcoin Price Using Bi-LSTM Network. *2021 International Conference on Computer Communication and Informatics (ICCCI)*. doi:10.1109/ICCCI50826.2021.9402427

Purbarani, S. C., & Jatmiko, W. (2018). Performance comparison of bitcoin prediction in big data environment. *2018 International Workshop on Big Data and Information Security (IWBIS)*. 10.1109/IWBIS.2018.8471691

Sattarov, O., Jeon, H., Oh, R., & Lee, J. (2020). Forecasting Bitcoin Price Fluctuation by Twitter Sentiment Analysis. *2020 International Conference On Information Science And Communications Technologies (ICISCT)*. 10.1109/ICISCT50599.2020.9351527

Sengupta, J., Kubendran, R., Neftci, E., & Andreou, A. (2020). High-speed, real-time, spike-based object tracking and path prediction on google edge tpu. *2020 2nd IEEE International Conference on Artificial Intelligence Circuits and Systems (AICAS)*. 10.1109/AICAS48895.2020.9073867

Serafini, G., Yi, P., Zhang, Q., Brambilla, M., Wang, J., Hu, Y., & Li, B. (2020). Sentiment-Driven Price Prediction of the Bitcoin based on Statistical and Deep Learning Approaches. *2020 International Joint Conference On Neural Networks (IJCNN)*. 10.1109/IJCNN48605.2020.9206704

Sheng, A., & He, J. (2020). Distributed Evolution Strategies Using TPUs for Meta-Learning. *2020 IEEE Symposium Series On Computational Intelligence (SSCI)*. 10.1109/SSCI47803.2020.9308334

Shin, M., Mohaisen, D., & Kim, J. (2021). Bitcoin Price Forecasting via Ensemble-based LSTM Deep Learning Networks. *2021 International Conference On Information Networking (ICOIN)*. 10.1109/ICOIN50884.2021.9333853

Sovbetov, Y. (2018). Factors Influencing Cryptocurrency Prices: Evidence from Bitcoin, Ethereum, Dash, Litcoin, and Monero. *Journal Of Economics And Financial Analysis*, 2(2), 1–27.

Wang, Y., Wang, Q., Shi, S., He, X., Tang, Z., Zhao, K., & Chu, X. (2020). Benchmarking the Performance and Energy Efficiency of AI Accelerators for AI Training. *2020 20Th IEEE/ACM International Symposium On Cluster, Cloud And Internet Computing (CCGRID)*. 10.1109/CCGrid49817.2020.00-15

Yang, S., Yu, X., & Zhou, Y. (2020). LSTM and GRU Neural Network Performance Comparison Study: Taking Yelp Review Dataset as an Example. *2020 International Workshop On Electronic Communication And Artificial Intelligence (IWECAI)*. 10.1109/IWECAI50956.2020.00027

Yao, W., Xu, K., & Li, Q. (2019). Exploring the Influence of News Articles on Bitcoin Price with Machine Learning. *2019 IEEE Symposium On Computers And Communications (ISCC)*. 10.1109/ISCC47284.2019.8969596

Yu, M. (2019). Forecasting bitcoin volatility: The role of leverage effect and uncertainty. *Physica A*, *533*, 120707. doi:10.1016/j.physa.2019.03.072

Chapter 5
Applications and Potential of Blockchain in Modern Technological Systems

Raghav Behal
Vellore Institute of Technology, India

Kalyanaraman Pattabiraman
Vellore Institute of Technology, India

ABSTRACT

Blockchain, a relatively recent technological breakthrough, has been spreading like wildfire with reports of adoption extending across industries. Although most popularly known in the context of cryptocurrencies, this technology's true strength lies in its ability to disrupt existing, traditional industries. It can be used to run any form of exchange, agreement (trust), or tracking processes. The capability to negotiate decentralized trust is key to its application in SCM. Holding the capacity to transform SC functionality, blockchain technology creates possibilities for security and redundancy mechanisms that are foundational for a revolutionary new approach to SCM. While aiming to explore the incumbent status, potential applications, and future directions of blockchain technology, this study iterates that its advent opens up new and exciting ways to implement a variety of different modern technological systems. The writers offer an overview of what an efficient SCM system should entail, with the aim to overcome any existent limitations via the application of blockchain technology.

INTRODUCTION

This academic work has been carried out with the objective to recognize and interpret the prevalent areas of research, challenges, and future scope concerning the Blockchain technology, particularly from the technical perspective. In other words, the purpose is to research, recognize, comprehend, and ask questions regarding blockchain as a technology. It aims to investigate the main topics and subjects of interest, in addition to the already implemented and ongoing projects trying to disrupt modern technological systems

DOI: 10.4018/978-1-7998-9274-8.ch005

like supply chain management. Overall, this research work intends to contribute to the knowledge and potential of blockchain applications in modern technological systems like supply chain management, in order to provide "a blueprint for such applications from the perspective of literature analysis".

With requirements of customers constantly evolving, rising competitive challenges, operations being run at separate geographical locations, plus the "adoption of new business models" make the current technological systems highly complex, and in some ways outdated. With the rising appeal for Environmental, Social, and Corporate Governance (ESG) businesses, it becomes more and more important to implement products that exhibit "autonomy, integrity, transparency, provenance, distributed and decentralized control, anonymity, and immutability". Blockchain technology particularly prioritizes such characteristics, and hence, holds the potential to disrupt all kinds of traditional as well as modern industries. "Prevalent and established technological systems are nowhere near versatile or transparent enough to accommodate the growing needs and demands of the future, leading to huge overheads in terms of error handling, costs, administration and fraud management."

Over the past decade, there has been a rising market appetite and desire for customized products, equipped with a methodical, lucid, and coherent experience for the user, in addition to being able to furnish "transparency about the value and provenance of products". These requirements, while delivering new opportunities, also impose "significant challenges to current technological implementations". For instance, "prevalent supply chains struggle to improve demand management, to provide data visibility for the entire flow, or to track goods from raw material to end consumer, all extremely complex procedures before blockchain". Furthermore, the technology in existent supply chains disappoints in impacting sufficient management of risk, reduction of costs, or even fulfillment of coping with ever-changing business requirements.

Thus, this study is written with the motivating force to shed light on what the main prevalent technological systems are where blockchain might be applied, what disruptions and challenges could arise in such systems due to "the adoption of blockchain", and what can be expected from "the future of blockchain".

BACKGROUND: WHAT EXACTLY IS BLOCKCHAIN?

Blockchain, as an idea, is said to have been first sketched out and drafted in 1991. Two researchers by the name of "Stuart Haber" and "W. Scott Stornetta" had the drive to achieve "a system where document timestamps could not be tampered with". However, as a concept that would actually be possible to execute, Blockchain only first emerged in 2008 as a technology to support transactions in the crypto-currency field (Nakamoto, 2008). Through a white paper, Satoshi Nakamoto (a presumed pseudonym or alias), proposed Bitcoin as "a purely peer-to-peer version of electronic cash that would allow online payments to be sent directly from one party to another without going through a financial institution".

(Risius & Spohrer, 2017) provide a more formal definition for blockchain technology. They refer to blockchain technology as a "fully distributed system for cryptographically capturing and storing a consistent, immutable, linear event log of transactions between networked actors". This, in essence, is very similar to "a consensually maintained distributed ledger that is validated by all the parties involved in the transactions within a network". "In such a network, blockchain technology enforces transparency and guarantees eventual, system-wide consensus on the validity of an entire history of transactions" (Risius & Spohrer, 2017). Despite this formal definition, study into this emerging research area over time has made it clear that blockchains also boast certain other distinctive characteristics like network security, data integrity, tamper-proof transactions, anonymity, and transparency, among the network members.

Maintaining integrity, availability, and confidentiality of all transactions and data involved makes blockchain a "highly disruptive, innovative, decentralized, and distributive state-of-the-art technology". "It is functionally an open, shared, and distributed ledger that can help store/record data and transactions backed by a cryptographic value (Tsan-Ming Choi, 2020) across a peer-to-peer network" (Shuchih Ernest Chang et al., 2019) (Tsan-Ming Choi et al., 2019). With committed blocks being immutable, blockchain can be considered to be "a distributed database organized as a list of ordered blocks" (Casino et al., 2019). "Blockchain is an immutable, tamper-proof distributed ledger technology (DLT) operated in a shared and synchronized environment where all the transactions are traceable and user-validated". As a result, a "decentralized environment" is created where all the network members can interact and communicate securely, without the requirement of a controlling, dominating entity. Moreover, distributed-consensus-enabled storage and validation further assist in eliminating the need for a central entity (Fu & Zhu, 2019) (Min, 2019).

"Blockchain is formed through a series of connected blocks, where transactions' history can easily be traced through previous blocks, thus making the technology transparent and trustworthy". Transaction security is ensured by the presence of a unique ID along with the previous block's hash with every block on the chain. Every transaction gets recorded and goes through validation conducted by the network users. The existence of time stamps and chronological arrangements in addition to every block's connection to its previous block makes them almost irreversible once inserted into the interconnected system. "This entire structure of blockchain makes it a trusted technology" (Queiroz, Maciel M et al., 2020). One of the most significant functionalities which establish blockchain as a "trusted, secure and transparent technology" is the "consensus mechanism". It is essential to abide by the consensus mechanism whenever new blocks are to be appended, records are to be embedded, linked through hash values, etc. This implies that to cause any alteration of an existing block, a participating member needs to compete with all other participating members to build the longer branch. Hence, the collaboration mechanism aids DLT in the maintenance and protection of historical data.

BLOCKCHAIN TECHNOLOGY: KEY ATTRIBUTES AND DISRUPTIVE POTENTIAL

(Iansiti & Lakhani, 2018) explain in clear terms the practical methods of the blockchain technology's functionality and operability, underscoring five basic principles as follows.

- *Distributed Database*: Every blockchain member has the ability to access "the entire ledger and its complete history". The technology while not giving any single node full control over the data, provides the capability to perform intermediary-less direct verifications.
- *Peer-to-Peer Transmission*: Blockchain members can directly communicate without "the need for a central authority or any intermediaries".
- *Transparency with Pseudonymity*: Entities with access to the blockchain's every single transaction can "stay anonymous" or give proof-of-identity via their "unique 30-plus-character alphanumeric address".

- *Irreversibility of Records*: The term 'chain' actually comes from the fact that all blocks hold the header's hash value of their previous block. The mining of a new block means that the alteration of records is no longer possible.
- *Computational Logic*: In essence. blockchain transactions can be programmed and be mapped to computational logic.

In simple words, the blockchain technology exhibits immense potential for future industrial applications. The characteristics that lend it its uniqueness incorporate being decentralized (multiple systems can access, monitor, store and update the available data), transparent (stored network data is fully traceable and available to view throughout its lifetime, with consensus), immutable (timestamps and controls furnished by the blockchain establish immutability), irreversible (each blockchain keeps indisputable and verifiable records for any transactions ever made), autonomy (without third party intervention, all blockchain nodes have the capability to safely retrieve, examine, store, transfer, modify and update data by themselves), open source (based on network hierarchy, open-source-access is given), anonymity (the nodes' individual identity can remain anonymous even as data transfer occurs between them), uniqueness and ownership (ownership records including "unique hash codes" for every possible exchange are stored), provenance (a digital record document certifying genesis/origin and authenticity is always present in the blockchain), smart contracting ("a small computerized program that helps to execute contract and replaces the need of a usual contract by providing better security and lower transaction costs"). "Smart contracts are typically coded to include conditions for rules, penalties, and actions that will be administered for all the parties involved in a certain transaction. Smart contracting reinforces quick response operations in supply chains as well" (Li et al., 2020).

(McKinsey & Company, 2017) refers to the blockchain as "a digital system for executing and recording transactions can be visualized as a building block constructed from intelligent algorithms and collected data, secured by cryptography. Whether in financial markets, healthcare, or the military, industries, and governments are using blockchain technologies to redesign hi-tech paradigms". Gartner ranks blockchain technology among "the top 10 strategic technology trends" for the years 2018 (Panetta, 2017), 2019 (Panetta, 2018), as well as 2020 (Panetta, 2019). In (Panetta, 2017), (Panetta, 2018), and (Panetta, 2019), the authors express that employing a public blockchain practically eliminates the requirement for trusted-central authorities to record transactions or arbitrate disputes. The very concept of "immutable" records on a "distributed ledger" means "trust" is built into the model of the blockchain technology. When viewed conceptually, "a blockchain is a distributed database comprising of records of transactions that are shared among participating members". "Since all transactions are confirmed by the consensus of a majority of the participating members, fraudulent transactions are unable to pass collective confirmation. A record, once generated and recognized by the blockchain, cannot get lost or be altered ever" (Efanov & Roschin, 2018). Some consider blockchain as the "most noteworthy invention after the Internet", as "it is widely anticipated to resolve trust issues through 'peer-to-peer networking' and 'public-key cryptography' solutions" (Efanov & Roschin, 2018). "Although it is commonly agreed that the technology still needs time to mature, mankind needs to be ready to adopt this unprecedented archetype. Hence, blockchain's potential to initiate radical transformations should impose fundamental questions for governments and even society as a whole" (Efanov & Roschin, 2018) (Wong et al., 2018) (IBM Institute for Business Value, 2017) (Iansiti & Lakhani, 2018) (Nallapaneni Manoj Kumar & Pradeep Kumar Mallick, 2018).

(J. AlJaroodi & Mohamed, 2019) highlights that the blockchain technology has emerged as a vanguard across industrial sectors with the capacity to straighten out bottlenecks and lay the foundations for future research work, addressing issues using "shared, secured, distributed and permissioned transactional ledgers" which have made it an inevitable choice for future product implementations.

According to (Peck, Morgen E, 2017) and deep-dive reports by Juniper Research (Kimmich, 2019), the majority of the world's largest corporations and businesses are studying and experimenting with blockchain technologies hoping to successfully integrate them into existing projects and enterprises. Projects are already underway that could "disrupt the management of healthcare records, property titles, supply chains, and even people's online identities". Experts believe that the entire digital ecosystem could potentially be remounted on the blockchain. (Peck, Morgen E, 2017) also provides a process map to decide if a certain product or project actually requires blockchain implementation or not.

Figure 1. Process map: do you really need a blockchain? (Peck, Morgen E, 2017)

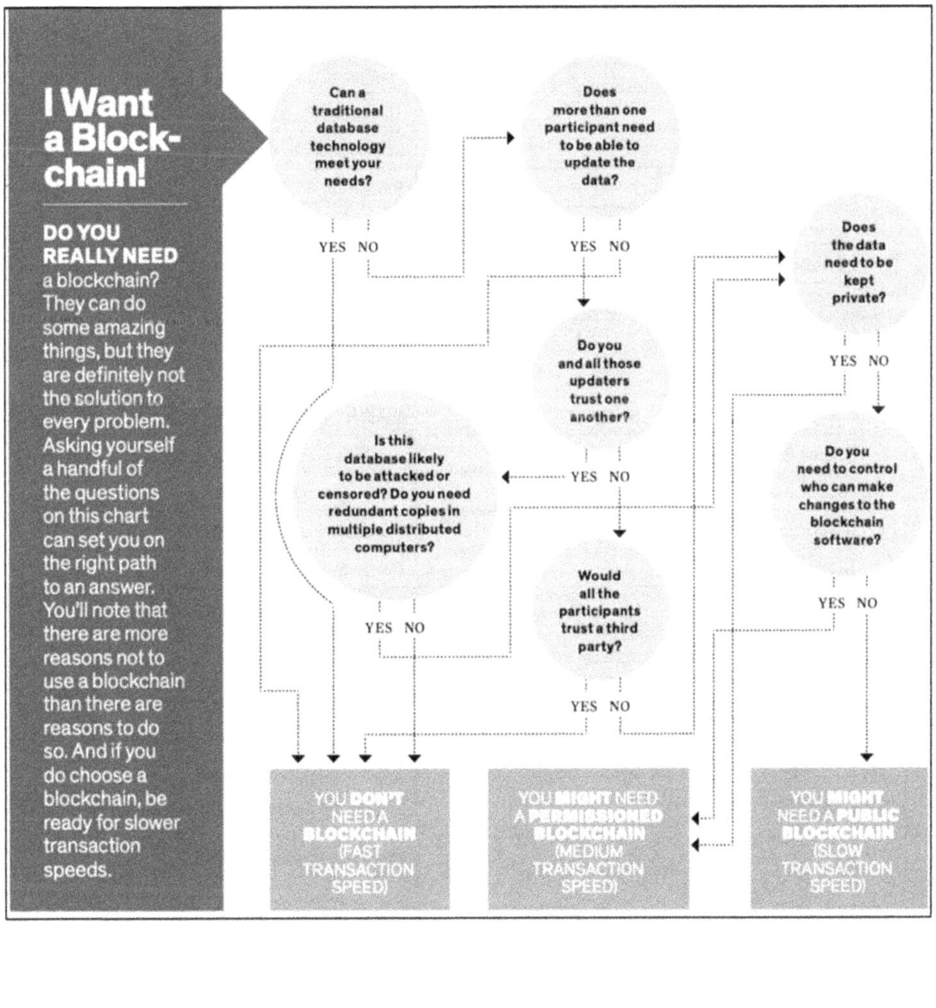

THE APPLICATIONS IN VARIOUS SECTORS

Table 1. Real-world applications of blockchain technology

Sector	Organization	Applications and Targeted Research
Financial Services	Major Central Banks Worldwide	Development of a CBDC (Central Bank Digital Currency; Bahamas, Ghana, Nigeria, and China have already begun trials and public rollouts.
	NASDAQ, Australian Securities Exchange	Record Keeping and Clearing
	Kingdom of Saudi Arabia	Value Added Tax (VAT) Structure
	Credit Union Alliance	Fractional Calculus Model for Bank Credit System
Manufacturing Sector	IBM-Watson IoT, IOTA	Protection of IoT Data
	Hijro	Decentralized Networks
Shipping Industry	IBM & Maersk Partnership	Improvement in Provenance and Transparency
	Provenance	Transparency in SC Product Lifecycle
	Fair Trade	Third-Party Product Information
Automotive Sector	Hyundai	SC Financing Ecosystem using Hyperledger Fabric
	MOBI Initiative	Blockchain Adoption
	Ford	Ethical Sourcing
Energy Sector	Korea Electric Power Corporation	Blockchain-based P2P Energy Transactions Platform
	Innogy	Automated Billing for EV Charging Stations
Construction	Briq, California	Ledger-based documentation
	HerenBouw, Amsterdam	Project Management Platform based on Blockchain
Healthcare Sector	National Health Service (NHS), UK	Medical Device Tracking with Automated Patientcare Monitoring
	MediLedger Project	Blockchain-based Pharmaceutical SC
	Estonian e-Health	Improved Storage of Health Records
Agriculture and Food	Walmart, Alibaba, JD.com	Food Safety and Traceability
	Etherisc	Decentralized Crop Insurance
	UPS	Better Handling of Logistical Challenges
	AgriChain, Ripe	SC Traceability
	AgriDigital	Commodity Management Platform for Grains
Aviation Industry	The Italian Airport Infrastructure	Airport Collaborative Decision Making (A-CDM)
	Norwegian Airways	Cryptocurrency Exchange
Tourism and Entertainment	SingularDTV	End-to-End Blockchain-based Transaction
Education Sector	Open-Source University	Distributed Database System
	LiveEdu, ODEM	Decentralized Marketplace for Educational Resources
E-Commerce	Amazon	DB, SC Traceability and AWS
	Alibaba	Logistics Platform based on Blockchain Technology
	OpenBazaar	Decentralized Marketplace
	Expedia, Overstock	Transactions via Cryptocurrency
Governments	Singapore, Denmark, Australia, Estonia	Keyless Signature Infrastructure (KSI)
	Russian Post	Crypto-based Digital Stamps used in Postal Services

Blockchain, when consciously applied, can drive innovation in all spheres of life. The decade of 2010-20 has been a testament to exactly that. Presently, multiple organizations throughout the globe are pursuing blockchain as a tangible means to solving key industrial and societal issues, thus contributing to global welfare. Some of these disruptive sectors include Financial Services (Fosso Wamba et al., 2019), Credit Accounting and Management, Asset Management and Trade Finance (O'Leary, 2017), Global Networks and Decentralized IoT Systems (Zelbst et al., 2019), Manufacturing Sector (Ko et al., 2018), Shipping Industry (Philipp et al., 2019), Automotive Sector (Sharma et al.,), Energy Sector, Construction and Real Estate (Işıkdağ, 2011), Digital Electoral Systems, Healthcare Sector (Burstall & Clarke, 2019) (Jamil et al., 2019), Agriculture (Kamble, Sachin S et al., 2020), Food Safety (Mao et al., 2018), Space and Aviation Sector (Di Vaio, Assunta & Varriale, 2020), Tourism and Entertainment, Education Sector, Postal Services and Online Advertising, E-Commerce Sector, Global Governance, etc.

BLOCKCHAIN IN SUPPLY CHAIN MANAGEMENT

Several studies in the past have reviewed blockchain technology with researchers landing on the conclusion that it holds the key to disrupt the way modern technological systems function. One such field is that of Supply Chain Management (SCM).

Supply Chain Management, being the enormous sector it is, can also be referred to as the main skeleton of every industry (Hughes et al., 2019). Nevertheless, "traditional SC systems are nowhere near versatile or transparent enough to be able to adjust to or handle the growing needs and demands of the future, thus resulting in huge overheads". These outdated supply chains struggle to boost performance or exhibit transparency at every step of the way. Furthermore, the SCM technology of old has failed to provide "ample risk management", to "scale down costs", or to keep up with the quickly changing market requirements.

Existing supply chains experience multiple challenges. In the recent past, it has come to be critical for SCs to tackle traceability, particularly concerning "customer service, planning, and forecasting in business operations". However, deploying a centralized system in an interconnected network becomes all the more complicated with trust being a limited resource among participants. Instead, there is a variety of "discrete systems" among "involved parties" composed of variegated databases (DBs) that hinder the process of tracking products all through the SC network (El Maouchi et al., 2018b). "Trust is an essential factor in SCM". It is believed that an unwavering foundation of trust is of the utmost importance when it comes to a high-functioning and effective SC network (Tyndall et al., 1998). Distrust amongst SC network participants is one of the greatest reasons hampering their improvement (Poirier, 1999). Consequently, most network stakeholders appoint third-party intermediaries to serve as "agents of trust" that can impartially perform verification on involved transactions. However, this causes the operational costs to dramatically rise, further decreasing process efficiency. The term "transparency" in the SC alludes to the extent of the stakeholders' acknowledgement for a "shared understanding of and access to reliable, precise and sufficient product information" (Frentrup & Theuvsen, 2009) (Pant et al., 2015). A transparent SC network augments trust amongst stakeholders by guaranteeing the "integrity of products and associated data". However, the centralized DBs in present SC networks extend "minimal transparency". Here, data and product transfer among stakeholders cause the loss of a lot of useful information. In addition to this, issues pertaining to inconsistent data sharing, excessive and unnecessary reliance on paper documentation, plus insufficient interoperability, worsen the situation even further. Despite many years of significant research investment, such critical challenges still exist. The "Chipotle Mexican Grill Outlets Crisis" (Kshetri, 2018) is one key instance of the prevalent SC systems' inefficiency. It goes to show that these systems are possibly unable to exhibit any degree of transparency when it comes to the merchandise's provenance and lifecycle.

Paper-based documentation to share data between organizations is still quite common when it comes to supply chains. "Often important documents, be it bills of lading, letters of credit, invoices, insurance policies, or other such certificates/credentials, must travel with their associated goods around the world" (Chang et al., 2020). For instance, to transport a single frozen goods' consignment from "Mombasa to Europe in 2014", the global transport and logistics company Maersk, needed approximately 200 communications (IBM, 2017). "These communications created a stack of documents about 25 centimeters in height" (Allison, 2016). Contrived by the inefficient and obsolete method of sharing information, airplanes, and cargo vessels are often slowed down at docks and harbours, all because of incoherent and inconsistent paperwork (Chang et al., 2020). In addition, there are compliance challenges with the way businesses currently implement and manage supply chains as they need to conform to an increasing extent of scrupulous and stringent regulatory standards to ensure product safety and satisfactory customer service. In the recent past, the U.S. F.D.A. and F.T.C. have adopted several standards to elevate food safety and offer "full visibility of food flows" in the SC. However, with current supply chain processes, one can only imagine how complex it would be to establish a system compliant with the latest regulations when the mere process of obtaining such detailed data from multiple stakeholders gets too complicated and difficult. Blockchain is one of the key innovative and disruptive technologies that has the potential to revolutionize digital SCM. As SCs become increasingly elaborate, grow complicated, cover more and more distinct stakeholders, and continue to rely on external intermediaries, blockchain easily emerges as a formidable contender with the ability to detangle communication exchanges and other data existing in the SC ecosystem.

In the Industry 4.0 era (Fernández-Caramés, Tiago M et al., 2019), the blockchain technology is proposed as a means to organize data and accounts in a "distributed manner through consensus mechanism" (Benčić et al., 2019) (Gao et al., 2018). With features such as "authenticity, transparency, disintermediation, trust, cost reduction, efficient, secure operations, and reduced waste", it has the potential to transform SCM altogether (Gurtu & Johny, 2019) (Philipp et al., 2019). The general conclusion is that blockchain-supported transactions are relatively more traceable, visible, anonymous, secure, efficient, economic, etc. (Queiroz, Maciel M et al., 2020). It is generally regarded as true that the blockchain's distributed nature helps it to "mitigate risks in the supply chain (Araz, Ozgur M et al., 2020) (Zhang et al., 2020) associated with piracy, hacking, vulnerability, expensive compliance with government rules and contractual disputes" (Min, 2019). Blockchain technology aids in facilitating order settlements in real-time in addition to automating manufacturing tasks via the implementation of smart contacts too (Sheel & Nath, 2019). It additionally ensures the "mitigation of ripple effect in SC", minimizing the "disruption caused through changing paradigms" (Ivanov et al., 2019).

In (Chang et al., 2020), the authors outline blockchain's use in global supply chain and trade procedures through scientific articles and industrial research reports along with the various "challenges, opportunities and future scope of research work". The work carried out focuses on the necessities for governments, industries, and academic entities to operate in harmony with each other in order to "fully harness the boundless potential of blockchain". (Gurtu & Johny, 2019) review the literature on "blockchain technology, its trends, implications and the potential value that can be generated in the field of SCM". (Juma et al., 2019) presents a "survey document on various applications of blockchain in trade

SC and its challenges in SC design while integrating blockchain", and (Khosla et al., 2019) present a "basic overview of blockchain technology in supply chain management and its ongoing popularity in supply chain applications".

(Wang et al., 2019) highlight that the technological breakthrough that is blockchain, possesses the potential to enhance and refine the processes of sharing data, managing records as well as role-based accesses. They draw attention to the corresponding/correlated research working on "designing new structural frameworks, models, architecture, etc.", however, "detailed documentation about actual or even prototype implementation is largely missing". Based on four pillars, namely "social, economic, environmental, and governmental", (Shen & Pena-Mora, 2018) discuss the pertinent implementation/application cases for smart and sustainable cities. The writers also state that the Hyperledger Fabric and Ethereum are the most commonly used platforms for applying blockchain technology currently. In (Tsan-Ming Choi et al., 2019), the writers review "the analytical literature using the mean-variance approach for responsive supply chain operations using air logistics" and also highlight blockchain's part/role in the same.

Authors in (Alick et al., 2017), (Chang et al., 2019), (C. Narayanaswami et al.), (Deloitte, 2017), (Infopulse, 2019), (Saberi et al., 2018), (Surjandy et al., 2019), and (Wang et al., 2019), list the following merits of a blockchain-based SC implementation.

- *Data management*: calibration of data/information real-time-capturing ensured across diverse SCs.
- *Improves transparency*: real-time-status-tracking of every item/task; data-analysis-automation; maintenance of permission-level-based end-to-end transparency.
- *Improves response time*: dynamic SC with efficient utilization of resources.
- *Smart contract management*: improved coordination via dedicated function and customization; helps design business operations; amplifies visibility and gets rid of the requirement for a central mediator.
- *Operational efficiency*: speed enhancement for end-to-end SCM; high robustness due to early debugging.
- *Disintermediation*: uninterrupted-transaction-chains made feasible with "improved speed and enhanced trust amongst process stakeholders".
- *Immutability*: "consensus mechanism for modifications means transaction security is ensured".
- *Intellectual property management*: IP-protection and registration.

IDENTIFIED CHALLENGES AND LIMITATIONS

It is important to note that blockchain technology too, in its current sense, has some technical challenges and limitations. (Swan, 2015) has identified and put forth 7 future technical challenges and hindrances for blockchain adoption.

- *Throughput*: Bitcoin network's prevailing-potential-throughput of issues is maximized to 7 transactions per second (tps). Other networks like Twitter and Visa process 5,000 tps and 2,000 tps respectively. To be able to handle transaction frequencies of such levels, the blockchain technology and networks will need to be refined and upgraded.

- *Latency*: At present, to complete a Bitcoin transaction with adequate security for the same, it takes about ten minutes, which is quite long when compared to a few seconds taken for a VISA transaction. To ensure security in an efficient manner, the network has to spend more time on a block, so that it can easily maneuver double-spending-attacks. "Double-spending is the result of successful spending of money more than once. Bitcoin protects against double-spending by verifying each transaction added to the blockchain, to ensure that the inputs for the transaction have not been spent previously" (Irreversible Transactions - Bitcoin Wiki, n.d.). Such a situation means that latency is a serious matter of concern for blockchain, especially when it comes to Bitcoin. Ideally, the process of block creation and transaction confirmation should be a matter of seconds, even with the requirement of maintaining security. So, in terms of latency, other prevalent networks do enjoy a relative advantage over blockchain technology.
- *Size and bandwidth*: The bitcoin blockchain size is subjected to continuous increase. As of February 2021, the bitcoin blockchain size is approximately 15 GB, growing at a rate of about 1 MB per hour (What is Bitcoin Blockchain and Bitcoin Blockchain Size? - Intellipaat, n.d.). However, on the other hand, to be able to handle transactions even comparable to the Visa network, the bitcoin blockchain size will need to rise at a rate of 1 GB per second. In the case of Bitcoin, another limitation is that a block of size 1 MB is created every 10 minutes (Bitcoin Wiki, n.d.), which further causes an issue for the number of transactions that can be handled in one block (Antonopoulos, 2014). In order to smoothly handle a growing number of transactions, resolution of these issues is vital.
- *Security*: The existing blockchains have a risk of 51% attacks. In such attacks, an entity manipulates the blockchain to take control of the majority, i.e. at least 51% of the network's mining hashrate. To overcome this, more research focused on blockchain network security is required.
- *Wasted resources*: Bitcoin mining squanders vast amounts of energy (approximately $USD 15M/day). The Proof-of-Work (PoW) effort is the main cause behind this. Here, the miner's work determines the likelihood of mining a block (Proof of work - Bitcoin Wiki, 2019). Alternatively, the industry has proposed alternatives like the Proof-of-Stake (PoS) effort where the amount of Bitcoin held by a miner is the resource that is compared (Proof of Stake - Bitcoin Wiki, n.d.). For instance, someone holding 0.5% ETH can mine 0.5% of the 'PoS blocks'. Solving this concern can lead to more efficient blockchain mining.
- *Usability*: Complex APIs at the moment raise the need for more developer-friendly ones for expanding the reach of blockchain technology.
- *Versioning, hard forks, multiple chains*: The possibility of a 51% attack becomes higher with smaller blockchains with a smaller number of nodes. Splitting of chains for administrative and versioning purposes further causes issues.

SCM: THE SHIFTING SCENARIO

In essence, SCM refers to "the management of the flow of goods and services and all activities that transform a raw material into a final product". Blockchain technology can enhance some key SC functions. These functional characteristics of SC include enhancement of security, SC resilience, SC provenance, SC reengineering, management of business processes and product, etc. In terms of resilience, blockchain provides multilayer protection for the supply chain network by administering "preventive and proactive" initiatives for managing risk. In the matter of provenance and reengineering, it provides data integrity and traceability of the product information, in addition to elevating the SC's "transparency and visibility". "This enables process automation, eliminates intermediaries, and facilitates tracking in real-time through traceability and data management techniques". As to the enhancement of security procedures, better "authentication and confidentiality" policies implemented through consensus mechanism and dynamic data storage intensify SC security. When it comes to the management of business processes and product, blockchain technology is a better means to implement supply chains because it involves smart contracting by "compiling the control flow and business logic of inter-organizational business processes", that further improve cycle time, productivity, and quality. This leads to a smoother product differentiation process for integration with SCs.

Figure 2. Traditional vs. blockchain implementation of supply chain management (Rooyen, 2017)

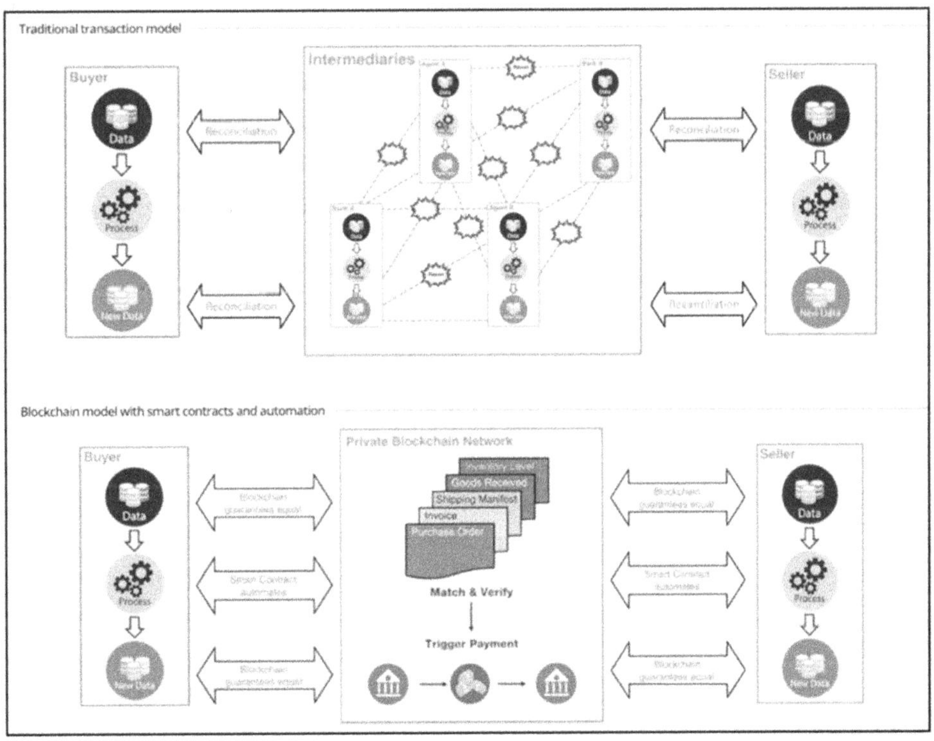

The majority of the world's largest corporations and businesses are studying and experimenting with blockchain technologies hoping to successfully integrate them. "Projects are already underway that could disrupt the management of healthcare records, property titles, supply chains, and even people's online identities. Experts believe that the entire digital ecosystem could potentially be remounted on the blockchain technology". In the Industry 4.0 era of automation, ML, AI, IoT & Cloud, the Blockchain Technology is proposed as a course of action to organize data in a "distributed manner through consensus mechanism". With features such as "authenticity, transparency, disintermediation, trust, cost reduction, efficient, secure operations, and reduced waste", it has the potential to transform SCM. All blockchain-technology-supported transactions are more secure, efficient, economic as well as transparent. The blockchain's distributed nature aids in SC risk mitigation. It not only helps facilitate real-time order settlement but also automates various tasks via the implementation of smart contracts. Furthermore, blockchain even ensures the "mitigation of ripple effect in SC", thus minimizing the "disruption caused through changing paradigms". SCM, being a highly expansive sector, is sometimes called "the main skeleton of all industries". However, traditional SC systems aren't anywhere near to be versatile or transparent enough to effectively handle the ever-growing future needs. This leads to enormous running costs. Recently, the desire for tailored products, equipped with a clear and comprehensive UI has been on the rise. Consumers are beginning to prioritize "transparency and provenance" again. However, with these prospects, there also arise significant technological challenges too complicated to implement before blockchain.

SOLUTIONS AND RECOMMENDATIONS: A PROPOSED FRAMEWORK

As the blockchain technology continues to evolve, the idea of a real and functional blockchain-based SCM system becomes clearer. The authors thus put forth a framework for exactly that. The supply chain designed to be independent and totally self-contained provides a simple mechanism for users to access and utilize various operations catering to them. For the purpose of application, a waste management cum e-commerce supply chain, operating over a blockchain network, is considered. The system records transactions on the Ethereum Blockchain while making use of technological frameworks like the Truffle Suite, Solidity Smart Contracts, Ganache Client, Metamask (Ethereum wallet), JavaScript, Web3.js, Node.js, HTML, CSS, Rinkeby Test Network, Mocha, Remix, Infura, IPFS, etc. Intended to be straightforward and effortlessly understood, the operation and functionality of the blockchain-based SCM are majorly based on three solidity smart contracts, namely AdminContract, NodeContract, and Migrations. The web interface interacting with the contracts involves the use of supply chain actors like admin, producer, retailer, consumer, recycling unit. Each role can interface with the blockchain in its own ways while maintaining the basic set of principles it operates on, i.e. transparency, traceability, immutability, etc. The goal behind a waste management supply chain is to make an effort towards mitigation of environmental threats while supporting and cultivating economic growth. This illustration not only exhibits the benefits of blockchain in supply chains but also propagates the need to practice waste management in a responsible manner. It helps nurture the environmental and ethical standards by giving each actor of the supply chain to participate in managing and recycling waste, all while running an e-commerce business.

The implementation opens doors to a multitude of possibilities through transparent monitoring of the supply chain and its pragmatic modules. Moreover, seamless interaction between various concerned parties with less fear of fraud makes this a valuable resource with an exciting future. In other words, this technology can provide accountability to the supply chain, by enabling free flow of information, with the purpose to eradicate all prevalent barriers to data interoperability, thus enabling transparency and access to respective information in real-time, supporting such implementation of the circular economy concept.

Figure 3. Interaction of SC Actors with the Blockchain DLT (Litke et al., 2019)

Architecture and Key Modules

Multiple experts have studied the blockchain architecture. (Min, 2019) and (Choi et al., 2020) explain the 5 modules governing the functioning protocols for blockchain applications.

- *Data source module*: "It supports to structure the blockchain as the 'distributed and shared DBs'. It ensures that the retrieved data stays unaltered, uncorrupted, and unchanged. Data immutability, tamper-proofed storage with any form, and shared data ledger through data APIs are the key aspects of blockchain."
- *Transaction module*: It manages, supervises, and supports the 'journey of a blockchain transaction'. It also assists in the validation process for any additions to the blockchain. Data that is explicitly visible to all participants is transferred through smart contracting transaction gates. Transactions are collected together in blocks to be transmitted to all nodes. Once executed, the transactions are almost impossible to delete or rollback in blockchains.
- *Block creation module*: Blocks, considered to be data structures developed by miners, hold important information. The block creation module replicates this information to all nodes of the network

and enables new blocks to be appended to an existing chain by supplying the previous block's hash values and links. The transaction sequences are stored in 'chronological blocks' so that the malicious or invalid blocks can be flagged more easily.

- *Consensus module*: To avoid the corruption of data, PoW and PoS algorithms are employed. The meticulous algorithms that confirm and validate all transactions maintain consistency in the decentralized and distributed system. Distributed consensus is good for transaction validation plus block-link-creation in the blockchain system.
- *Connection and interface module*: It keeps track of all transactions and gives real-time data on smart contracts. This module synchronizes all the IT platforms, algorithms, and software needed for applying blockchain technology. Depending on use cases, multiple DLT programmes that offer consensus algorithms for the blockchain system, no matter the blockchain type (public, private, permissioned, permissionless/non-permissioned, or hybrid), can be made available in the market.

Figure 4. Architecture diagram for proposed system: a typical SC process using blockchain

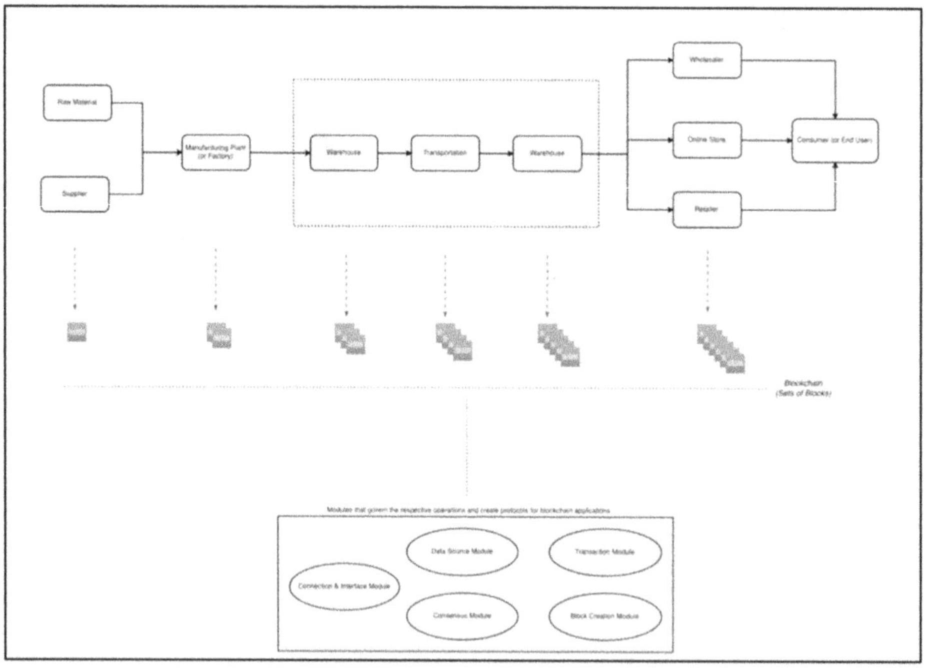

DISCUSSION

Blockchain, once just an idea of a system with tamper-proof document timestamps, finally became a reality in 2008, with the emergence of Bitcoin. Although blockchain is most popular for its role in the cryptocurrency boom, the actual underlying technology is far beyond the concept of digital currencies. It is an "immutable, tamper-proof DLT executing in an environment that is shared and where the transactions are transparent and validated through consensus".

Blockchain technology's real potential lies in its ability to disrupt existing, traditional industries. With an unrivalled union of properties like decentralized structure, distributed notes and storage mechanism, consensus algorithm, smart contracting, and asymmetric encryption, in order to ensure network security, privacy, anonymity, transparency, visibility, and data integrity, it establishes a decentralized environment where all the network members can securely interact without the requirement of a dominant controlling authority.

Experts believe that blockchain holds the key to disrupt the way modern technological systems and industries operate. Supply chain management is at the very core of any business. Moreover, the prevalent supply chain systems in recent times have failed to live up to the standards of the evolving market. Blockchain technology's attributes of transparency and provenance align perfectly with the core properties that are needed today for transforming supply chain management into a successful traceability system.

Although traceability might not be a priority for corporations, reducing their overheads by managing risk and scaling down costs definitely is. Hence, the application of blockchain technology in modern technological systems is an appealing venture for both corporate organizations and their customers.

It is apparent that blockchain technology is the most befitting approach to revolutionize industries. As a result, huge research projects and investments into this domain have already been initiated. However, it is important to note that there do arise points where science and practicality might not totally coincide. Issues of latency, throughput, bandwidth, security, resource stage, etc. are a concern in the pursuit towards high operational efficiency.

CONCLUSION AND FUTURE RESEARCH DIRECTIONS

There is huge momentum and driving force for the technological breakthrough known as blockchain, especially in the field of SCM. This is majorly due to distinct features like decentralization, end-to-end visibility, transparency, enhanced decision making, traceability, augmented sharing and security of data, full-fledged integration, etc. The application of automated computer programs called smart contracts is being used to propel humanitarian aid. "Blockchain-based supply chain helps curb social and political issues like corruption via quicker, transparent and low-cost transactions". Another benefit includes the enforcement of "labour and environmental standards" in the corporate organizations throughout the world employing crowdsourced accountability, thus upholding health, safety, and inspectability standards of the industry. Traceability and provenance further ensure that only legally sourced raw materials are used at all times. Reduction of unnecessary paperwork or old systems makes it possible for these new SCM systems to be much more efficient. Also, an adequate amount of risk management coupled with better customer service, lower costs, planning, and forecasting in business operation, makes blockchain application preferentially economical as well.

Since the concept of blockchain is not vividly known to many, despite extensive research, a large section of the industry is still not ready or comfortable enough to shift its operations and applications from a traditional database to a blockchain network. A lot of the time, one feels better sticking with an existing technology that works, rather than adopting a totally new one with very little commercial awareness or success. Hence, learning and understanding a concept to make essential decisions is an elemental factor pertaining to software engineering projects. Analyzing the requirements is a decent method to better comprehend whatever the target problem is.

The developed system must possess attributes allowing users to use any of the functionalities both a supply chain and blockchain have to offer. Some of these include trust, transparency, immutability, anonymity, integrity, efficiency, reliability, and portability. In order to successfully deploy the system developed, several pre-requisites need to be fulfilled. To begin with, a truffle environment needs to be developed where solidity smart contracts can be tested, compiled, and migrated. A metamask wallet also needs to be initiated for the purpose of all blockchain transactions, including functions as they require gas to be executed. Once the contracts are ready to go, both the front-end and back-end systems are run. Based on the system requirements and availability of ethers in one's metamask wallet, deployment can be made locally on the host system, on an Ethereum test network like Rinkeby, or on the main Ethereum network as well. The developers involved should all be well-versed in software engineering practices and principles while working on the system. Documentation is an essential aspect of the development life cycle. In addition, the specific SCM system under consideration requires knowledge and experience in web development, networking, and blockchain technologies and tools. One should have in-depth knowledge about how to work with Ethereum, truffle, solidity, etc. Only with the proper expertise can a high-performing, secure, and efficient blockchain-based supply chain be achieved.

Furthermore, one of the most intriguing events in the future that could practically revolutionize the blockchain world, and specifically the blockchain-based SCM systems, will be the launch of Ethereum 2.0. ETH2.0 is what many call the expected upgrade to the Ethereum network. This new version aims to enhance the network's security and scalability by shifting the current mining model to a staking model. With this, transaction costs, as well as energy effects, could be managed judiciously.

SUMMARY

This academic work provides an outline of the design/architecture as well as features/characteristics of blockchain technology, in addition to few extremely essential background surveys on blockchain in supply chain management and the intricacies of the prevalent handiwork. It also puts forth the huge potential applications of blockchain in various industries. The authors propose a blockchain-based SCM system. It involves a demonstration of how multiple actors throughout the supply chain can directly interact with one another, easily track everything happening in the supply chain, all while keeping overheads to a minimum. The proposed system is developed on the Ethereum blockchain using the truffle suite and solidity smart contracts. Smart contracts refer to "transactional scripts on the top of the blockchain technology" exhibiting extensive prospects in assuring SC disintermediation, including the possibility to diminish costs related to implementation, execution, enforcement, and transactions. The exploration and analysis exhibit that the application of smart contracts with blockchain technology strengthens the effectiveness of entrepreneurial collaborations and cross-organizational business processes, thus leading to smart supply chains.

In addition, readers can get a synopsis of the recently emerging blockchain developments along with the potential disruptive applications of this technology in various sectors. The research work has made an effort to report all relevant studies and projects published in the concerned field, with an intention to facilitate industries as well as academics towards an improved perception of the related studies. Various theories, challenges, and impacts discussed can provide further benefit to concerned stakeholders by acting as valuable references. Overall, this academic work contributes to the knowledge and potential of blockchain's implementation prospects in modern technological systems like supply chain management, to provide a blueprint for such applications.

REFERENCES

Alick, K., Davies, A., Leopoldseder, M., & Niemeyer, A. (2017). *Blockchain technology for supply chains--A must or a maybe?* McKinsey & Company. https://www.mckinsey.com/business-functions/operations/our-insights/blockchain-technology-for-supply-chainsa-must-or-a-maybe

J. AlJaroodi, & Mohamed, N. (2019). Blockchain in Industries: A Survey. *IEEE Access, 7*, 36500–36515. doi:10.1109/ACCESS.2019.2903554

Allison, I. (2016, October 12). Shipping giant Maersk tests blockchain-powered bill of lading. *International Business Times UK*. https://www.ibtimes.co.uk/shipping-giant-maersk-tests-blockchain-powered-bills-lading-1585929

Antonopoulos, A. M. (2014). Mastering Bitcoin: Unlocking Digital Cryptocurrencies. In *Google Books*. O'Reilly Media, Inc. https://books.google.com/books?hl=en&lr=&id=IXmrBQAAQBAJ&oi=fnd&pg=PR4&dq=Antonopoulos+AM.+Mastering+Bitcoin:+unlocking+digital+cryptocurrencies.+%E2%80%9C+O%E2%80%99Reilly+Media

Araz, O. M., Choi, T., Olson, D. L., & Sibel, S. F. (2020). Role of Analytics for Operational Risk Management in the Era of Big Data. *Decision Sciences, 51*(6), 1320–1346.

Benčić, F. M., P. Skočir, & Žarko, I. P. (2019). DLTags: DLT and Smart Tags for Decentralized, PrivacyPreserving, and Verifiable Supply Chain Management. *IEEE Access, 7*, 46198–46209. doi:10.1109/ACCESS.2019.2909170

Burstall, R., & Clarke, B. (2019, January 14). *Blockchain, IP, and the Pharma Industry – How Distributed Ledger Technologies can help Secure the Pharma Supply Chain – Blockchain*. blockchain.bakermckenzie.com https://blockchain.bakermckenzie.com/2019/01/14/blockchain-ip-and-the-pharma-industry-how-distributed-ledger-technologies-can-help-secure-the-pharma-supply-chain/

Casino, F., Dasaklis, T. K., & Patsakis, C. (2019). A systematic literature review of blockchainbased applications: Current status, classification and open issues. *Telematics and Informatics, 36*, 55–81. doi:10.1016/j.tele.2018.11.006

Chang, S. E., Chen, Y., & Lu, M. (2019). Supply chain reengineering using blockchain technology: A case of smart contract based tracking process. *Technological Forecasting and Social Change, 144*, 1–11. doi:10.1016/j.techfore.2019.03.015

Chang, Y., Iakovou, E., & Shi, W. (2020). Blockchain in global supply chains and cross border trade: A critical synthesis of the stateoftheart, challenges and opportunities. *Null, 58*(7), 2082–2099. doi:10.1080/00207543.2019.1651946

Choi, Wen, Sun, & Chung. (2019). The mean-variance approach for global supply chain risk analysis with air logistics in the blockchain technology era. *Transportation Research Part E: Logistics and Transportation Review, 127*, 178–191.

Choi, T., Guo, S., Liu, N., & Shi, X. (2020). Optimal pricing in ondemandserviceplatformoperations with hired agents and risksensitive customers in the blockchain era. *European Journal of Operational Research*, *284*(3), 1031–1042. doi:10.1016/j.ejor.2020.01.049

Choi, T.-M. (2020). Creating all-win by blockchain technology in supply chains: Impacts of agents' risk attitudes towards cryptocurrency. *The Journal of the Operational Research Society*, *0*, 1–16. doi:10.1080/01605682.2020.1800419

Deloitte. (2017, September 20). *Using Blockchain to Drive Supply Chain Transparency and Innovation*. Deloitte United States. https://www2.deloitte.com/us/en/pages/operations/articles/blockchain-supply-chain-innovation.html

Di Vaio, Assunta, & Varriale. (2020). Blockchain technology in supply chain management for sustainable performance: Evidence from the airport industry. *International Journal of Information Management, 52*, 102014.

Efanov, D., & Roschin, P. (2018). The AllPervasiveness of the Blockchain Technology. *8th Annual International Conference on Biologically Inspired Cognitive Architectures, BICA 2017 (Eighth Annual Meeting of the BICA Society), 123*, 116–121.

El Maouchi, M., Ersoy, O., & Erkin, Z. (2018b). TRADE: A Transparent, Decentralized Traceability System for the Supply Chain. *ERCIM - The European Research Consortium for Informatics and Mathematics*. doi:10.18420/blockchain2018_01

Fernández-Caramés, T. M., Blanco-Novoa, O., Froiz-Míguez, I., & Fraga-Lamas, P.Fernández-Caramés. (2019). Towards an autonomous industry 4.0 warehouse: A UAV and blockchain-based system for inventory and traceability applications in big data-driven supply chain management. *Sensors (Basel)*, *19*(10), 10. Advance online publication. doi:10.339019102394 PMID:31130644

Fosso Wamba, S., Kala Kamdjoug, J. R., Epie Bawack, R., & Keogh, J. G. (2019). Bitcoin, Blockchain and Fintech: A systematic review and case studies in the supply chain. *Production Planning and Control*, *31*(2-3), 115–142. doi:10.1080/09537287.2019.1631460

Frentrup, M., & Theuvsen, L. (2009). Information technologies and transparency in agri-food supply chains: empirical results from the German pig and dairy production. *Georg-August University Goettingen*. Department of Agricultural Economics and Rural Development.

Fu, Y., & Zhu, J. (2019). Big production enterprise supply chain endogenous risk management based on blockchain. *IEEE Access: Practical Innovations, Open Solutions*, *7*, 15310–15319. doi:10.1109/ACCESS.2019.2895327

Gao, Z., Xu, L., Chen, L., Zhao, X., Lu, Y., & Shi, W. (2018). CoC: A Unified Distributed Ledger Based Supply Chain Management System. *Journal of Computer Science and Technology*, *33*(2), 237–248. doi:10.100711390-018-1816-5

Gurtu, A., & Johny, J. (2019). Potential of blockchain technology in supply chain management: A literature review. *International Journal of Physical Distribution & Logistics Management*, *49*(9), 881–900. doi:10.1108/IJPDLM-11-2018-0371

Hughes, A., Park, A., Kietzmann, J., & ArcherBrown, C. (2019). Beyond Bitcoin: What blockchain and distributed ledger technologies mean for firms. *Business Horizons, 62*(3), 273–281.

Iansiti, M., & Lakhani, K. (2018, March 6). The Truth About Blockchain. *Harvard Business Review*. https://hbr.org/2017/01/the-truth-about-blockchain

IBM. (2017, March 5). *Maersk and IBM Unveil Supply Chain Solution on Blockchain*. https://www-03.ibm.com/press/us/en/pressrelease/51712.wss#feeds

IBM Institute for Business Value. (2017, January). *Building trust in government*. IBM. https://www.ibm.com/thought-leadership/institute-business-value/report/blockchain-for-government

Infopulse. (2019, October 7). *Blockchain in Supply Chain Management: Key Use Cases and Benefits*. Medium. https://medium.com/@infopulseglobal_9037/blockchain-in-supply-chain-management-key-use-cases-and-benefits-6c6b7fd43094

Irreversible Transactions - Bitcoin Wiki. (n.d.). *En.bitcoin.it*. https://en.bitcoin.it/wiki/Double-spending

Işıkdağ, Ü. (2011). Barriers to E-Procurement in Turkish AEC Industry. *CIB W078 2011 Conference*. https://www.academia.edu/5469207/Barriers_to_E_Procurement_in_Turkish_AEC_Industry

Ivanov, D., Dolgui, A., & Sokolov, B. (2019). The impact of digital technology and Industry 4.0 on the ripple effect and supply chain risk analytics. *Null, 57*(3), 829–846. doi:10.1080/00207543.2018.1488086

Jamil, F., Hang, L., Kim, K., & Kim, D. (2019). A Novel Medical Blockchain Model for Drug Supply Chain Integrity Management in a Smart Hospital. *Electronics (Basel), 8*(5), 505. doi:10.3390/electronics8050505

Juma, H., Shaalan, K., & Kamel, I. (2019). A Survey on Using Blockchain in Trade Supply Chain Solutions. *IEEE Access*, 1. doi:10.1109/ACCESS.2019.2960542

Kamble, S. S., Gunasekaran, A., & Sharma, R. (2020). Modeling the blockchain enabled traceability in agriculture supply chain. *International Journal of Information Management, 52*, 101967.

Khosla, D., Sharma, M., Sharma, A., Budhiraja, A., & Singh, S. (2019). Blockchain based supply chain management: An overview. *International Journal of Control and Automation, 12*, 424–430.

Kimmich, M. (2019, November 5). *Devices & Technologies - Blockchain*. https://www.juniperresearch.com/researchstore/devices-technology/blockchain-research-report

Ko, T., Lee, J., & Ryu, D. (2018). Blockchain Technology and Manufacturing Industry: Real-Time Transparency and Cost Savings. *Sustainability, 10*(11), 4274. doi:10.3390u10114274

Kshetri, N. (2018). Blockchain's roles in meeting key supply chain management objectives. *International Journal of Information Management, 39*, 80–89. doi:10.1016/j.ijinfomgt.2017.12.005

Kumar, N. M., & Mallick, P. K. (2018). Blockchain technology for security issues and challenges in IoT. *Procedia Computer Science, 132*, 1815–1823. doi:10.1016/j.procs.2018.05.140

Li, G., Li, L., Choi, T.-M., & Sethi, S. P. (2020). Green supply chain management in Chinese firms: Innovative measures and the moderating role of quick response technology. *Journal of Operations Management, 66*, 958–988.

Litke, A., Anagnostopoulos, D., & Varvarigou, T. (2019). Blockchains for Supply Chain Management: Architectural Elements and Challenges Towards a Global Scale Deployment. *Logistics, 3*(1), 5. Advance online publication. doi:10.3390/logistics3010005

Mao, D., Wang, F., Hao, Z., & Li, H. (2018). Credit Evaluation System Based on Blockchain for Multiple Stakeholders in the Food Supply Chain. *International Journal of Environmental Research and Public Health, 15*(8), 1627. Advance online publication. doi:10.3390/ijerph15081627 PMID:30071695

McKinsey & Company. (2017). *Artificial intelligence takes shape*. McKinsey & Company. https://www.mckinsey.com/quarterly/the-magazine/2017-issue-4-mckinsey-quarterly#

Min, H. (2019). Blockchain technology for enhancing supply chain resilience. *Business Horizons, 62*(1), 35–45. doi:10.1016/j.bushor.2018.08.012

Nakamoto, S. (2008). *Bitcoin: A Peer-to-Peer Electronic Cash System.* https://bitcoin.org/bitcoin.pdf

Narayanaswami, Nooyi, Govindaswamy, & Viswanathan. (n.d.). Blockchain anchored supply chain automation. *IBM Journal of Research and Development, 63*(2/3), 7:1–7:11. doi:10.1147/JRD.2019.2900655

O'Leary, D. E. (2017). Configuring blockchain architectures for transaction information in blockchain consortiums: The case of accounting and supply chain systems. *Intelligent Systems in Accounting, Finance & Management, 24*(4), 138–147. doi:10.1002/isaf.1417

Panetta, K. (2017, October 3). *Smarter With Gartner.* https://www.gartner.com/smarterwithgartner/gartner-top-10-strategic-technology-trends-for-2018/

Panetta, K. (2018, October 15). *Smarter With Gartner.* https://www.gartner.com/smarterwithgartner/gartner-top-10-strategic-technology-trends-for-2019/

Panetta, K. (2019, October 21). *Smarter With Gartner.* https://www.gartner.com/smarterwithgartner/gartner-top-10-strategic-technology-trends-for-2020/

Pant, R. R., Prakash, G., & Farooquie, J. A. (2015). A Framework for Traceability and Transparency in the Dairy Supply Chain Networks. *Operations Management in Digital Economy, 189*, 385–394.

Peck, M. E. (2017). Blockchain world - Do you need a blockchain? This chart will tell you if the technology can solve your problem. *IEEE Spectrum, 54*(10), 38–60. doi:10.1109/MSPEC.2017.8048838

Philipp, R., Prause, G., & Gerlitz, L. (2019). Blockchain and Smart Contracts for Entrepreneurial Collaboration in Maritime Supply Chains. *Transport and Telecommunication Journal, 20*(4), 365–378.

Poirier, C. C. (1999). Advanced Supply Chain Management: How to Build a Sustained Competitive Advantage. In *Google Books*. Berrett-Koehler Publishers. https://books.google.co.in/books?hl=en&lr=&id=66XhLR3Ih_MC&oi=fnd&pg=PR7&dq=C.+C.+Poirier

Proof of Stake - Bitcoin Wiki. (n.d.). *En.bitcoin.it.* https://en.bitcoin.it/wiki/Proof_of_Stake

Proof of work - Bitcoin Wiki. (2019). *Bitcoin.it*. https://en.bitcoin.it/wiki/Proof_of_work

Queiroz, M. M., Telles, R., & Bonilla, S. H. (2020). Blockchain and supply chain management integration: A systematic review of the literature. *Supply Chain Management*, *25*(2), 241–254. doi:10.1108/SCM-03-2018-0143

Risius, M., & Spohrer, K. (2017). A Blockchain Research Framework. *Business & Information Systems Engineering*, *59*(6), 385–409. doi:10.100712599-017-0506-0

Saberi, S., Kouhizadeh, M., Sarkis, J., & Shen, L. (2018). Blockchain technology and its relationships to sustainable supply chain management. *International Journal of Production Research*, *57*(7), 2117–2135. doi:10.1080/00207543.2018.1533261

Sharma, P. K., N. Kumar, & Park, J. H. (n.d.). BlockchainBased Distributed Framework for Automotive Industry in a Smart City. *IEEE Transactions on Industrial Informatics, 15*(7), 4197–4205. doi:10.1109/TII.2018.2887101

Sheel, A., & Nath, V. (2019). Effect of blockchain technology adoption on supply chain adaptability, agility, alignment and performance. *Management Research Review*, *42*(12), 1353–1374. doi:10.1108/MRR-12-2018-0490

Shen, C., & Pena-Mora, F. (2018). Blockchain for Cities—A Systematic Literature Review. *IEEE Access: Practical Innovations, Open Solutions*, *6*, 76787–76819. doi:10.1109/ACCESS.2018.2880744

Shuchih, Chen, & Lu. (2019). Supply chain re-engineering using blockchain technology: A case of smart contract based tracking process. *Technological Forecasting and Social Change, 144*, 1–11.

Surjandy, M., Hidayanto, A. N., & Prabowo, H. (2019). The latest adoption blockchain technology in supply chain management: A systematic literature review. *ICIC Express Letters*, *13*, 913–920. doi:10.24507/icicel.13.10.913

Swan, M. (2015). Blockchain: Blueprint for a New Economy. In *Google Books*. O'Reilly Media, Inc. https://books.google.com/books?hl=en&lr=&id=RHJmBgAAQBAJ&oi=fnd&pg=PR3&ots=XRrEJ-WOc6&sig=Kn7hlqSVIsLoQSnvbLHAgAD6bxI

Tyndall, G., Gopal, C., Partsch, W., & Kamauff, J. (1998). *Supercharging supply chains. New ways to increase value through global operational excellence.* John Wiley & Sons, Incorporated. https://trid.trb.org/view/672145

van Rooyen, J. (2017, May 3). *Blockchains for supply chains - part I*. Resolve Solution Partners. https://resolvesp.com/blockchains-supply-chains/

Wang, Y., Han, J. H., & Beynon-Davies, P. (2019). Understanding blockchain technology for future supply chains: A systematic literature review and research agenda. *Supply Chain Management*, *24*(1), 62–84. doi:10.1108/SCM-03-2018-0148

What is Bitcoin Blockchain and Bitcoin Blockchain Size? (n.d.). *Intellipaat Blog*. https://intellipaat.com/blog/tutorial/blockchain-tutorial/what-is-bitcoins-blockchains/

WikiB. (n.d.). *En.bitcoin.it*. https://en.bitcoin.it

Wong, M. C., Yee, K. C., & Nohr, C. (2018). Socio-technical consideration for blockchain technology in healthcare. *Studies in Health Technology and Informatics*, *247*, 636–640. doi:10.3233/978-1-61499-852-5-636 PMID:29678038

Zelbst, P. J., Green, K. W., Sower, V. E., & Bond, P. L. (2019). The impact of RFID, IIoT, and Blockchain technologies on supply chain transparency. *Journal of Manufacturing Technology Management*, *31*(3), 441–457. doi:10.1108/JMTM-03-2019-0118

Zhang, J., Sethi, S. P., Choi, T.-M., & Cheng, T. C. E. (2020). Supply chains involving a mean-variance-skewness-kurtosis newsvendor: Analysis and coordination. *Production and Operations Management*, *29*, 1397–1430.

Chapter 6
Developing the Real Estate Rental Sector in Third World Countries Using Blockchain Technology:
Iraq as Case Study

Muthana Alisawi
https://orcid.org/0000-0003-0257-1811
College of Computer Science and Information Technology, University of Kirkuk, Iraq

Aras Al-Dawoodi
College of Computer Science and Information Technology, University of Kirkuk, Iraq

Yousif Mohammed Wahab
College of Computer Science and Information Technology, University of Kirkuk, Iraq

Layth Hammood
College of Computer Science and Information Technology, University of Kirkuk, Iraq

Asmaa Yaseen Nawaf
College of Computer Science, University of Anbar, Kirkuk, Iraq

Alaan Ghazi
Department of Computer Technical Engineering, Al-Qalam University College, Iraq

ABSTRACT

Blockchain technology represents an attractive factor for sectors that are interested in managing and keeping their transactions on a database that records all the movements that take place on each transaction. Therefore, to keep pace with the digital transformation taking place in this field of the real estate sector, this chapter aims to submit a proposal in the Iraqi real estate sector to implement this technology. In addition, the chapter clarified the current mechanism adopted in Iraq in completing real estate rental transactions and the problems that accompany this method. Then, the authors explained the advantages provided by the application of blockchain technology through the smart contract, in addition to eliminating the problems of tax evasion in this sector. Despite the negative points that this technology suffers from, it still represents a safe environment for preserving, managing, and exchanging information.

DOI: 10.4018/978-1-7998-9274-8.ch006

INTRODUCTION

Countries seek to achieve digital transformation for all their sectors in order to facilitate procedures and ensure accuracy in their completion, and Iraq is one of the countries lagging behind the rest of the countries in the digital sector, and the digital transformation is intended to invest technology to convert traditional work into electronic, as the real estate sector produces a huge number of transactions and information. Which needs technologies to facilitate the process of managing it digitally and the ability to complete transactions electronically. Blockchain technology is one of the most important modern technologies that allow the exchange of value and the completion of transactions digitally and ensure the correctness and accuracy of transactions, as it is characterized by transparency and reliability, so it works to manage and secure transactions through the Internet, and at the same time the blockchain technology contributes to the management of information, as it works To preserve information in a decentralized manner, and to organize it in a succession linking information with different transactions of the same thing; meaning that the information of a plot of land can be retrieved since the establishment of its deed to the last person who owned it in a coherent and sequential manner, through which this information is also exchanged between the network parties, and it allows access to and retrieval, and thus it is interested in information from its inception, through its organization and preservation to its availability, so it can be considered as one of the techniques that helps in managing information. Also, the use of this technology in the real estate rental sector contributes to achieving digital transformation, as well as provides many features that facilitate the digital management of the process.

The study problem revolves around two basic points: The first is that the real estate rental sector lacks a link between it and the providers of services such as: electricity, water, telecommunications, and others, as linking the various parties to create secure digital assets which represents one of the goals of the digital transformation. The second is not to dispense with the existence of a real estate broker between the owner and the tenant to complete the process in its current form, and to solve this problem and create a kind of regulation and link between the concerned authorities as well as between the owner and the tenant without an intermediary, blockchain technology can be applied to create a unified and reliable database that brings together the parties. The importance of the study stems from the importance of employing technologies in all sectors to achieve the digital transformation of Iraq, and the real estate rental sector is one of the important sectors in which many relevant parties contribute, such as the Electricity Company, the Water Company, the Ministry of Justice, the Ministry of Housing and others, so finding a network that allows the implementation of transactions digitally and linking those parties Together, they contribute significantly to achieving digital transformation in Iraq. To achieve the goals of this study, it answers several questions, the most important of which are:

- What is blockchain technology, and what are its most prominent applications in the real estate field?
- What is the mechanism currently used in Iraq in the rental of real estate?
- What are the problems of the current mechanism and the opportunities to solve it by using rental blockchain technology?
- How can blockchain technology be applied in the real estate rental sector in Iraq?
- What are the challenges of applying blockchain technology in the real estate rental sector in Iraq to achieve digital transformation?

STUDY APPROACH

The study adopted the case study approach, which falls under the descriptive approach, whereby the current situation of the real estate rental sector in Iraq will be studied, a detailed study and then a proposal for the application of blockchain technology in this area to develop the current system in a way that bridges the gaps in it and ensures greater advantages and benefits including in line with the digital transformation of the rest countries of the region.

TERMINOLOGY OF STUDY

- **Blockchain**: A decentralized (peer-to-peer) network that manages information or processes; It allows for preservation, exchange and retrieval (Sudha Sadasiuvam, 2021).
- **Block**: The container that holds the data that is stored in the blockchain (Nakamoto, 2008).
- **Nodes**: Are the devices that make up the network, and which keep a copy of the database and ledger inside (Guegan, 2018).
- **Distributed Database or Distributed Ledger Record**: The place where blocks and transactions are collected, and every device in the network has a copy of it, so copies distributed in more than one device have given it the distribution feature (Zivic et al., 2020).
- **Miner node workers**: Every node that a person works on who verifies and confirms transactions and data, they can be called workers, participants or validators, as well as prospectors in some blockchain cases that require mathematical operations to obtain a block symbol to add to the chain, and the process of searching for a symbol The block can be called prospecting, so they are called prospectors (Sudha Sadasiuvam, 2021).
- **Hash**: Hash is a type of encryption that converts data into a fixed-length code called an encryption token. It is as unique as a fingerprint (Gomathi et al., 2021).
- **Encryption symbol**: It is the code produced by the (hash) function, which works to link the blocks and distinguish the block from other blocks (Ma, 2021).
- **Timestamp**: It is the time when the block was created and added to the chain. Here it represents the time and date of birth of the block (Nakamoto, 2008).

Previous Studies

Blockchain technology is considered one of the modern technologies, as there are few related studies that are branched out in different fields. Previous studies will be reviewed according to topics relevant to the current study:

1. **Studies in the field of real estate**: Davidson County in the United States of America is an example of the use of blockchain technology in the real estate sector, as Spielman's study showed the benefits that Davidson County, USA, when applying blockchain technology in registering real estate ownership, and the problems of the current system that can be solved through technology; One of the most important of these is solving fraud cases resulting from identity theft by providing security for information through blockchain encryption technology (Spielman, 2016).

The study (Corluka & Lindh, 2017) confirmed that blockchain technology will affect the real estate market, as it will change it significantly, and the study (Kirit, N., & Sarkar, 2017) provided a model for using blockchain technology to transfer real estate properties and maintain transaction records through smart contracts. Using the Ethereum blockchain.

The study (Martijn, 2017) aimed to define the stage in which blockchain technology could be used in real estate management, and it reviewed the opportunities and challenges to apply technology in order to improve the current real estate management process, the most prominent of which are: technical challenges, lack of standards, the need to organize the process at the level Government.

The study (J. Michael Graglia, 2018) confirmed that the role of blockchain in providing a transparent decentralized network is one of the reasons that made the use of blockchain logical in the real estate field. The study (Kopylash, 2018) developed a blockchain-based real estate system for a real estate company that relies on the Ethereum platform for smart contracts, and (Wetzel Livia Verena, 2018) presented a blockchain-based model for a company specializing in real estate management in Switzerland, focusing on a process. The lease agreement and the settlement of fees, and she confirmed that the model achieves integration in the property management process so that the process becomes more efficient, transparent, and traceable. Also, The study (Park et al., 2021) presented Ziraffe framework on the Ethereum platform, which is based on protecting user privacy while supporting the authentication of origin for external data and showed that the model is cost-effective due to the low cost of verification complexity.

2. **Studies in e-government**: The study (lnes, S., Ubacht, J., & Janssen, 2017) reviewed blockchain technology and its impact on government operations, and concluded that the technology will bring potential benefits in e-government, and it will change government operations, but a lot of research and experiments must be done so that the technology can be understood in depth before it is applied, Their strategies must also be identified.

The study (Hou, 2017) discussed the application of blockchain technology in Chinese e-government, and identified the difficulties and challenges facing the application of technology at the present time, and how technology contributes to the development of government services, and concluded that technology achieves many benefits in e-government, including: Improving government services Significant transparency, easy access to information, and the development of the information exchange process. Moreover, a study (Alketbi, A., Nasir, Q., & Abu Talib, 2018) after reviewing the literature to identify potential cases of using blockchain applications in government services confirmed that there is a huge potential for using the technology.

3. **Studies in smart contracts**: A study (Scherer, 2017) compared public and private blockchains (licensed or restricted with terms of use) by performance, applicability, and scalability, and provided a blockchain-based smart contract model capable of development. On the other hand, the study (Mohanta, B. K., Panda, S. S., & Jena, 2018) focused on the different elements of smart contracts and the principle of working in them and reviewed their various uses in blockchain and the challenges of their application.

BLOCKCHAIN TECHNOLOGY

The Genesis of Blockchain

Blockchain technology was associated with digital currencies, so that at first some believed that it was not possible to create a blockchain without a digital currency, and the reason was that in 2008 Satoshi Nakamoto presented a scientific paper explaining cash transactions between peer-to-peer without a central system (Mohanta, B. K., Panda, S. S., & Jena, 2018), proposed a payment system called Bitcoin without the need for a central system, as the trusted third party (The bank in payment cases) replaced with cryptography, which is explaining the concept of the blockchain and not the name (Mulár, 2018), this was the first blockchain appearance.

The Concept of Blockchain Technology

The naming of the blockchain with this name is due to the nature of its work and the method of recording and storing transactions. It records every transaction that takes place within the network in a block and links the blocks with each other, so it is called the blockchain or the blockchain. The idea of blockchain technology lies in the possibility of exchanging value between two parties without a central system, and the value here can be money, or properties such as: real estate ownership, car ownership, intellectual property, and others, anything that has a value that is usually exchanged between two parties in There was still a central system that verified, documented, and approved this exchange, but the blockchain relied on cryptographic techniques, agreement algorithms, a decentralized network, and the blockchain protocol to make this exchange without a central system (Seebacher & Schüritz, 2017).

Definition of Blockchain

Until the time of preparing this study, there was no single agreed definition of blockchain, as some defined it on the basis of its first concept, which is the Bitcoin transaction system, among these definitions: Blockchain is a digital ledger that stores transactions after verification by nodes or network devices (Mohanta, B. K., Panda, S. S., & Jena, 2018), and in another definition: it is a distributed and shared ledger that provides authentication and the ability to verify transaction (Daniel, F., & Guida, 2019), and some of them considered it a common database of these definitions: Blockchain is a distributed database that works in a peer-to-peer network, each peer in the network owns a complete or partial copy of the database (Kopylash, 2018). Also, defined by Deloitte as: just another type of database for recording transactions that are replicated in all computers participating in the network (Deloitte, 2016).

Based on the foregoing, a comprehensive procedural definition can be given to say a blockchain: it is an information network that contains a group of devices or nodes, each device in which represents a database and a ledger, as it saves all transactions that take place within the network, and every transaction between two devices is subject to verification. And the confirmation of its validity by the rest of the network devices.

Blockchain Components

1. **Nodes (Network devices):** They are the parties to the network, and each blockchain network contains many devices called nodes, and transactions or data are saved in them and it is the distributed ledger of the blockchain (Huhmo, 2018). The workers on these machines are called miners, workers, participants, or prospectors, and their job is to verify operations and agree on the correct date of the block (Corten, 2018), and this is done through agreement or consensus algorithms, which also called decentralized consensus; They are algorithms that allow nodes or devices within the blockchain to reach consensus about which blocks are added to the blockchain, and there are many of these algorithms, the most famous of which are:
   a. **Proof-of-Work (PoW) algorithm:** It is one of the most famous of these algorithms used in Bitcoin, as it relies on granting a financial reward to verify the correctness of transactions (Kalis, 2018). After confirming the validity of the transaction data, network devices work to guess the block code by solving arithmetic transactions, and when one of the devices finds a solution The block is created within the blockchain and the prospector gets a financial reward, so it is called the process of prospecting or mining, this process requires great computing power and processors who specialize in solving mathematical problems so that the prospector can create the block and get the reward (Wetzel Livia Verena, 2018).
   b. **Proof-of-Stake (PoS) Algorithm:** This algorithm requires the presence of an amount of money in the wallets of the network workers, and they are not called prospectors here, except that they do not carry out exploration for a block code, but after the process of verifying and confirming the transactions, the block is raised to the network and everyone gets a financial reward, but in the event of tampering One of the devices with data, the blockchain withdraws digital currencies from its wallet as a fine, and this algorithm does not require huge computing power or processors (Nguyen, G.-T., & Kim, 2018).
2. **Encryption**: Encryption is one of the advantages of blockchain, so we find that there are two types of encryptions:
   a. **Hash**: The hash function encodes inputs of different lengths and converts them into fixed-length outputs expressed in unique and fixed-length symbols, not all data in the block are displayed, but the block is encoded and a unique code is issued for each block, this block code and each block for which the cipher code is generated based on the cipher code in the previous block (Wang, W., Hoang, D. T., Hu, P., Xiong, Z., Niyato, D., Wang, P., Kim, 2019), except for the first block, it differs from the rest of the blocks, as there is no previous encryption code (Bruyn, 2017). There are many types of the hash function, including ***SHA-256*** that was used in Bitcoin, and if someone wanted to change the data in a block, all the cryptographic symbols for the previous blocks would change (Kalis, 2018). When creating the code for the block, a timestamp is added to it, and the time of its creation is specified.
   b. **Key encryption**: Key encryption is known as symmetric, analog, or analog encryption, and it has many forms, as files are encrypted using keys and the idea of this encryption can be illustrated in general in Fig.1. This encryption has been used in the blockchain to issue digital signatures on transactions, so we find that every user has two keys, Public key, and private key:
      i. **Public key**: It is a key consisting of numbers and letters. It is used to identify the identity of the user in the transmission and reception process, and it is visible to all.

ii. **The private key**: It is a type of password, and it is considered personal and confidential, and it also consists of numbers and letters, and it is longer than the public key, and this is one of the differences between them and is used for digital signature (Huhmo, 2018).

Figure 1. Encryption using keys in general

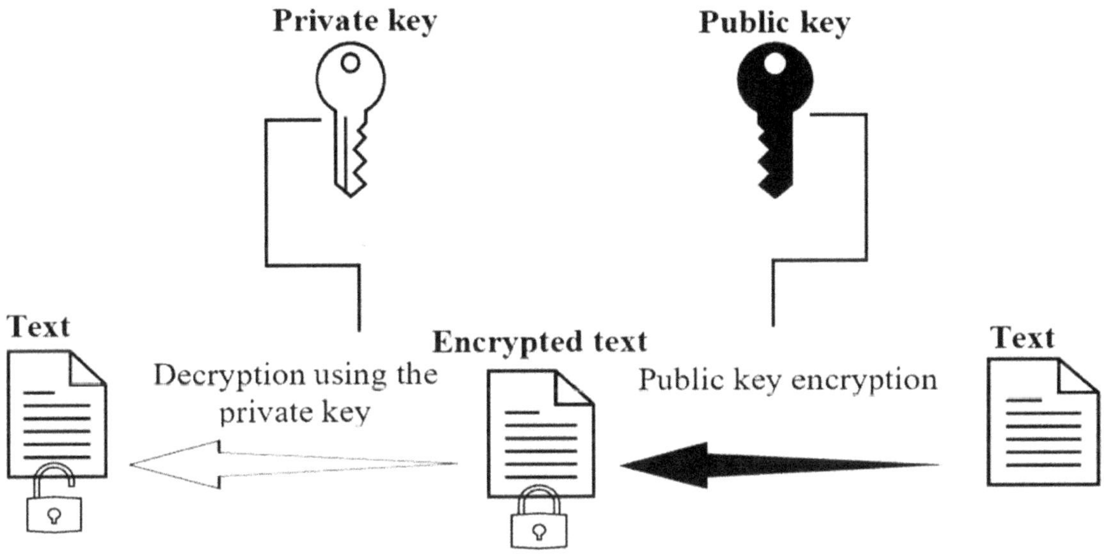

If a person we represent with (A) wants to send a transaction whether it is (monetary value, smart contract, etc.) to a person we represent with (B), then the transaction is first encrypted using the hash function and the encryption code is issued for it, and then this code is encrypted using the person's private key (A), and at this stage results in a transaction signed by the first party, after which it is sent to the other party (B), in addition to the transaction itself (Kikitamara, 2017). When the transaction reaches Party (B), it confirms the transaction through the public key of the sending party intended in this example (A) and the encryption code using the hash function, then decrypts the transaction using its private key, and the process is verified as the digital signature is The result of this process is available for any node within the network to verify and confirm it (Thompson, 2017).

3. **The Block**: The block contains the transaction data, and the block consists of two parts which are:
   a. **Block header**: which is the top of the block and contains: block number, previous block cipher symbol, timestamp; Which is the time of block generation, data agreement algorithms (Bruyn, 2017).
   b. **Its content**: which is the lower part of the block, and contains transaction data such as: amounts, addresses of parties, and the cipher code for the current block (Deloitte, 2016).
4. **Distributed ledger or distributed database**: Each node within the blockchain maintains a copy of the transaction log, because the process of verifying new transactions and adding blocks to the blockchain requires the complete record of previous transactions, so each node or device within the network has a complete copy of the database or ledger (Corten, 2018). In order to clarify the difference between the two terms, as the term ledger is used in accounting systems, and it is a

record to which all financial operations that have been recorded in the journal are transferred (Al-Husban, 2014), and it shows the account balance, its history, the amount and the statement of the operations that have taken place, and the use of this term in the blockchain is considered It makes sense if the goal of the blockchain is to create and process financial transactions between two parties, where the data that is stored inside the blockchain and distributed in the network devices are financial transactions that make up the distributed ledger record. If the blockchain is not intended to carry out financial transactions or does not use a digital currency, then it is called a distributed database, because it maintains the data in distributed devices, and each device contains a copy of the database.

5. **Digital currency**: In some uses of blockchain, digital currency is one of its basic components, and digital currencies appeared for the first time in history in the eighties, specifically in 1983, as *David Chaum* presented the idea of digital currencies at that time, but it did not succeed, perhaps one of the reasons for its failure was the lack of a central structure for the network and its lack of standards Regulatory (Kopylash, 2018). Then came *Satoshi Nakamoto* in 2008 and introduced *Bitcoin*, which is the most famous digital currency, and after that many digital currencies appeared, such as *Ether*, which works on the *Ethereum* platform, the second largest blockchain platform after Bitcoin and other digital currencies, also called the cryptocurrency or cryptocurrency This is because it is based on the principles of encryption, and it is electronic money that is traded through a decentralized network (Iskakova et al., 2020).

CREATING A BLOCK IN BLOCKCHAIN

We need digital identities, a peer-to-peer network and blockchain protocol, and these are briefly reviewed below:

- **Digital identities**: In general, identity refers to a set of characteristics and characteristics of an entity, through which it can be identified and distinguished from other entities. In the digital world, the identity is a set of digital records that represent the user, and the data and information in the identities of the parties are determined according to the nature of the blockchain and the work that they do, and a national digital identity can also be created in the blockchain that is managed by the relevant authorities (Kikitamara, 2017), and in blockchain digital identity depends on keys, and as mentioned previously; The public key and the private key together produce a digital signature.
- **Peer-to-peer network**: It is a group of different distributed resources that are connected to the network, which is the opposite of the idea of the server / client, where the participants in the network share some of their resources such as processors, storage space, network bandwidth, etc., and these resources provide the server and content, and allow access to them from directly accept network participants without an intermediary (Uurtsaikh Jamsrandorj, 2017). These devices save the transactions that take place in the network and participate in the verification and confirmation of these transactions.
- **Blockchain Protocol**: The blockchain protocol preserves the block, prevents information from being tampered with, adds block timestamps and digital signature, as well as gives it cipher codes and links it to the antecedent using agreement or consensus algorithms (Huhmo, 2018).

The digital identity is represented by the private and public key, and the peer-to-peer network works to validate this block using one of the proof-of-work algorithms provided by the blockchain protocol, which then adds a timestamp to the block, thus creating the block and the block can be represented in its final form in Fig. 2.

Figure 2. The form of block (Bauerle, n.d.)

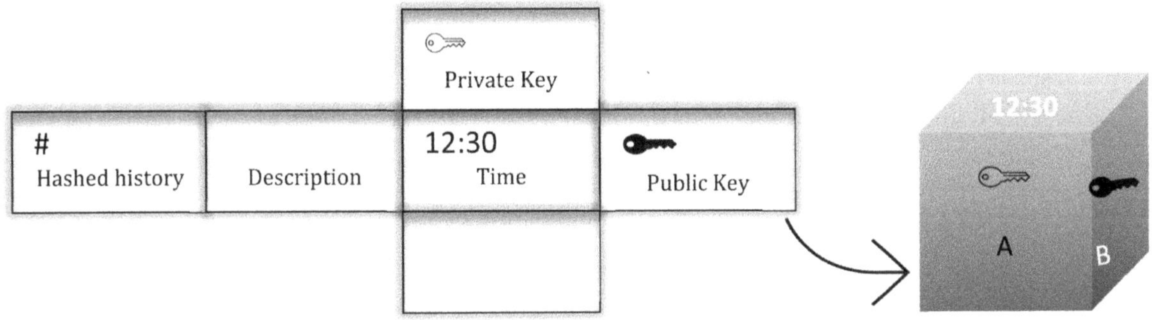

TYPES OF BLOCKCHAIN

It can be categorized into four types:

- **Public Blockchain**: It allows everyone to access data, verify transactions and add blocks, so anyone can access the network, see the data, perform the verification process, and verify transactions that take place within it without conditions, such as Bitcoin.
- **Public Blockchain with Authority (permission):** It allows everyone to access the data but in order to verify operations and add blocks, needs to some requirements such as an ID card.
- **Private Blockchain**: It pre-identifies the entities, devices, and people who have access to the data, creating and verifying transactions, and adding blocks.
- **Private Blockchain with Authority (permission)**: in which devices and people are identified and given full powers, and it is considered the most limited of the types of blockchain, as it can be used within companies and only allow users within the company to read and store data (Huhmo, 2018).

Private blockchains allow public access to data and access to it, but they do not have the authority to create, verify and add blocks, and this is what distinguishes it from blockchain with (permission) powers, and the powers are in multiple levels: (permission to connect to the network, permission to build and create assets, Permission to send transactions, permission to receive transactions, permission to confirm and verify transactions, permission to view transactions (Uurtsaikh Jamsrandorj, 2017).

Based on the foregoing, the types of blockchain can be classified into two parts according to:

- **Verifications and Confirmations**: Public blockchains do not need permission to create or verify a transaction, and private blockchains need permission to verify and confirm transactions.

- **Data accessibility**: Blockchain is public, data can be accessed by anyone, and blockchain in particular is restricted to users within the network.

FEATURES OF BLOCKCHAIN TECHNOLOGY

Although there is more than one type of blockchain, all types share the following features (Deloitte, 2016):

- Distribution and decentralization.
- Consensus on the validity of transactions.
- The use of digital signature and proof of identity.
- Transparency.
- Programmability according to their use.
- The difficulty of modifying data and transactions.

PROBLEMS OF BLOCKCHAIN TECHNOLOGY

Most of the studies that dealt with blockchain problems were from the side of the flaws and problems of Bitcoin and Ethereum, which are public platforms that handle huge transactions and use proof-of-work algorithms, and this is one type of blockchain, as well as one type of algorithm that cannot be generalized to all types of blockchains and their algorithms, Among these defects (Gatteschi, V., Lamberti, F., Demartini, C., Pranteda, C., & Santamaria, 2018):

- High energy consumption, as one transaction in Bitcoin costs six dollars the energy consumed by the network nodes.
- Huge computing power in the process of mining the block symbol, the process here is similar to competition for the financial reward, and in order for the node to get the reward, it is keen to solve mathematical problems very quickly, which requires the presence of huge computing power.
- High storage space for each node within the network to be able to store a full copy of the blockchain.
- The previous defects do not exist when using other algorithms, for example the validation or voting algorithm, they do not require huge computing power to solve mathematical problems, as well as in the use of special blockchains to process limited transactions, so they do not need huge spaces to store a large amount of crisis transactions.

From the above, the problems and disadvantages of blockchain technology in general are represented in two issues:

- **First: The security issue:** There are many attacks that affect the blockchain, and they differ according to the algorithm that was used. Among these attacks are related to the PoW algorithm and PoS, and these attacks (Golosova, J., & Romanovs, 2018):
    - Double spending attack, which is when the same amount is spent twice.
    - DDos attack, it is denial of service attacks in distributed networks.

- ◦ 51% attack.

The attack (51%) is one of the defects of the blockchain in general, because one of the fundamentals of the blockchain is the agreement of most network nodes to add, modify, or delete a block, and this means the change will done when most of the network nodes (more than half of the existing nodes) agree to change, even if the change is not correct. On this basis, they can control and manage the network and inflict damage on transactions, and this is one of the most and fastest attacks on the blockchain (Ye, C., Li, G., Cai, H., Gu, Y., & Fukuda, 2018).

- **Second: Privacy:** Every node in the network that stores a copy of the blockchain that can access the content and thus may violate privacy (Gatteschi, V., Lamberti, F., Demartini, C., Pranteda, C., & Santamaria, 2018).

USES OF BLOCKCHAIN

Blockchain has been used in many areas, including bank remittances, in keeping records such as patient records and intellectual property rights, in supply chains, as it achieves transparency, in e-government and in the real estate market, as well as it is used to conclude and implement smart contracts and other uses (Johansson & Nilsson, 2018).

- **Smart contracts:** Smart contracts in the blockchain are a type of contract that can be executed automatically without the need for a reliable third party (Mulár, 2018). In general, they are contracts concluded between two parties digitally using codes, perhaps the most famous of which is the Ethereum platform (Kalis, 2018). Also, the smart contract processes the value of transactions on the basis of digital assets. When a smart contract is entered into with the blockchain, a code is created that works to manage and implement the operations according to the preconditions (Sultan et al., 2018), for example when determining home rental payments, the contract is based By completing automatic transfer of funds between parties as soon as the payment is due, and there are many applications of smart contracts on the blockchain, including: What *Visa* and *DocuSign* have done to create a car rental project that manages transactions between multiple parties, including lessee, lessor and insurance companies (Docusign, n.d.).
- **Blockchain applications in real estate:** Blockchain technology has been widely applied in the real estate field, as we find that it has been applied in land registration, transfer of real estate properties, long-term and short-term real estate leasing, real estate investment, and the field of digitizing contracts and agreements. There are many applications that are working in the process of being created, among the most prominent applications that have been implemented and worked with:
    - ◦ In the field of transferring real estate properties (*SIMMST for Handover Application 2017 - Propy in 2017*), in investing in real estate assets (*ATLANT 2016 - Meridio*), in digitizing contracts and agreements (*BUILDIN 2016 - Agent Not Needed in 2017*), and in leasing real estate (*Rentberry 2015 - Bee Token 2017 - The LendingCoin 2017 - Etheera 2017*) (Mire, 2018).

THE REAL ESTATE SECTOR IN IRAQ (CURRENT MECHANISM)

Lease agreement in Iraq in the property sector is regulated based on the Real Estate Lease Regulation *Law No. 184 of 1978*, according to the aforementioned law, the entities regulating the lease contract for the real estate sector are: The Land Registry Department, the Real Estate Tax Department, and the Mediator (*Real Estate Office*) (Iraqi Laws and Legislation, 1984). The rental system in the real estate sector in Iraq is devoid of any digital automation, as it is still dependent on the paper contract, which contains many problems, including false information and tax evasion.

The necessary steps to complete renting a property in Iraq are simple steps, going to the real estate broker's office and then concluding a contract between the two parties, the lessor, and the lessee, when finding the right property for the lessor. Despite the simplicity of these steps, there are many problems that occur during this process. Among these problems:

1. **Centralization of the real estate broker**: The real estate broker controls most of the steps from the stage of finding the property until signing the contract, as the broker is responsible for the process of entering data into the contract, and he is the documenter and the certified one for the validity of the data. In addition, there is no control over the real estate broker to verify the accuracy of the data. Moreover, the mechanism of renting the real estate sector in Iraq lacks transparency as it is done through a single real estate broker.
2. **The process of searching for units and real estate available for rent and data reliability**: The process of searching for real estate units is represented by the real estate broker's office or by the landlord himself. Furthermore, this information may be incorrect or misleading.
3. **Collecting the amount of rent**: The amount of rent is paid face to face, and this process takes place either through the real estate broker or between the landlord and the tenant directly.
4. **Connecting the concerned authorities or those providing services such as water and electricity**: The current mechanism lacks the ability to obtain the amounts of services provided to the real estate unit through the concerned authorities from the tenant due to the lack of any digital process, which results in problems at the end of the contract and the tenant's exiting.

Solutions to the above problems can be provided by using blockchain technology, as we will show in the next section.

Applying Blockchain Technology to Develop Appropriate Solutions to the Above Problems:

- To solve the real estate broker and transparency problem represented by Problem No. 1, the Blockchain technology eliminates the role of the real estate broker and achieves transparency. The Blockchain technology is a distributed network, meaning that there are a certain number of devices that confirm and save transactions, and this feature ensures the validity of the data and the difficulty of modification, this is due to the multiplicity of copies in more than one device. In addition, the blockchain technology is a (peer-to-peer) network, the completion of a process within it is between the two parties without the presence of a third party in the process, dependent on the devices in the network. Among these devices are the concerned authorities to confirm the validity of the entered data, and the operations that take place within the network such as deals, conclusion of contracts and financial transfers, and because it requires confirmation of 51% of the total devices within the network, this feature ensures the validity and transparency of operations without the presence of a third party that approves and documents this process. Consequently, the blockchain operates without an intermediary between the lessor and the lessee, and it is transparent, so the entered data cannot be modified or deleted, and if one of the devices tries to amendment the data, the time and the device that did so will be determined, which facilitates the process of monitoring the operations that take place within the blockchain (Zhang et al., 2019).

By looking at the real estate sector rental business models for some companies that rely on blockchain technology and digital currencies, a suitable solution to the previous problem can be developed, for example, the Rentberry company which based on Ethereum, and Bitcoin (© Rentberry, 2018), (Shanker, 2019):

- Creating digital identities for each of the lessor - the tenant - the housing unit and managing the data of digital identities by controlling availability and encryption. Some data is made available from the digital identity of the housing unit, such as: the age of the property, unit evaluation, rental price, and other data that the tenant needs when searching for units and encrypting data that does not need to be publicly displayed, such as: data of the leased owner, data of the instrument, data of the identity of the tenant, and others.
- The tenant searches for and selects units, then sends a request to rent and negotiate, and after approval of the request, he visits the site, then the identity of the unit, the property, the lessor and the tenant is fully displayed to both parties, then a request is sent to sign a lease contract according to the agreement between them, the blockchain devices are verify the identities and contracts, and the terms of the contract are determined by the lessor, and the items can be specified in the form

Figure 3. Completing a rental process without a real estate broker through the blockchain

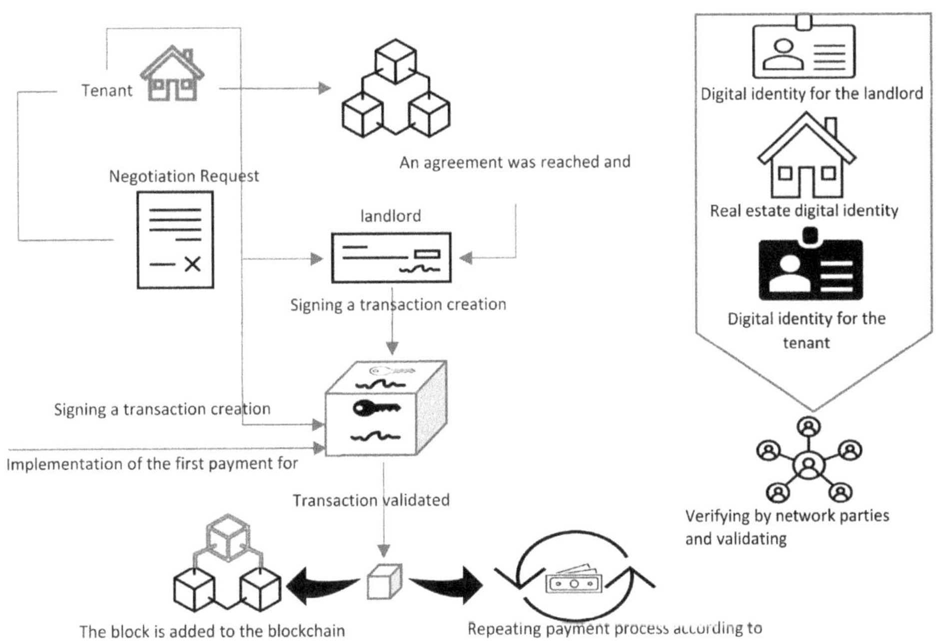

of a list of choices. Then, the contract is sent to both parties to view and sign it, and the blockchain devices verify the signatures, after that the smart contract is created, the block is added to the blockchain, and the payment process begins. Thus, a rental process can be completed without a real estate broker through the blockchain, as in Fig.3.

- As for problem No. 2, which is related to the search for real estate units while ensuring confidence in the displayed data, the Blockchain technology allows adding the units' data by the lessor in a way that allows managing that data and controlling access to it while ensuring confidence and credibility, as the data that is provided can be controlled. It is initially shown to the tenant before the contract, while ensuring the credibility of the displayed data, such as the age of the property, address, etc., and since one of the properties of the Blockchain is that it does not allow modification to the data, it shows to the tenant only the real data about the unit. After reviewing the business model of a number of companies selling and renting real estate based on blockchain technology such as (Atlant) (ATLANT, n.d.), a processing method was developed to solve the previous

Figure 4. The process of adding real estate units in the blockchain and making it available for search

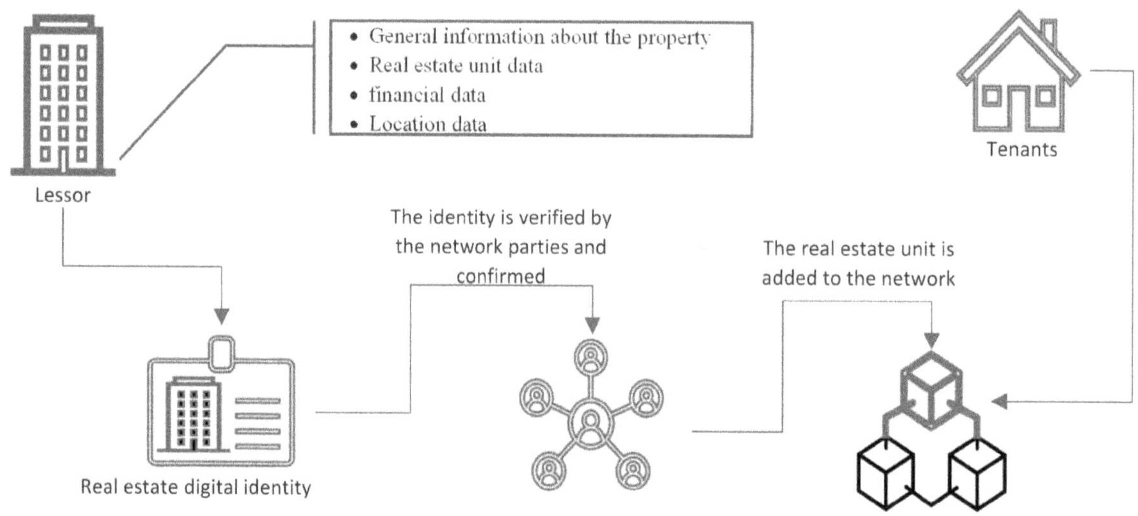

problem, where the platform allows the lessor to enter the units' data, and then the entered data is verified and approved, and the tenant is allowed to initially search for the units, and preliminary data for the unit that is not linked to a contract is displayed, and some features are available in the search process such as: search limitations and sorting according to the age of the property, the region, unit size, etc., in Fig.4, we show the process of adding real estate units in the blockchain and making them available for search (Allessie et al., 2019).

- We come to problem No. 3, which is related to obtaining rent allowance, and we will add to it here a solution to the problem of the insurance amount, depending on the blockchain technology, the rent amount can be obtained automatically. According to the blockchain technology, a lease contract can be created that allows the automatic collection of funds according to the agreement contained in the smart contract (The rent value, payments, and payment dates), so the amounts are automatically transferred from the tenant to the lessor. Moreover, not only the rent amounts, but the insurance amounts can be managed, so that the insurance amount is collected from the tenant with a private key and the amount is kept as a guarantee, and at the end of the contract, if there is no damage to the unit, the ban is lifted and returned to the tenant, and in the event of any damage, it is deposited For the lessor, and in the event of renewing the contract, the amount is kept. The process here is similar to the process of blocking amounts in credit cards (Ante, 2020). Putting appropriate solutions to this problem is to create smart contracts based on blockchain technology, that is, the first payment of rent and the amount of insurance or guarantee is withdrawn from the tenant and deposited in the escrow account with a private key, and then based on the agreement to pay the amounts in the contract, the amounts are transferred from the tenant to The lessor periodically according to the dates of the specified payments as in Fig.5 (Karamitsos et al., 2018). Then at the end of the contract, the ban on the insurance amount is lifted, whether for the lessor or the lessee. The performance of this process requires the presence of a digital currency, and alternative solutions can be used: such as issuing certified checks through the blockchain based on the experience of Emirates NBD "Cheque Chain" in issuing and settling Checks through blockchain technology, by printing a unique QR code on checkbooks (Peyton, 2018), or relying on digital currency to collect amounts in real estate operations through smart contracts using the Blockchain without the presence of a digital currency used in it, where contracts contract directly with banking systems for processing payments in real fiat currencies (US dollars, Euros, etc.) (Midasium Contracts, n.d.) Fig.6.

- With regard to problem No. 4 of linking the concerned parties to the rental process of real estate units, blockchain technology allows automatic linking of utility bills. After examining the work of a number of companies' models (*Like Rentberry*) (© Rentberry, 2018) that link utility bills to the tenant based on Blockchain technology, a solution to this problem was developed. That is, the concerned parties are added to the network as devices that confirm the information and verify the data, then the usage is calculated from the signing of the contract until the date of completion, and the invoices are linked to the identity of the tenant on the blockchain network, and bills are issued on a monthly basis on the identity of the tenant based on the smart contract (Qi-Long et al., 2019).

ADVANTAGES THAT WILL COME FROM APPLYING SMART CONTRACTS IN IRAQ

1. **Digitizing identities for individuals, units, and institutions**: Many companies have developed appropriate solutions to this issue, such as *Atlant and Deloitte* (ATLANT, n.d.), (Deloitte, 2016), which deals with digital identities to complete the processes of buying, selling and renting real estate. According to the models of the two mentioned companies, the data of real estate units can be registered, then the data entered by the devices in the network is confirmed, and through the Real Estate Department database for "Building Title" data, as well as through the municipality for "Building Permit" data that determined the number of units Its area, number of floors, age of the

property, electricity and water meters, then a digital identity is created for the real estate units, as well as the creation of a digital identity for the lessor and the tenant, with emphasis on data through government agencies such as the Directorate of National Card Affairs, and working to update the digital identity with every transaction that takes place through the network and linking the results of the evaluation process to the digital identity of the unit, the lessor and the tenant(Junis, 2019).

2. **Tax administration**: Based on the Deloitte tax collection company's business model, the tax value is calculated after the lease contract is completed and the tax value is deducted from the landlord's account or the rental payment, and then the tax value is transferred to the tax authority account automatically (Deloitte, 2016).
3. **Add evaluation process**: At the end of the contract, each party must evaluate the other, as well as the tenant's evaluation of the real estate unit, adding the evaluations and linking them to the digital identities of the unit, the lessor and the tenant.
4. **Accelerate and facilitate the rental process**: Registering a digital identity for the real estate unit provides opportunities for leasing without the need for a real estate broker to bring the tenant, as it allows the search process and facilitates the selection process, as well as allowing the conclusion of contracts through the network reduces the transaction time and the amounts of money paid to the real estate broker (Norta et al., 2018).
5. **Information Security and Cyber Security**: The presence of two types of public or private blockchain reduces security risks to information, as the powers of verification and verification differ between public and private, which achieves security for this information, and the presence of a private and public key for each party in the process, and the presence of the encryption feature for documents increases and enhances security . In addition, the multiplicity of devices in the network reduces the process of data manipulation (Dai et al., 2017). Especially in the public blockchain, we find that the process of controlling the availability of data and the encryption process for identities enhances the security of this information (Zyskind et al., 2015), while in the private blockchain, we find that identifying devices in the network and requesting permissions enhances security in them (Jayachandran, 2017).

THE PROPOSED MECHANISM FOR APPLYING BLOCKCHAIN TECHNOLOGY TO MANAGE THE RENTAL PROCESS IN IRAQ

We suggest creating a private network or using a private Ethereum blockchain platform to implement lease contracts, which allows creating a smart contract, collecting payments automatically, and applying all the opportunities mentioned earlier, the reasons for choosing a private blockchain type are:

- The nodes within the network are predetermined, which ensures security for this network.
- There is no need for proof-of-work algorithms, or validation algorithms.
- Provides authorization mechanisms (or (permissions)) that allow greater control over them.

Later, we will show who the network representatives are (ie the nodes and their roles), and then explain the data that digital identities must have.

Network Representatives and Their Roles

- **Nodes or external devices:** They are the accounts of the lessor and the lessee as external parties. These accounts are controlled using the private key. The network allows these accounts to send requests, create smart contracts, create a transaction to transfer money and the evaluation process, each of these accounts must have a digital identity, in addition to the creation of a digital identity for the real estate unit by the lessor.
- **Ministry nodes or devices**: These devices are predetermined, and these nodes are controlled through their code, and they are allowed to display data and store the Blockchain on their devices, as well as create smart contracts or new blocks, send them to all devices and receive them from all devices, then confirm and verify of transactions made within the network.

These nodes or devices can be limited to the Ministry of Housing, and it is preferable that the service providers and the Ministry of Interior participate in a number of devices. If they are not involved, other representatives will be added to the network, namely:

- **Nodes and devices of service providers**: They are predetermined, and are controlled through a special code. These devices do not verify and confirm transactions and allow them to display transactions and store the Blockchain in their devices. In the event of creating a new contract, they add unit counter data to the contract and linking the counters to the identity of the tenant from the beginning of the contract until its end, issuing bills on the network, in addition to the process of broadcasting and receiving blocks from all nodes or devices.

Digital Identities

Before starting to review the proposed mechanism, we review the digital identities within the blockchain network and its contents:

- The digital identity of the lessor and tenant and contains:
    - Personal data: name, nationality, identity type, number and version, mobile number, e-mail.
    - Financial data: digital currency wallet or bank account number, contracted unit service bills.
    - Evaluation data within the network.
- The units' digital identity and contains:
    - Ownership data: identity of the lessor or owner, title deed number and date, issuer, place of issuer.
    - Real estate data: address, type of real estate building (villa or building), type of real estate use (family or individuals), number of floors and units, number of elevators and parking, facilities data such as: (swimming pool, guard).
    - Units data: unit type (apartment or villa), unit number, floor number, furnished, furnishing status (new or used), fitted kitchen cabinets, number of rooms, room area, air conditioning type and number.
    - Services data: electricity meter number, current reading of the electricity meter, water meter number, current reading of the water meter.

- ◦ Financial data for the units: the annual rent amount, the guarantee amount, in the case of shared service meters (the monthly electricity fee, the water monthly fee).
- ◦ Pictures and videos of the units.

CHALLENGES FACING THE APPLICATION OF BLOCKCHAIN TECHNOLOGY IN IRAQ

We can say in general that the main challenge is the lack of seriousness or ability of the government to implement digital transformation in the joints of the state, but specifically, there are a number of challenges summarized in the following:

- The absence of a strategy and policies for the use of blockchain technology in Iraq until the date of conducting the study, as well as the lack of technology with international standards and standards.
- Lack of full awareness of the advantages of blockchain technology and the revolution that it may cause in the field of transaction and information management by institutions in their various sectors.
- Not adopting the digital currency in general.
- Educational institutions have not opened the way to learn technology, whether in a special section or as a curriculum, and this is considered a challenge, as the Blockchain is a very complex technology, and requires high skills in construction, programming and use.
- The use of blockchain technology to manage government operations requires a link between the relevant authorities, and most government agencies work on separate and mostly non-digital systems, and there is no strategy for linking.
- Legal challenges, as Iraq needs, before applying the technology, to amend and enact new laws in the field of cybersecurity in line with international laws.
- The biggest challenge facing the real estate rental sector to implement blockchain technology is the lack of connection between the relevant authorities and the non-use of digital systems.

RESULTS OF STUDY

1. The use of blockchain technology in general represents the beginning of achieving digital transformation in Iraq.
2. The current mechanism in managing the real estate rental sector in Iraq is centralized in the presence of a real estate broker who manages this process, and the proposed mechanism for using the blockchain technology eliminates the role of the real estate broker and ensures the validity of data and maintains its security.
3. The use of blockchain in managing the real estate rental sector provides a decentralized database on vacant and non-vacant real estate units, and allows the ability to search for these units and send an agreement request, which contributes to the digitization of these assets and agreements and increases confidence in the information displayed on the units.

4. Blockchain technology provides a complete record of real estate leasing operations, and ensures the validity of operations and information security from any change, deletion or modification.
5. The use of the blockchain to conclude smart contracts provides the ability to collect amounts automatically, and manage the flow of funds between the lessor and the lessee, which provides immediate collection and settlement of payments faster, and the management of insurance amounts.
6. Smart contracts through the blockchain allow linking utility bills to the actual user of the real estate unit, whether he is a renter (in case the unit is not vacant) and the tenant (in case the unit is vacant), and that linking is done automatically, which achieves transparency and safety, and solves the problem of payment of dues on real estate units. It allows accurate documentation of invoices and payments.
7. The evaluation process for real estate units, the lessor and the tenant through the blockchain achieves the principle of transparency, and creates a spirit of competition to ensure the provision of services with efficiency and high quality.
8. Blockchain reduces fraud by digitizing identities, increases transparency by verifying information and data in more than one stage, and enhances information security through encryption.
9. The possibility of managing taxes using the Blockchain faster by listing the real estate units that have been rented and their tax value, and collecting the value of taxes automatically.
10. The use of the blockchain facilitates the rental process, reduces the necessary operations, and reduces the time to complete the transaction.
11. The ability to analyze big data resulting from the use of the blockchain to obtain real-time information that contributes to improving the quality of decisions related to managing the rental process, and knowing the trends and patterns of current and future users.

STUDY RECOMMENDATIONS

1. The government should start directing its departments with the necessity of digital transformation, and linking the relevant authorities together. To achieve the principle of e-government.
2. Interest in conducting more studies on blockchain technology in all fields.
3. Adopting educational curricula for creating, programming and using Blockchain technology in Iraq.
4. Legislate the policies, procedures and laws necessary for the use of blockchain in Iraq.
5. Establishing the national digital identity of the Directorate of National Card Affairs to facilitate its use in government blockchain platforms.
6. Creating a digital currency for the government sector in Iraq to facilitate payments through the Blockchain.
7. Establishing blockchain platforms to manage real estate operations in Iraq represented in the rental process, digitization of assets, transfer of ownership and land registration.

REFERENCES

Al-Husban, A. A. (2014). Accounting Information Systems. Al-Yazuri Publishing House.

Alketbi, A., Nasir, Q., & Abu Talib, M. (2018). Blockchain for government services — Use cases, security benefits and challenges. 15th Learning and Technology Conference (L&T). doi:10.1109/LT.2018.836849410.1109/LT.2018.8368494

Allessie, D., Sobolewski, M., & Vaccari, L. (2019). Blockchain for digital government. JRC Science for Policy Report: Vol. EUR 29677. 10.2760/93808

Ante, L. (2020). Smart Contracts on the Blockchain – A Bibliometric Analysis and Review. SSRN Electronic Journal, 10, 1–48. doi:10.2139/ssrn.3576393 doi:10.2139srn.3576393

ATLANT. (n.d.). Atlant CTO Attends Cityscape's Real Estate Summit in Dubai. https://blog.atlant.io/denis-donin-atlant-cto-attends-cityscapes-real-estate-summit-in-dubai-e65a47e6a79f

Bauerle, N. (n.d.). How does blockchain technology work? https://www.coindesk.com/learn/blockchain-101/how-does-blockchain-technology-work

Bruyn, A. S. (2017). Blockchain an introduction Research paper. https://beta.vu.nl/nl/Images/werkstuk-bruyn_tcm235-862258.pdf

Corluka, D., & Lindh, U. (2017). Blockchain - A New Technology That Will Transform the Real Estate Market. Thesis, 74.

Corten, P. A. (2018). Implementation of Blockchain Powered Smart Contracts in Governmental Services (Issue March). https://repository.tudelft.nl/islandora/object/uuid:87709465-b9a1-48da-9ba5-eba98bc263d7

Dai, F., Shi, Y., Meng, N., Wei, L., & Ye, Z. (2017). From Bitcoin to cybersecurity: A comparative study of blockchain application and security issues. 2017 4th International Conference on Systems and Informatics, ICSAI 2017, (61471129), 975–979. 10.1109/ICSAI.2017.8248427

Daniel, F., & Guida, L. (2019). A service-oriented perspective on blockchain smart contracts. IEEE Internet Computing, 23(1), 46–53. doi:10.1109/MIC.2018.2890624 doi:10.1109/MIC.2018.2890624

Deloitte. (2016). Blockchain Opportunity Contents. Deloitte.

Docusign. (n.d.). eSignature Plans and Pricing. https://www.docusign.com/products-and-pricing

Gatteschi, V., Lamberti, F., Demartini, C., Pranteda, C., & Santamaria, V. (2018). To blockchain or not to blockchain: That is the question. IT Professional, 20(2), 62–74. doi:10.1109/MITP.2018.021921652 doi:10.1109/MITP.2018.021921652

Golosova, J., & Romanovs, A. (2018). The advantages and disadvantages of the blockchain technology. 6th Workshop on Advances in Information, Electronic and Electrical Engineering (AIEEE). doi:10.1109/AIEEE.2018.859225310.1109/AIEEE.2018.8592253

Gomathi, S., Soni, M., Dhiman, G., Govindaraj, R., & Kumar, P. (2021). A survey on applications and security issues of blockchain technology in business sectors. Materials Today: Proceedings. Advance online publication. doi:10.1016/j.matpr.2021.02.088 doi:10.1016/j.matpr.2021.02.088

Graglia. (2018). Blockchain and property in 2018: At the end of the beginning. Academic Press.

Guegan, D. (2018). The Digital World : II – Alternatives to the Bitcoin Blockchain ? Halshs.

Hou, H. (2017). The application of blockchain technology in E-government in China. 26th International Conference on Computer Communication and Networks (ICCCN). doi:10.1109/ICCCN.2017.803851910.1109/ICCCN.2017.8038519

Huhmo, M. (2018). Blockchain technology, Bitcoin as a case. In Oulu University of Applied Sciences. Oulu University of Applied Sciences.

Iraqi laws and legislation, Pub. L. No. 184 (1984). http://wiki.dorar-aliraq.net/iraqilaws/law/18200.html

Iskakova, A., Kumar Nunna, H. S. V. S., & Siano, P. (2020). Ethereum blockchain-based peer-to-peer energy trading platform. PECon 2020 - 2020 IEEE International Conference on Power and Energy, 327–331. 10.1109/PECon48942.2020.9314591

Jayachandran, P. (2017). The difference between public and private blockchain. IBM. https://www.ibm.com/blogs/blockchain/2017/05/the-difference-between-public-and-private-blockchain/

Johansson, J., & Nilsson, C. (2018). How the blockchain technology can enhance sustainability for contractors within the construction industry. Chalmers University of Technology. https://publications.lib.chalmers.se/records/fulltext/255155/255155.pdf

Junis, F. (2019). A Revisit on Blockchain-based Smart Contract Technology. Academic Press.

Kalis, R. (2018). Using blockchain to validate audit trail data in private business applications. Universiteit van Amsterdam.

Karamitsos, I., Papadaki, M., & Al Barghuthi, N. B. (2018). Design of the Blockchain Smart Contract: A Use Case for Real Estate. Journal of Information Security, 09(03), 177–190. doi:10.4236/jis.2018.93013 doi:10.4236/jis.2018.93013

Kikitamara, S. (2017). Digital Identity Management on Blockchain for Open Model Energy System. Radboud University.

Kirit, N., & Sarkar, P. (2017). EscrowChain: Leveraging ethereum blockchain as escrow in real estate. International Journal of Innovative Research in Computer and Communication Engineering, 5(10).

Kopylash, V. (2018). An Ethereum-based Real Estate Application with Tampering-resilient Document Storage. University of Tartu.

Lnes, S., Ubacht, J., & Janssen, M. (2017). Blockchain in government: Benefits and implications of distributed ledger technology for information sharing. Government Information Quarterly, 34(3), 355–364.

Ma, F. (2021). Design of running training assistance system based on blockchain technology in wireless network. Eurasip Journal on Wireless Communications and Networking, 2021(1). 10.1186/s13638-021-01897-4

Martijn, D. (2017). Blockchain: Towards Disruption in the Real Estate Sector: An exploration on the impact of blockchain technology in the real estate management process. https://repository.tudelft.nl/islandora/object/uuid%3Ab6ec7ece-e879-4ae3-8232-d8144ac2642d?collection=education

Midasium Contracts. (n.d.). The Blockchain of Real Estate. http://midasium.herokuapp.com/smart-contracts

Mire, S. (2018). 37 Startups using blockchain to transform real estate [market map]. https://www.disruptordaily.com/blockchain-market-map-real-estate/

Mohanta, B. K., Panda, S. S., & Jena, D. (2018). An overview of smart contract and use cases in blockchain technology. 9th International Conference on Computing, Communication and Networking Technologies (ICCCNT). doi:10.1109/ICCCNT.2018.849404510.1109/ICCCNT.2018.8494045

Mulár, O. (2018). Blockchain technology in the enterprise environment. Masaryk University. http://search.ebscohost.com/login.aspx?direct=true&db=izh&AN=131823111&lang=pt-br&site=ehost-live

Nakamoto, S. (2008). Bitcoin: A Peer-to-Peer Electronic Cash System. 10.1162/ARTL_a_00247

Nguyen, G.-T., & Kim, K. (2018). A survey about consensus algorithms used in blockchain. Journal of Information Processing Systems, 14(1), 101–128. 10.3745/JIPS

Norta, A., Fernandez, C., & Hickmott, S. (2018). Commercial Property Tokenizing with Smart Contracts. Proceedings of the International Joint Conference on Neural Networks. doi:10.1109/IJCNN.2018.848953410.1109/IJCNN.2018.8489534

Park, J., Kim, H., Kim, G., & Ryou, J. (2021). Smart contract data feed framework for privacy-preserving oracle system on blockchain. Computers, 10(1), 1–12. 10.3390/computers10010007

Peyton, A. (2018). Emirates NBD goes blockchain for its "Cheque Chain." https://www.fintechfutures.com/2018/04/emirates-nbd-goes-blockchain-for-its-cheque-chain/

Qi-Long, C., Rong-Hua, Y., & Fei-Long, L. (2019). A Blockchain-based Housing Rental System. CTISC, 184–190. 10.5220/0008097201840190

Rentberry. (2018). A Decentralized Global Home Rental Platform Whitepaper. https://rentberry.cryptonomos.com/#whitepaper

Scherer, M. (2017). Performance and Scalability of Blockchain Networks and Smart Contracts. UMEA. http://www.diva-portal.org/smash/record.jsf?pid=diva2%3A1111497&dswid=8567

Seebacher, S., & Schüritz, R. (2017). Blockchain technology as an enabler of service systems: A structured literature review. Lecture Notes in Business Information Processing, 279(April), 12–23. doi:10.1007/978-3-319-56925-3_2 doi:10.1007/978-3-319-56925-3_2

Shanker, M. (2019). Use Case: Smart Contract for Lease Agreements using Blockchain Technology. International Journal of Scientific Research in Computer Science and Engineering, 7(6), 1–09.

Spielman, A. (2016). Blockchain: Digitally Rebuilding the Real Estate Industry. Vanderbilt University.

Sudha Sadasiuvam, G. (2021). A critical review on using blockchain technology in education domain. In Handbook of Deep Learning in Biomedical Engineering. Elsevier Inc. doi:10.1016/B978-0-12-823014-5.00005-3 doi:10.1016/B978-0-12-823014-5.00005-3

Sultan, K., Ruhi, U., & Lakhani, R. (2018). Conceptualizing blockchains: Characteristics & applications. Proceedings of the 11th IADIS International Conference Information Systems 2018, IS 2018, 49–57.

Thompson, S. (2017). The preservation of digital signatures on the blockchain. The University of British Columbia ISchool Student Journal, 3(Spring).

Uurtsaikh Jamsrandorj. (2017). Decentralized Access Control Using Blockchain. University of Saskatchewan. https://harvest.usask.ca/bitstream/handle/10388/8087/JAMSRANDORJ-THESIS-2017.pdf?sequence=1&isAllowed=y

Wang, W., Hoang, D. T., Hu, P., Xiong, Z., Niyato, D., Wang, P., & Kim, D. I. (2019). A survey on consensus mechanisms and mining strategy management in blockchain networks. IEEE Access: Practical Innovations, Open Solutions, 7, 22328–22370. Advance online publication. doi:10.1109/ACCESS.2019.2896108 doi:10.1109/ACCESS.2019.2896108

Wetzel. (2018). A use case of blockchain technology in the real estate industry for the rental agreement and accessory charge settlement processes. Academic Press.

Ye, C., Li, G., Cai, H., Gu, Y., & Fukuda, A. (2018). Analysis of security in blockchain: Case study in 51%-attack detecting. 5th International Conference on Dependable Systems and Their Applications (DSA). doi:10.1109/DSA.2018.0001510.1109/DSA.2018.00015

Zhang, R., Xue, R., & Liu, L. (2019). Security and privacy on blockchain. ACM Computing Surveys, 52(3), 1–34. Advance online publication. doi:10.1145/3316481 doi:10.1145/3316481

Zivic, N., Kadusic, E., & Kadusic, K. (2020). Directed Acyclic Graph as Hashgraph: An Alternative DLT to Blockchains and Tangles. 2020 19th International Symposium INFOTEH-JAHORINA, INFOTEH 2020 - Proceedings, 2019–2021. 10.1109/INFOTEH48170.2020.9066312

Zyskind, G., Nathan, O., & Pentland, A. S. (2015). Decentralizing privacy: Using blockchain to protect personal data. Proceedings - 2015 IEEE Security and Privacy Workshops. SPW, 180–184. Advance online publication. 10.1109/SPW.2015.27

Chapter 7
Blockchain Technology in E-Healthcare

Abhinav Yadav
Vellore Institute of Technology, Vellore, India

Amrit Kanoi
Vellore Institute of Technology, Vellore, India

Ramani Selvanambi
Vellore Institute of Technology, Vellore, India

ABSTRACT

Blockchain is one of the technologies that is gaining significant attention within the healthcare industry. In order for blockchain to reach its full potential, it's essential that governments and public institutions map the blockchain ecosystem and establish a blockchain framework for coordinating early adopters. The authors have proposed a web-based application using blockchain with security algorithms to store patients' and other users' details who sign up on the website. Blockchain is highly secure as any tampering with data in any block is not possible. Blockchain attributes ensure smooth interaction between different applications and systems along the availability chain. They also reward the patient with their own implemented bitcoin on their transaction. Ethereum library Web3 is used to make a complete decentralized peer-to-peer chat service without any involvement of a third party. The application will be able to send encrypted messages both securely and anonymously. Blockchain technology has been seeing widespread interest as a means to ensure the CIA triad in fewer environments.

INTRODUCTION

What exactly is the blockchain, and why is it being spoken about so much? Immutable and distributable data records that are shared peer to peer between networked database systems are created using this technique. The technique captures digital events in such a way that the data cannot be altered or recognized until it reaches the intended recipient. As a result, the most major benefit of blockchain is that the data is theoretically secure and protected from data breach threats. (Hölbl et al., 2018) The banking sector

DOI: 10.4018/978-1-7998-9274-8.ch007

has already recognized the benefits of blockchain technology. Will medical institutions be prepared to benefit from cryptography? There are numerous blockchain applications in healthcare, and in this article, we'll look at the technology's long-term potential as well as existing applications.

The beginning of the twenty-first century saw significant advancements in digital technology, which are reshaping the global healthcare sector. Healthcare systems are undergoing a continuous and efficient transition from paper to electronic records, resulting in a revolution in the business. By providing a stage that efficiently exchanges healthcare data among different stakeholders and entities in the system, such signals of progress bring great efficiency and flexibility to healthcare services. Paper-based records were transformed into computerized E records, such as Electronic Medical Records, as a result of this evolution. Medical histories, statistical data, medicincs, immunization status, laboratory test reports, and other sensitive information are all included in these records. EHD systems offer advantages over traditional paper-based recordkeeping. EHR requires less people, time, and physical storage than paper-based records. EHRs provide several advantages, including quicker and faster access to clinical data, the ability to maintain an active medical workflow, the correction of medical errors, greater patient care, lower medical expenses, and a better and stronger provision for experimental decision making.

Digitization and outsourcing of data, on the other hand, are easily available to a wider range of users. Due to the sensitive and secret nature of personal data, such a kind of data availability poses additional issues in terms of security and privacy; it also introduces new challenges in terms of scalability, reliability, and integrity (Feng et al., 2018). In this chapter, we conducted a complete search and plan to use blockchain and encryption to address the security and privacy concerns associated with E-healthcare data with patient portals, in order to protect against various assaults and challenges.
Healthcare as an industry stands to profit most from blockchain integration. Here's how:

- Identity Management: A blockchain-based patient identification system would improve the most unsafe and hardest problems.
- Data Protection: The healthcare industry is often targeted to the highest number of data breaches. This data includes the personal information of patients, physicians, and medical records. A distributed system protects data from outbreaks or failure of local nodes.
- Preventing Fraud: this issue is common in almost every industry. In the healthcare industry, fraud often refers to untrue medical records, claims, and proof of work. However, a tamper-proof ledger of hashes can solve this problem.

LITERATURE REVIEW

Table 1. Summary of Key Concepts and Limitations of Blockchain in Healthcare

S.No.	Paper Title	Author(s)	Journal Details and Date	Key Concepts	Highlights	Limitations
1	Systematic Review of the Use of Blockchain in Healthcare	Marko Hölbl, Marko Kompara, Aida Kamišali´c, Lili Nemec Zlatolas	MDPI, 10 October 2018	Use of blockchain in healthcare. Acceptance in different sectors of the healthcare system, advantages and difficulties in implementing blockchain in the healthcare	Blockchain acts as a secure, transparent, decentralized network for the security and privacy of sensitive data. Using it in healthcare would provide better services and privacy for the user and the hospitals	Lack of Infrastructure and other acceptance difficulties especially related to supply of items.
2	A Survey on Privacy Protection in Blockchain System	Qi Feng, Debiao He; Sherali Zeadally; Muhammad Khurram Khan; Neeraj Kumar	Journal of Network and Computer Applications, 29 October 2018	Privacy in the blockchain. Advantages of using blockchain-like transparency, reliability, robustness, etc. Security and data privacy in any kind of implementation	Techniques used for breaching privacy in a public blockchain network, overcoming techniques. Benefits of replacing the current system with blockchain in terms of privacy	Privacy of private systems unaddressed. How even private systems can fail the trust model and lead to unauthorized gains.
3	Supply Chain Management based on Blockchain: A Systematic Mapping Study	Youness Tribis, Abdelali El Bouchti, Houssine Bouayad	MATEC Web of Conferences, 20 January 2018, International Workshop on Supply Chain Engineering	Supply Chain and healthcare	Blockchain in the healthcare supply chain and its advantages.	Security and authenticity of the network
4	Real-time supply chain A blockchain architecture for project deliveries	Petri Helo, A.H.M. Shamsuzzoha	ELSEVIER, September,2019	Blockchain in supply chain	Uses and advantages of supply chain and blockchain	Privacy issues in the public network
5	Protecting Patient Data the New Scope Cybersecurity for Healthcare	Marci J. Swede, Vincent Scovetta, Marie Eugene-Colin	Elsevier Journal of Allied Health, Summer 2019 June 2019	Proposal of the partnership of different sectors and use there a potential area for better security	Use various sector key concept using a certain pattern	The pattern can be an issue once known can result in applying of blended attack which can result in data leakage
6	Blockchain anchored supply chain management	C Narayanaswami, R Nooyi,R Viswanathan, SR Govindsawmy	IBM Journal of Research and Development, vol. 63, March-May 2019	Integrating blockchain technology helps with providing visibility and recording provenance, as demonstrated through a use case in production.	Consumer demand for verifiable authenticity, ethical sourcing and manufacturing, regulatory compliance, and after-sales services has increased.	To meet the increasing complexity of customer demand, supply chains have evolved into multi-enterprise networks with many production flow pathways.
7	Bitcoin: A simple explanation of Bitcoin and Blockchain technology	Richard Lee Twesige	Conference: Bitcoin Cryptocurrenct, Volume: 1,Project: Decentralized Computing	BlockChain features Fast CheapEasily Available & PublicOpen Source & Programmable Transparent Distributed	Can be used in different fields, innovation lets the clients move esteem comprehensively	-
8	A Decentralized Connectivity Service for private social communication apps	Christos Aslanoglou; Michalis Konstantopouls; Nikos Chondros ; Mema Roussopoulos	IEEE International Conference, 08 July 2020	Decentralized Connecting Service Privacy Preserving	Without the use of a centralised party, blockchain-based networking is possible.	Lack of monitoring in the transactions

Continued on following page

Table 1. Continued

S.No.	Paper Title	Author(s)	Journal Details and Date	Key Concepts	Highlights	Limitations
9	Using Blockchain for Electronic Health Records	Ayesha Shahnaz; Usman Qamr; Ayesha Khalid	IEEE Access (Volume 7) Aug 2019	Implementing Blockchain Technology for Electric Health Record (EHR)	Highly Secure, Tamper – Proof, and UnBreachable	EHR notion of preserving data is still work under progress
10	Blockchain- Enabled Smart Contracts: Architecture, Applications, and Future Trends	Shuai Wang; Liwei Yong Yuan;	IEEE Transactions On Systems, Man And Cybernistics Nov 2019	Implementing smart contract using Ethereum and Hyperledger	Automatic Fascilate, verify and enforce the transaction	Lack of central hub to resolve conflict and handle any mis- happenings
11	Blockchain-based Electronic Patient Records for Regulated Circular Healthcare Jurisdictions	Sofia Alexaki; George Alexandris; Nikolaos E., Vasilis Katao;	IEEE23rd Workshop on ComputerAided Modelling and Design of Communication Links and Networks NOV 2018	Circular Data Driven Healthcare	Provision of health care cycle by provid in medical data integrity and privacy to patient data	Business Logic not clear
12	Blockchain Technology and Cryptocurrencies	Siddhart Rajput; Archana Singh; Tushar Bansal; Sanyukta Shrestha	International Conference on Artificial Intelligence FEB 2019	Advancements of Blockchain in Bitcoin	Use of Blockchain headway consolidates splendid transactions	Nothing Mentioned about corporate financing and interest on bitcoin
13	Performance analysis of ethereum transactions in private blockchain	Sara Rouhani; Ralph Deters;	IEEE International Conference on Software Engineering and Service Science APRIL 2018	Ethereum based Technology in Blockchain Transactions	Analysis of two ethereum client Geth and Parity on a private Blockchain	Reentrancy issue not covered
14	Blockchain-based payment collection supervision system using pervasive bitcoin digital wallet	P. Chen, B. Jiang and C. Wang	IEEE 13th International Conference on Wireless and Mobile Computing, Networking and Communications (WiMob), 2017	Crypto currencies inpayments	Usage and advantages of bitcoin	No comparison with traditional methods.
15	Blockchain in FinTech: A Mapping Study	Fernandez-Vazquez S.,Rosillo,R. David De La Fuente,Priore P.	Sustainability 2019	Fintech and adoption of blockchain in fintech	Use of blockchain and its benefits	Disadvantages or challenges related to adoption

A blockchain can be thought of as an interlinked chain of a data structure (blocks) that carry timestamps and are linked together via cryptographic hash code. These blocks behave like only readable entities that mean they are secure, sealed and immutable. The chain acts as a never ending progressive chain that appends new blocks at the end, every new block included contains the hash code of the previously connected block. System of these interconnected blocks is considered as a network and the responsible entities or the shareholders are considered as nodes of a network that send and receive data. The network functions similarly to a peer-to-peer (P2P) network. (Tribis et al., 2018) Each node has two keys, which are analogous to the public and private keys used in asymmetric cryptography. The public key encrypts the communication sent to the network, while the private key decrypts it. This ensures consistency, non-reputability and irreversibility these also are the key features of blockchain. One field where blockchain can have immense impact is of e healthcare. Blockchain can serve multiple purposes for this field, whether it is updating data in Real-time to interconnect different healthcare systems. It can be used for

drug authentication (supply chain), managing audit trails or to support access control. Blockchain can help to make healthcare more patients centric and could increase accessibility, reliability and security of patients' data as it can be seen by the patients' themselves. Similarly, the trailed report data can be used for disease prediction purposes that can transform the overall experience of the patient. Other than these blockchain have immense potential to become a security and privacy system for a private or internal network because it is very robust and reliable, moreover the access control mechanisms are an added advantage. As a result, blockchain is an excellent framework for electronic health-care systems. On the basis of managed data, data availability, and user permissions, blockchains are classified into many categories.
These are:

- Public permission less
- Consortium (public permissioned)
- Private

When comparing the three, the public permissionless network is the most dangerous in terms of network security because it is open, transparent, and available to anyone (public), allowing anyone to act as a simple or miner node in the network. These are commonly used in cryptocurrency to provide the user with additional privacy.

The consortium blockchain network only allows only selected nodes in the process. It is like a private blockchain but for a group of participants that are predefined. It is best used for multi organizational networks where the network stays partially centralized with public use. Only selected members can join the network as nodes on the private blockchain. In this manner the network stays centralized and distributed at the same time. All the processes are permission based. This network can be used for supply chain purposes. For E healthcare public permission less and consortium are of no use as it can be a source of disruption of the anonymity of the system. Private Blockchain is best suitable for e healthcare. Seeing the current medical trends of dependent data systems which are very vulnerable and costly or healthcare networks. Private Blockchain network will act as the best possible interconnected, distributed and centralized network that is immutable and reliable.

Privacy

Privacy of information stands for protection of the information from any unauthorized or unwanted access. With blockchain system privacy becomes a huge issue as the information is generally very sensitive and in the eyes of invaders on the network. In blockchain network privacy can be insured by two steps:

1) The links of the transactions must not be available or discoverable to others.
2) The content of the transaction should only be visible to the shareholders. (Public blockchain is considered in the analysis. In a private blockchain network for access control permissions are needed which satisfies the need of privacy of the system but complete transparency is at cost in the following process. Privacy requirements can be considered on two factors:

Identity Privacy: This is referred to as the intractability of personal information of the shareholder and the transactional scripts between the real identities of the shareholders. With all the measures for privacy protection, identity privacy is not completely secured. Moreover on an unencrypted network some behavior identity analysis techniques are used to identify the customers and attackers (Helo & Shamsuzzoha, 2020). Transaction Policy: It is referred to the safety of the transactional content of the network. The transactional content includes the amount or pattern of transaction. This information can be accessed only by specific users. A block in the blockchain contains transactional information, such as the participants' addresses, trade values, timestamp, and sender's signature. By using data mining on a public blockchain, it is feasible to extract the stakeholder's personal information from transactional data. Techniques used for privacy breach:

1. De anonymization: It is possible to perform a static analysis of the blockchain or extract the required information from the network data in a blockchain network. Methods used for de anonymization:
    a. Network Analysis
    b. Address Clustering
    c. Transaction Fingerprinting
    d. Dos Attack
    e. Sybil Attacks
2. Transaction Pattern Exposure The transactional information flows on the network and can be accessed by which the transactional patterns can be determined and the secure information can be at threat.

Methods used for TPE:

a. Transaction graph analysis
b. AS level deployment analysis

Supply Chain

Supply chain in logistical terms means a system used for supplying goods from a source to a destination. Now this system contains intermediaries and these intermediaries play a huge role in completion of the chain by providing necessary routes and medium to transport the goods. But at times these intermediaries can also be the reason for huge losses, as the major problems in supply chains include lack of trust and information. The intermediates are often not trusted by the manufacturer as they are generally a third party company. Similar to the trust factor lack of information also plays a huge role in the processing of the manufacturer as if the information of the goods is not received on time; the manufacturer loses money and commitment. (Protecting Patient Data the New Scope Cybersecurity for Healthcare, 2019) The third problem is the mixture of this lack of trusted information; this causes the manufacturer loss of reputation in the long term, as wrongly passed information shatters the image of the organization. Generally, a supply chain whether it be for a product or a service needs two things, firstly least amount of inventory for producer and/or retailers and true information about the service or goods. Supply chains were created to facilitate the producers by connecting them to new markets and the consumers also by providing quality products or introducing new products in their lives. The main purpose of these chains was also, to provide maximum satisfaction of requirements by giving minimum efforts. Blockchain technology with its key features like reliability, traceability, privacy and security can be a boon for supply chain systems. With the blockchain technologies in hand companies can create their own networks (whether being private or consortium), they can easily trace the goods or services that are being supplied since all of the blocks in the network are connected a trail of information can be easily associated with the block which gives unharmed, reliable and Real-time information to the producer.

The major factors that will get an impact of the blockchain in supply chain management are:

Traceability tracing the product for its route or the source of the product is an important factor in the supply chain. Products like medicines or foods have been reported to be replaced in the chain and this causes huge losses to the companies, in fact promoting illegal supplies of these products in unknown places. The products with risks involved like medicines can cause threat to the lives of the users. Blockchain introduces time stamped registration of the product and each intermediate trail is also constantly validated, so product tampering is not possible. Moreover, the information is only readable so this also ends the possibility of wrong information being supplied to the system.

Distribution Blockchain networks are distributed independent of the type of network being used. With the advantage of the distributed nature, supply chains benefit a lot from the network. Each chain does not need a central authority to operate and all the nodes are not mandatory for the operations. So, if one of the nodes loses its functionality others can be used for references, as the data is being stored and cross validated at each of the nodes. Moreover, due to a node independent system the producer can cut middlemen and reduce cost because now the producer can be directly connected to the consumer via the network.

Immutability: The information that is once stored on the system can be changed or removed; this helps the traceability directly because without an immutable system traceability can't be achieved.

Globally the industries related to the supply chain are facing a tough time to create a trusted Real Time view of the supply chain. Tracing and tracking of the products or shipment becomes a huge task as the information used for the task depends on multiple entities (or companies) in the real world. Nowadays the logistical systems try to use positioning information to alter the next actor of their chain but generally the methods used in this process are GPS, GTIN or other different positioning systems but all of these systems have their own flaws and which increases the concern of security and authenticity for the companies. As the tracking devices generally have too many vulnerabilities' which can be easily exploited to change the information on the systems, which leads to a bottled neck situation for the companies where they are expecting payments at a certain time but the delivery has not been done which delays all the other tasks. (Narayanaswami et al., 2019) Blockchain having the features of traceability and Immutability have a very high potential to be a part of the positioning services for companies. Companies can decentralize their task without the worry of authentication and still achieve the same amount of reliability from the system. The blockchain technology with all the other positioning systems works as a reliable framework for positioning not only for the producer but also for all the stakeholders in the scene. These advantages of the Blockchain system facilitate its usage in the drug industries for the supply chain management of the drugs. The key advantages of blockchain that can be relied upon for completing the current are:

- Decentralized management Shareholders of the system share a common need to access immutable and transparent data.
- The data include detailed trails of transaction data.
- Reliable and trustworthy due to transparency and fraud detection

A shared and secure network with the participants sharing common interests. Blockchain when combined with the current technologies like AI and Iota gives unmatchable results in the form of data that cannot be generated otherwise. This data can be used for prediction modeling that can even enhance the system more. With the use of Iota devices the products are always in track of the producer and gives extra control over the product in real time. With the help of these systems the transportation of the goods has become a hustle free task and the same can be implemented in transportation of the healthcare supplies to mark the authenticity and integrity of the supply

Partnership

Partnership is proposed here. Partnerships with community hospitals, human resources and IT professionals in an advisory capacity are proposed here. The partnerships will help inform the specific challenges and needs of their local healthcare providers. This gives crucial ground for development, depending on the individual department's approach and challenges, as well as local facilities' clear awareness of the regulations and policies that these institutions are trying to implement owing to a lack of knowledge and current issues.

The IT provides with latest protection from latest attacks that comes in notice which aren't in knowledge with the other sectors while the human resources gives the details about which data and how they need to be secured for better utilization and keeping safe from forgery and data leak. The hospital provides the latest data and policies they need and use to enforce better working and keeping in knowledge the state of each patient. (Twesige, 2015)

- Cyberspace Laws and Regulations –HIPPA or HITECH Privacy and Security rules are proposed to be used for encryption and privacy protection.
- Cloud computing being used to store the hospital data while having different limitations related to data in motion and data at rest, which are worked upon by the IT sector.
- IT departments secure patient data given by the hospitals that are stored in the cloud, working on privacy and security rules based on the policies keeping in mind the latest trend of attacks to provide a safe and secure way of preserving and using data.
- Cross-departmental partnerships will result in better understanding of the real world and working with it.

Digital Technology in Supply Chain

Digital technology advancements are having a significant impact on the world around us, resulting in significant changes in customer behavior. Many supply chains, on the other hand, are unable to keep up because they are designed in a very traditional and sequential manner. Traditional supply chains, we believe, must evolve into linked, intelligent, and highly efficient supply chain ecosystems with layers to address custom solutions and digital operations companies capable of setting up and running client solutions and digital operations. Ecosystems empower themselves to proactively manage and efficiently fulfil customer needs across multiple channels, quickly identifying changes in demand and organizing their supply chain accordingly. This allows them to turn their supply chain into a competitive advantage and use it as a platform for new digital business models. (Aslanoglou et al., 2020)

Direct link between demand and manufacturing. At the same time, an updated demand plan is shared with all relevant suppliers, who immediately incorporate the new requirements into their material planning and adjust their manufacturing priorities accordingly. Meanwhile, the customer can track the order's progress, staying informed about when it's being manufactured and when it'll be delivered.

Blockchain Features

Fast: Communication between peers on the system is swift, which is beneficial in situations where time is a critical factor for chores and exchange.

Cheap: Using the Block Chain for purposes of installation and value exchange is less expensive than using any other framework now in use. Easily accessible and public: Anyone with a Bitcoin customer, such as Armory, Bitcoin Center, Bitcoin, and others, can use the Block Blockchain innovation anywhere in the world.

It is quite likely that it can be customized to fulfil a specific purpose because it is open source and programmable. Consider the necessity for a monetary payment system, which might be adjusted by anyone with programming development skills to fulfil a unique need.

Transparent: This is one of the most appealing features of the Blockchain; its simplicity makes it visible to everybody on the network. It's a bit like when everyone opens a record. It is not only visible to everyone, but it is also time stepped.

Distributed: The way that the Blockchain uses a disseminated arrangement, this makes it safer than some other incorporated organization like budgetary banks and the rest alike. This is on the grounds that imperative information like cryptographic keys utilized for online budgetary installments and exchange are not kept in a similar spot. On the off chance that an aggressor was to bargain the Block Chain, pursues are thin that he could have an effective assault.

Decentralized Privacy-Preserving Blockchain

We associate with many new people in our day life. Daily communication has become an integral part of our life. This is our basic social need and has become our social identity. But, do we really care what we share, what do we text, whether our message is read by someone else. Hardly have we paid any attention to this detail and as a result our data is stolen and breached. In this paper we read about a decentralized chat application that will run without a server. Whenever a user sends a message, it will be sent to each and every node but only the user with the desired key can access it. This is called Darkness. Though it has one major drawback of increased latency and increased bandwidth. The messages will take time to travel from user to recipient. But it is highly secure and safe. Each user will be given a key which will be their own unique identity and generated through SHA-256. (Shahnaz et al., 2019)

The data will be safe and attackers cannot steal it. Nowadays, people rely on a variety of chat apps to collaborate, socialize, inform and chat with each other. There is a huge emergency of such a Decentralized central application. The first and foremost priority is user's security and user's data privacy which we as security students must pay very high attention to in making any application. The Internet is becoming decentralized day by day. With increasing and advancements of cloud-based technology, the health care sector can lag behind and resist this change. Patients and doctors are very unlikely to communicate with each other once they visit the clinic. But this should be improved. Patients need doctor advice each time so such a decentralized app will really fascinate the intercommunication of patient and doctor. Using a third party application means loss of data privacy. Every day we hear news of big giants being blamed for data loss and data sales. Data is the modern currency of this 21st century. It is this data on which the market runs.

- The authors proposed DCS, a decentralised privacy-preserving blockchain-based connectivity service for supporting point-to-point connections between users of social apps like VoIP, chat, instant messaging, video conferencing, and file sharing.
- DCS separates the ownership and management of a user's contacts list from the use of applications and third-party services, giving the user authority over the "social" component of her identity (with whom she interacts and when and how frequently). Prior decentralised identity management solutions haven't addressed this crucial aspect of a user's identity.
- In networked apps, the user is treated as a first-class citizen. DCS authenticates user identities and recognises the target programme, the available device, and the IP address when users connect with their friends using a simple nickname.
- DCS has no single point of failure or trust, while also offering monetary incentives for end users to participate in the service by acting as DCS nodes.

Blockchain Use-Case

Blockchain's use-case is now not a hidden thing. The safest way to store and transfer data is blockchain. Blockchain are immutable which makes them perfect for any sensitive data. Health records of patients are highly sensitive data. No one in today's generation would like to spread his health details across the globe and let others misuse it. It must be securely guided and prevented from any miss happenings. So that is how Blockchain came into picture. We need some system where data can be protected from any hacker or any spyware or any malware. Blockchain consists of a chain of blocks where each block contains one hash value which is generated through SHA-256. (Wang et al., 2019) Any tamper with the blockchain is not possible. For storing health records, blockchain makes a perfect choice. Software developers need to ensure that whatever we create, any application should be data secure and data privacy.

The main aim of the chapter is to propose a paper for a blockchain based framework to store electronic health records. But the electronic health record notion of storing is relatively a progress and not that famous. The work still needs many advancements and technology still needs to develop. With so many advancements in cloud-based technology it won't be wrong to say that this blockchain technology is a future for health records which is more safe and secure to use and highly efficient against any type of attack be it malware, spyware or any Trojan. Moreover, the main issue is the scalability of the proposed framework. Whether it will be able to handle a large user database, not only patient details also we are talking about doctors' details, retailers and manufactures who work on the ground level providing the necessary goods and services. Supply chain is also an important use case of blockchain. Any forfeit of goods or medicine must be eliminated anyhow.

If a patient takes any defected medicine or any nutrient that is defected, instead of getting healed he will fall sick. This type of case is on a daily surge. We also need to address this issue such issues regularly hamper the growth of the billion worth market of the ever-growing industry of healthcare. Whatever we can do to maintain data security, data privacy, data integrity, data safety, all safety choices should be strictly adhered to. Any lack in the approach can risk someone's life and unexpected events happen every now and then. They do, however, have certain concerns about the security of medical records, user ownership of data, data integrity, and so on. The implementation of a revolutionary technology, such as Blockchain, could be the answer to these problems. This technology promises to provide a safe and secure platform for keeping medical records and other health-related data. The healthcare industry used a paper-based system to preserve medical records before the arrival of modern technology, i.e., a handwritten method. This tamper-proof, paper-based medical record system was inefficient, insecure, and unstructured. It also had to deal with data duplication and redundancy because the patient's medical records were duplicated and redundancy was an issue. It is high time that we realize that data is the new currency and any loss of data is like bringing more trouble and more risk in our lives. Blockchain technology is not going to stop here. It has a long way to go in improving the health security, health identity of individuals all across the globe.

Smart Contract

Smart Contract is really a buzzing word in today's market. But what exactly is it? What is so smart about them? Why should we really care? Programming code represents any type of contract we will call them as smart contracts. (Wang et al., 2019) Lets take an example, now if you talk about a lot of online websites which are e-commerce websites, most of them are the websites which we don't trust. If I want to buy something online, I have some specific website in mind, for example in India Amazon, Flipkart, Snapdeal are quite famous but I don't even try other websites. Let's say there's a website named abc.com where you can buy stuff online. Maybe I will not buy from them due to lack of trust and security. So how do you build that trust and how do you make it secure, when you say I don't trust them? What's wrong with them? So let's say if I order something on abc.com as a prepaid order, what if the product is not delivered, there's a transaction from my side but what if abc.com can be trusted but not the delivery service. Let's say company XYZ is responsible for taking that product from ABC and delivering it to me. What if they are doing some fraud? And if the product is not delivered then xyz are using ABC's money as commission and delivery payment included in the order amount. There's a trust issue between three different parties right from my side to ABC from ABC to XYZ. Okay, I should trust myself right because I belong here to solve that problem.

Blockchain provides a solution to this problem. We can use smart contracts here so the entire payment system will be automated. For example I will buy something from ABC Company which will be prepaid but then ABC will not receive that money they will receive the money only when I receive the product so it is something like if-then in programming. There is a decentralized authority involved that ensures credibility and security in the payment process. When I am paying for a product it will not be received by ABC it will be received only when I get delivery. XYZ will receive the commission from ABC only when they deliver the product right. There are certain benefits of this, it is fast, it is secure because they're using blockchain here and it builds trust which is very important in terms of ecommerce. It excludes the middlemen in the systems and automates the process fully, as any of the parties involved will straightaway receive the payment if they have met the conditions stated in the contract, because everything is on a blockchain there is a major advantage of conflict resolution and protection from masquerading. The smart contracts in simple terms are plans of codes which runs automatically when certain conditions are met. Only in the mere future real power of blockchain and its tremendous use case will be unleashed, currently we are just exploring with the tools we have. Blockchain power is not still underway in the health sector. Still many areas are left to explore.

We know one of the most famous use case for blockchain is cryptocurrency. Bitcoin, Ethereum, alt coins are well known terms used with blockchain but blockchain and healthcare has an ocean of opportunities in terms of improvisation and improvement. Blockchain can be a better medium to exchange information between hospitals. There are other issues which can be taken care of with the help of blockchain. A similar use case of smart contracts can be applied in the field of supply chain mainly for the medicines that are sold in different parts of the world and have to go through a long chain of middlemen, exporters and importers (Alexaki et al., 2018). Inclusion of multiple middlemen creates a threat to the system as there are plenty of chances of forgery. There have been cases where fake medicine can be replaced in the original shipment. No one can be sure if the medicines they are taking are legitimate and the price they are paying because of the middlemen is correct. These two issues can be solved using blockchain in the supply chain. So how can we use blocks in here, first of all just to reiterate blockchain is a distributed ledger so basically, all the information will be stored in a decentralized way, so no one will own your data, everything will be distributed between different between nodes between entities and it is immutable.

Blockchain of course is decentralized and it is immutable so what if all the hospitals share the same information. There are certain issues here: the first issue is even if you have a centralized database where all the hospitals are sharing the information about a patient. Instead of saving the information in a central database which can be corrupted, they can have a distribution of responsibility so the hospital has a power to not share my information as health records are owned by a hospital. I don't have the power to access my own records, that's the issue, so what if you can have a centralized database where every patient will get an identifier and they will hold the information, so there is no power for the information which will be in the central database. Every hospital you visit they will request for your information from you so you can provide the access so if you say maybe you can have an app on your mobile phone and say hospital will ask you for your information and that information can be shared using the unique identifier. Privacy is also maintained of course not everyone will know about your health records. Health records will be on the distributed data it will be on the decentralized internet but it will be anonymous right no one will know your records of course we can implement that with public or private blockchain, it makes more sense to have this type of scenarios for private blockchain those things can be can be solved with the help of the option here so you as a patient you can provide the access for your information so you basically own your data and the amazing thing is your data will be always updated so let's say you got some disease now so you go to a hospital maybe a small clinic even those people can update your data here this is what they you have been diagnosed with now.

Bitcoin

What is Bitcoin and why is there so much hype regarding it? The current price of Bitcoin is approximately fifty-seven thousand dollars and if we convert that into Indian currency it comes out to be 4.3 million rupees. In March 2018, the price was approximately one thousand dollars and so we can imagine the exponential growth from 1 to 57 thousand dollars. Bitcoin is basically a currency which is digital and which uses cryptography and that's why we also term it as a digital cryptocurrency. Now when we think about cryptography we always think about encryption. In cryptography we use asymmetric cryptography where we have two keys - a private and a public one. We use this public key to encrypt data so that we can use a private key to decrypt. But then we don't use cryptography in Bitcoin to achieve secureness, we use cryptography in Bitcoin to achieve authentication (Digital Signature). How do we know if someone transfers money, then who is transferring it and to whom it is being transferred to. That's why we use this cryptographic concept. Bitcoin is open-source and that means the volunteers worldwide are contributing to the Bitcoin technology. It's not just a currency, it's a whole set of new technologies we can use.

(Rajput et al., 2019) Bitcoin technology has four parts - the first one is a decentralized peer-to-peer network which is also the Bitcoin protocol, the second one is it maintains a public ledger to maintain all the transactions, then we have some set of consensus rules and finally the fourth one is the proof of work to make sure that the transaction is complete and authenticate it. Bitcoin is a peer - peer network which also means that we don't have any central body. For example, in the real world, to do an online transaction we have a bank in between and then we do a transaction from one bank to another. Bitcoin says no more banks. We can transfer money from one place to another place by using a peer-to-peer network. With Bitcoin we can buy products, we can sell products, we can convert that Bitcoin into real currency and we can convert a real currency into a Bitcoin. The moment a transaction is done, it will be assigned to a blockchain and that block will be broadcasted throughout the network for all the miners. But this transaction is not in plain english, it is encrypted and the miner who verifies this transaction gets a reward in return.

(Rouhani & Deters, 2017) Ethereum is a platform to easily build decentralized applications using blockchain technology. Ethereum does what Bitcoin did for money. It removes the need for a trusted third party. Ethereum blockchain is different in a sense that not just transactions are stored in blocks but we can also store code snippets in those blocks. Ethereum is a programmable blockchain while Bitcoin isn't. Bitcoin does have its own scripting language but it's not complete or capable as an Ethereum one. In a decentralized application stack, we have a blockchain at the bottom and a blockchain which allows for decentralized consensus over application level since there are so many things in an app that different parties need to agree upon such as a valid username, a valid identity, a valid address, a valid reputation score, a valid owner of data. In a decentralized application we need consensus and the blockchain was really that missing ingredient to achieve consensus. So, in a distributed decentralized way at the bottom of the stack is the Ethereum blockchain or any blockchain that allows for smart contracts. In the ethereum platform if we want to use a cryptocurrency we have ether because if we want to run this decentralized application we do need the cryptocurrency there and that is ether. One can simply go to Ether Network and can build their own application or we can use the existing applications.

E-Commerce

In an e-commerce environment payments are essentially needed and the payment methods must be beneficial for not only the end consumers but also the suppliers and distributors as well. One of the major concerns of supply chains is also the payment methods. Generally for a payment of a very huge amount of sum from entity A to entity B there is a bank involved which facilitates the payments and ensures that the transaction is safe and secure. The bank therefore takes a fixed amount or percentage of the payment as the transactional fees. This system is used in the real time application of e-commerce but there are many inbuilt issues in this system. The issues related to conventional banking systems are:

1. Banking business can be performed when the bank is open or in a certain period of time with certain guidelines to be followed.
2. International payments take almost 4-5days to get processed by the bank and a huge amount of documentation is needed.
3. There can be failure of the banking online systems at times.
4. Lack of trust in the system by users.

Blockchain payment system or crypto currencies is an one stop solution for all the banking related issues as the system is robust, secure, immutable and decentralized it has a edge over the traditional system. In a supply chain involving companies from different countries the process time of the payments has always remained an issue. (Chen et al., 2017) With the use of crypto-currencies not only the speed of processing improves but it also increased anonymity of the users in the transaction and decreases the chances of cyber-attacks. Crypto currencies uses peer to peer connection which ensures decentralization has the user in the transaction not only works as a client but can also work as a server to serve the network. It gives the currency an edge over the traditional centralized server. This reduces the risk of denial of service attack as the user is also serving the network and increases anonymity at the same time. There are no changes of failure of the system as the failure of one node does not affect the working of a decentralized network. If blockchain is used in e-commerce systems for transactions there will be no need for manual data entering and the transaction records will be transparent to the users, this will ensure there are no problems in the cash flow. The customers can trust the system more as the data cannot be changed by any means. This will in fact reduce the auditing work of the banks and governments all the data related to a single user can be retrieved easily.

Finance Industry

The finance industry has a huge importance in the global economy and the traditional methods used in these industries by the back are generally outdated or too slow for progressive world. The banks only used traditional methods until a few years back which created a lack of credibility in users. The methods used by banks were too slow and were more process oriented; they rely on the method more than the result for a given task. For example if money is to be transferred from America to Europe it took days as there was a lot of paper and transactional work involved within multiple banks to fulfill the consumers' needs. (Fernandez-Vazquez et al., 2019) After The global recession in 2008 people had no trust in the banks and started to find new solutions for finance. Seeing this opportunity a new industry started to grow known as fintech which is financial technologies that aimed to use new technologies in finance. The main focus of fintech was to provide all the necessities the banks lagged. The main preference of these fintech companies was customer and hence they created new means of finance methods using newer technologies which created user friendly, fast and reliable methods for cash flow in comparison of the old slow and unreliable methods involving a number of third parties. These companies created new methods for transactions that suited the customer needs and focused on the following aspects of transactions:

1. Improved transparency
2. Security
3. Minimization of risks

Therefore blockchain became a solution for these needs as it is secure, immutable, reliable, etc. The companies now use smart contracts which is termed as a digital contract that the two parties agree to while performing any transaction, this ensures the authenticity and trust between the parties, none of the parties can back out of the transaction or claim it to be false after it is complete due to this, which happened a lot in the traditional banking system. (Hasselgren et al., 2020) The Contracts share:

1. Digital Signatures- This ensures verification of the parties in the transaction

2. Oracles- This is the external system that acts as a middleware between the parties it contains the data on which both parties rely on.
3. Self-Execution- When the conditions of the contract are met the execution takes place automatically without any action by the parties.

Proposed Methodology

Blockchain are immutable which makes it super powerful in dealing with sensitive data. We will use Cryptography Algorithms to give each block a new hash, which will allow the chain to be safe and prevent from unauthorized access. We will be using Mongo DB to store our chain after validating each block that we are going to add is valid. This will achieved with proof of work function. We are going to use supply chain management for drugs in our project. We will be creating a smart contract for the retailers where they can register their medicines and the end user can directly see them. We will be creating a database for the drugs (medicines) which can be read or written by the appropriate user. The main advantage of this system will be the system will show a trail of intermediates and the end user can directly redirect to buy the drugs, this helps in creating a safe and trusted environment for the end user to buy medicines with the guarantee of the on demand supply chain information stored.

Figure 1. Block diagram of Supply Chain Management

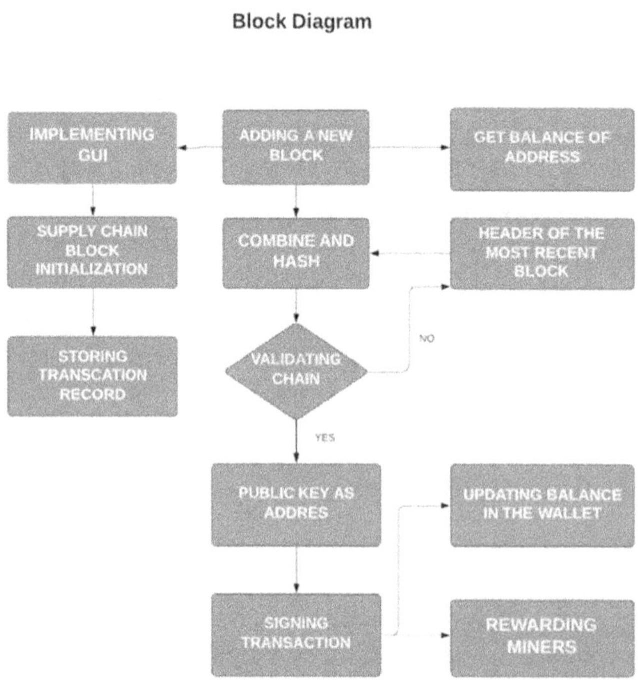

We will be creating a personalized cryptocurrency totally for the purpose of exchange in medical context. We will make a simple cryptocurrency. We created a block that can handle multiple transactions.

We will have a new transaction class. Transaction receives to Address, Forwarders and a certain amount. All the transactions that are made in between blocks are temporarily stored in this pending Transaction array. Users will be assigned a public and private key. In that way, we will ensure that a user can only spend coins in the wallet if he has the private key of it. For key generation we have imported a library – elliptic. (Hasselgren et al., 2020) Before we sign a transaction, public key should be equal to from address. At each class, various proof of work is called to make sure that the user always has valid transactions.

Novelty

In our proposal, not only we are focusing to provide security and tamper proof of healthcare data of the user, but also going forward our small idea of using blockchain in supply chain industries and rewarding users with Bitcoin can revolutionize the health care sector. Blockchain gives all industries a separate inventory network with access to the same information, potentially reducing communication or transfer data errors. Our idea can be implemented into a startup with proper funding and execution of the model.

RESULTS

The target hash is generated after mining loads of SHA256 hashed codes and that hash is associated to the block, the mining of these hashes makes the hash un-predictable so that no unauthorized user can guess it. These hashes are then sending to the previous block associated in the chain. So this creates a chain of blocks connected by unknown hashes. When the block is mined in the server the miner who gets the block has to submit a proof of work if the hash in the proof-of-work and block matches the block is added to the chain. This ensures no delicacy and additional security to the chain.

Figure 2. Hash generation 1560 times and random hash selection

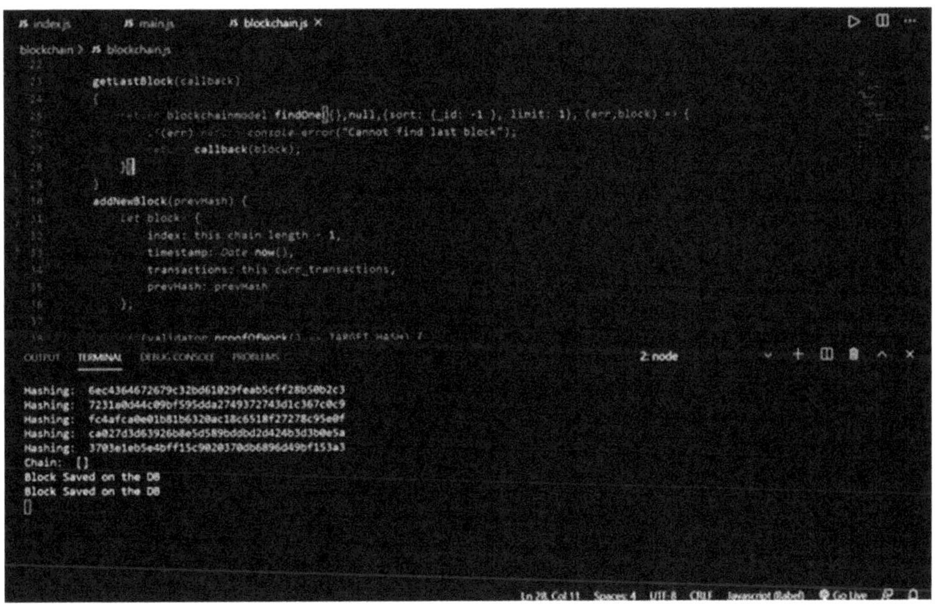

Figure 3. Blocks in the database.

Figure 4. Transaction and prevHash chain formation

Figure 5. Representation of the transaction details for each transaction.

Website Summary

This is a prototype for a website that can be used to connect doctors to patients throughout the world. The website has doctors and patients according to the patient the doctor can be assigned to him or her. Once the doctor is assigned they can be added to a private chat group where the patient can consult the doctor. Besides this there is a record of medicines used by the patient and their prize (score) which is provided to the blockchain supply chain.

In full working there will be analysis of patients' time of wait and the severity of disease can be accessed by the score of medicines he/she has been using in the past. So there can be dynamic doctor and patient pairings to facilitate customers quickly who have more value in terms of medical score.

Figure 6. Sign-in portal

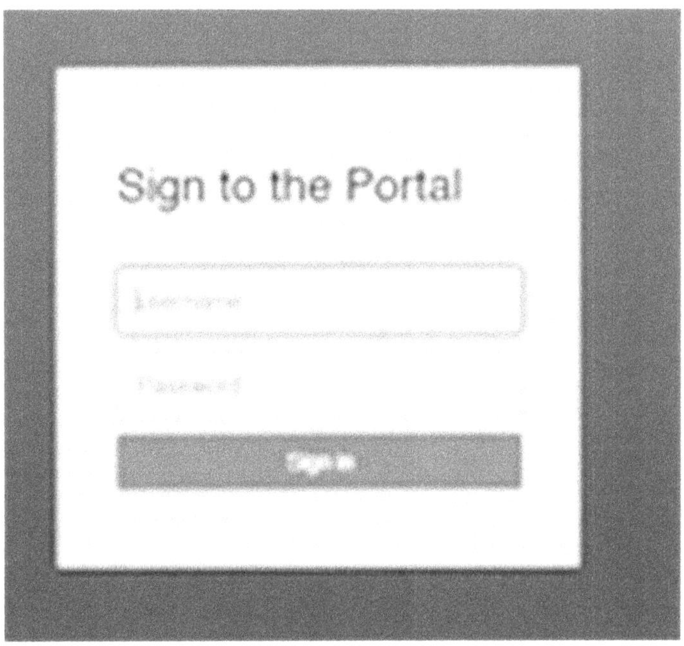

Figure 7. Data view

Figure 8. Adding patient details.

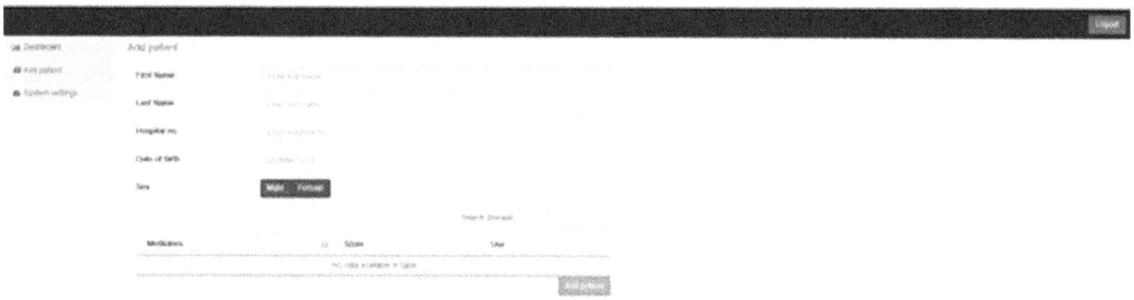

Figure 9. Representation of the medicine database

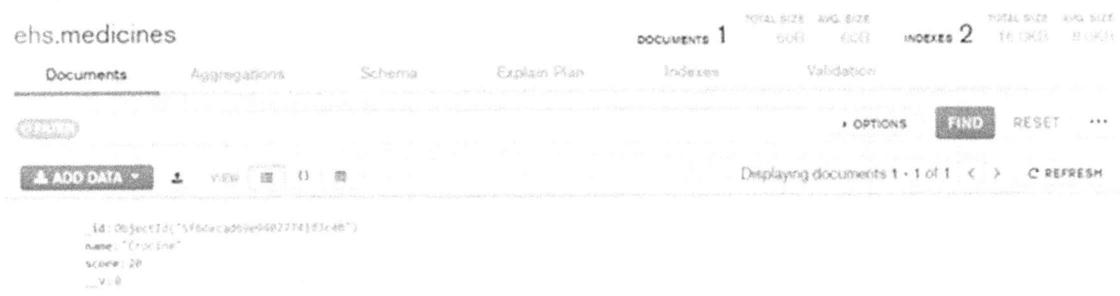

Figure 10. Representation of patient database.

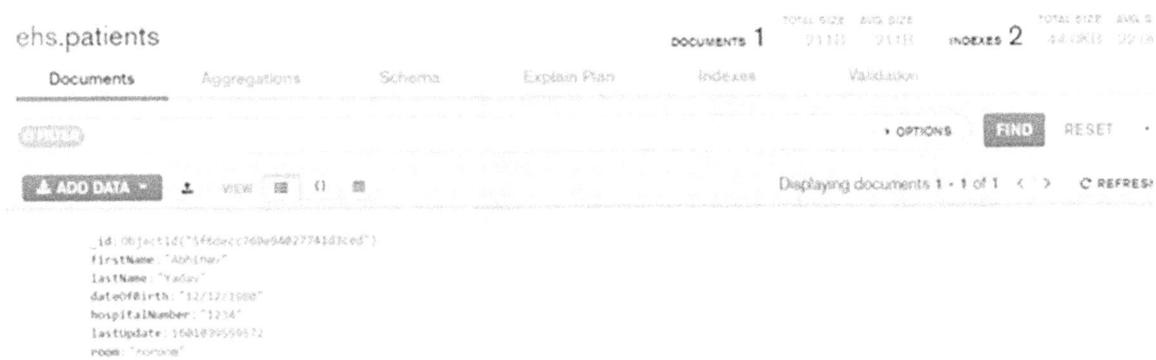

Figure 11. Entire history of our blockchain

```
PS C:\Users\amritk49\Desktop\BLOCKCHAIN> node main.js
{
    "chain": [
        {
            "index": 0,
            "previousHash": "0",
            "timestamp": "01/01/2017",
            "data": "Genesis block",
            "hash": "4373c7fb1437035365d9228c77eca2cfd240523e274163e78c1eba11effd8b38"
        },
        {
            "index": 1,
            "previousHash": "4373c7fb1437035365d9228c77eca2cfd240523e274163e78c1eba11effd8b38",
            "timestamp": "10/07/2020",
            "data": {
                "amount": 10
            },
            "hash": "6523313982255b58bdfb66cadacb22ce754e6b94ce647a270fbe1782f744515b"
            },
            "hash": "6523313982255b58bdfb66cadacb22ce754e6b94ce647a270fbe1782f744515b"
        },
        {
            "index": 2,
            "previousHash": "6523313982255b58bdfb66cadacb22ce754e6b94ce647a270fbe1782f744515b",
            "timestamp": "10/07/2020",
            "data": {
                "amount": 20
            },
            "hash": "1a1d023f61712edd04ed6ddb1a97ac342c5bc40886b6e61595579309994c4194"
        }
    ]
```

Figure 12. Proof of work

```
PS C:\Users\amritk49\Desktop\BLOCKCHAIN> node proof_of_work.js
Mining Block 1...
Block mined : 0000dbf229d568bb20cd0e6afd9cdca6832a08e9141144c0c8064ec6d0871313
Mining Block 2...
Block mined : 0000c497e0f2cab126e4bf4efb3dedd956c9990eec17d36df21089ea6b0e7b80
{
    "chain": [
        {
            "index": 0,
            "previousHash": "0",
            "timestamp": "01/01/2017",
            "data": "Genesis block",
            "hash": "c62b8e4162c75e1fe45dad6af8fcd2672a258941af50ee87f19eacfc190c2ce3",
            "nonce": 0
        },
        {
            "index": 1,
            "previousHash": "c62b8e4162c75e1fe45dad6af8fcd2672a258941af50ee87f19eacfc190c2ce3",
            "timestamp": "10/07/2017",
            "data": {
                "amount": 4
            },
            "hash": "0000dbf229d568bb20cd0e6afd9cdca6832a08e9141144c0c8064ec6d0871313",
            "nonce": 82139
        },
        {
            "index": 2,
            "previousHash": "0000dbf229d568bb20cd0e6afd9cdca6832a08e9141144c0c8064ec6d0871313",
            "timestamp": "12/07/2017",
            "data": {
                "amount": 10
            },
            "hash": "0000c497e0f2cab126e4bf4efb3dedd956c9990eec17d36df21089ea6b0e7b80",
            "nonce": 35660
        }
```

Figure 13. Mining process

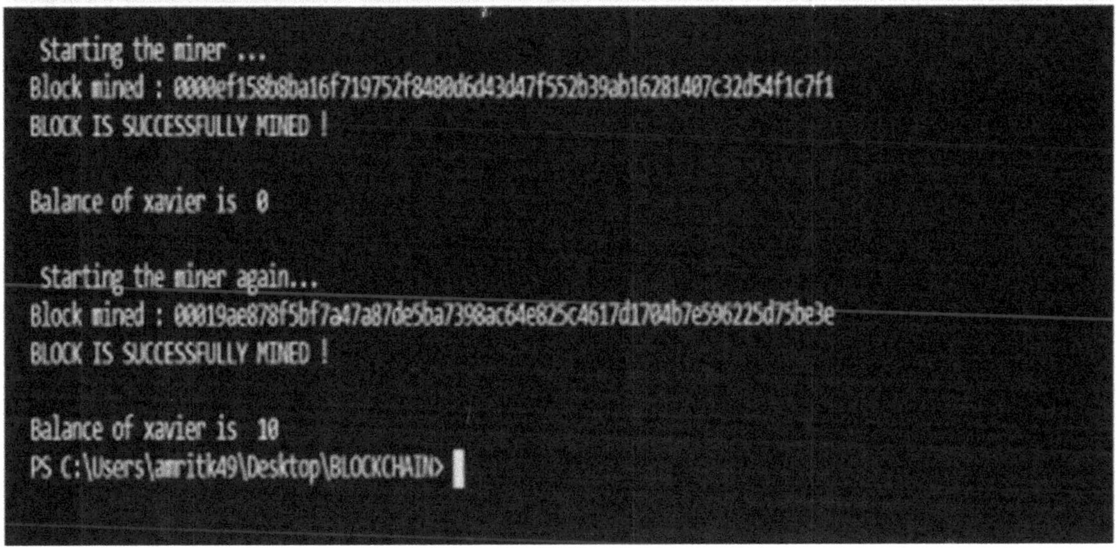

Figure 14. Generating keys - every user will have its public key and private key.

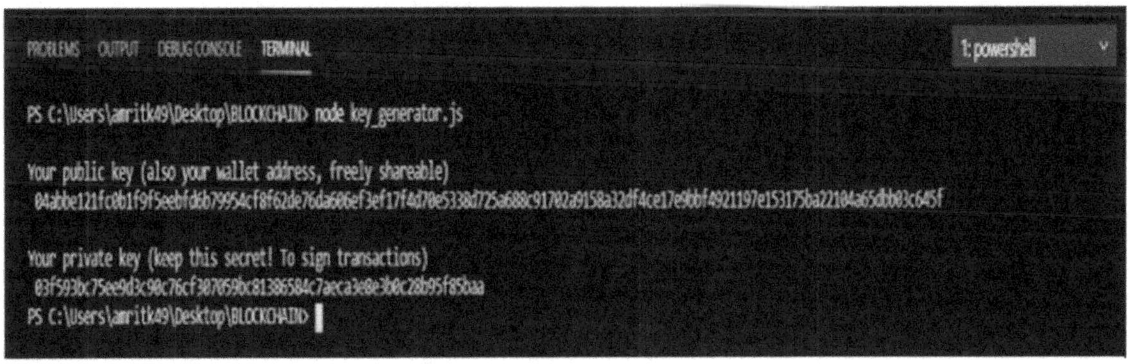

Figure 15. Chain validity

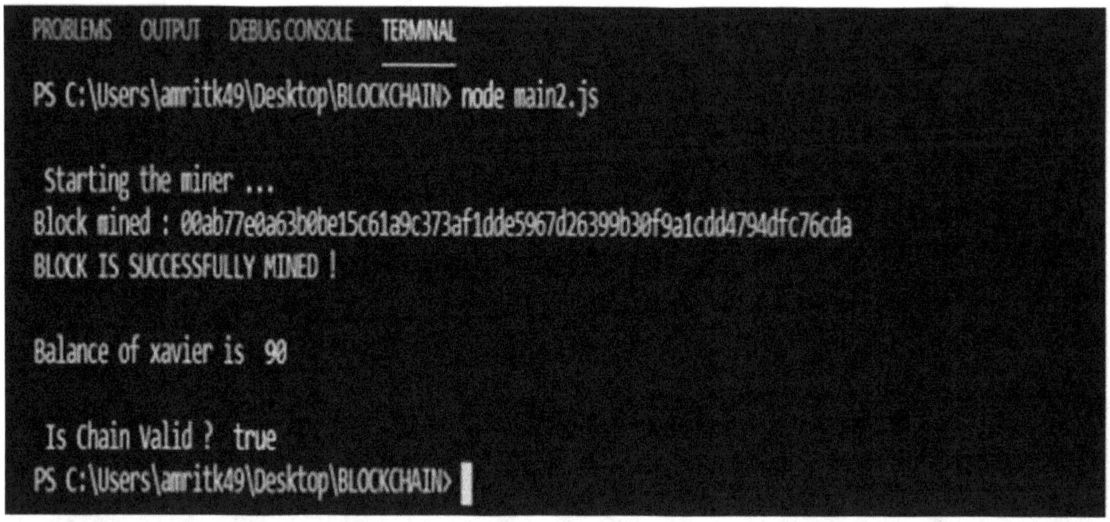

Figure 16. Incorrect validation

CONCLUSION

Blockchain development is still at early periods of commercialization and remembering that there are various industry masters who acknowledge there is a promising future for the utilization of this advancement across industry. Blockchain is in high demand in a few specific industries, including banking, insurance, and logistics, healthcare, and government administration. While Bitcoin and blockchain have taken the globe by storm, India is also making great progress in adopting the technology. With India's rising focus on digitization, the potential for Blockchain in India appears to be expanding. The use of blockchain in financial transactions will not only save time and money, but it will also make transaction processing and authentication much more frictionless. Furthermore, because all transactions are permanently saved on the Blockchain network, Blockchain can be an ideal instrument for monitoring money laundering and black money accumulation. Double spending and unlawful expenditure are further problems that blockchain can solve. Throughout this project, we explored various use cases of blockchain that can be harnessed in health-care security. Though our product is still a prototype but our project can be scaled to bring revolution in the health care sector.

REFERENCES

Alexaki, S., Alexandris, G., Katos, V., & Petroulakis, N. E. (2018). Blockchain-based Electronic Patient Records for Regulated Circular Healthcare Jurisdictions. *IEEE 23rd International Workshop on Computer Aided Modeling and Design of Communication Links and Networks (CAMAD),* 1-6. 10.1109/CAMAD.2018.8514954

Aslanoglou, C., Konstantopoulos, M., Chondros, N., & Roussopoulos, M. (2020). Take Back your Friends with DCS: A Decentralized Connectivity Service for private social communication apps. *2020 IEEE International Conference on Decentralized Applications and Infrastructures (DAPPS)*, 133-138. 10.1109/DAPPS49028.2020.00017

Chen, P., Jiang, B., & Wang, C. (2017). Blockchain-based payment collection supervision system using pervasive Bitcoin digital wallet. *IEEE 13th International Conference on Wireless and Mobile Computing, Networking and Communications (WiMob),* 139-146. 10.1109/WiMOB.2017.8115844

Feng, Q., He, D., Zeadally, S., Khan, M. K., & Kumar, N. (2018). A survey on privacy protection in blockchain systems. *Journal of Network and Computer Applications.*

Fernandez-Vazquez, S., Rosillo, R., De La Fuente, D., & Priore, P. (2019). Blockchain in FinTech: A Mapping Study. *Sustainability*, *11*(22), 6366. doi:10.3390u11226366

Hasselgren, A., Kralevska, K., Gligoroski, D., Pedersen, S. A., & Faxvaag, A. (2020). Blockchain in healthcare and health sciences—A scoping review. *International Journal of Medical Informatics*, *134*, 104040. doi:10.1016/j.ijmedinf.2019.104040 PMID:31865055

Helo, P., & Shamsuzzoha, A. H. M. (2020). Real-time supply chain— A blockchain architecture for project deliveries. *Robotics and Computer-integrated Manufacturing*, *63*, 101909. doi:10.1016/j.rcim.2019.101909

Hölbl, M., Kompara, M., Kamišalić, A., & Nemec Zlatolas, L. (2018). A Systematic Review of the use of Blockchain in Healthcare. *Symmetry*, *10*(10), 470. doi:10.3390ym10100470

Khan, F. A., Asif, M., Ahmad, A., Alharbi, M., & Aljuaid, H. (2020). Blockchain technology, improvement suggestions, security challenges on smart grid and its application in healthcare for sustainable development. *Sustainable Cities and Society*, *55*, 102018. doi:10.1016/j.scs.2020.102018

Narayanaswami, Nooyi, Govindaswamy, & Viswanathan. (2019). Blockchain anchored supply chain automation. *IBM Journal of Research and Development, 63*(2-3), 7:1-7:11.

Protecting Patient Data the New Scope Cybersecurity for Healthcare. (2019, Summer). *Elsevier Journal of Allied Health*.

Rajput, S., Singh, A., Khurana, S., Bansal, T., & Shreshtha, S. (2019). Blockchain Technology and Cryptocurrencies. *2019 Amity International Conference on Artificial Intelligence (AICAI)*, 909-912. 10.1109/AICAI.2019.8701371

Rouhani, S., & Deters, R. (2017). Performance analysis of ethereum transactions in private blockchain. *8th IEEE International Conference on Software Engineering and Service Science (ICSESS)*, 70-74. 10.1109/ICSESS.2017.8342866

Shahnaz, A., Qamar, U., & Khalid, A. (2019). Using Blockchain for Electronic Health Records. *IEEE Access: Practical Innovations, Open Solutions*, *7*, 147782–147795. doi:10.1109/ACCESS.2019.2946373

Tribis, Y., El Bouchti, A., & Bouayad, H. (2018). Supply Chain Management based on Blockchain: A Systematic Mapping Study. *MATEC Web of Conferences, 200*, 20. 10.1051/matecconf/201820000020

Twesige. (2015). *Bitcoin. A simple explanation of Bitcoin and Block Chain technology*. GRIN Verlag.

Wang, S., Ouyang, L., Yuan, Y., Ni, X., Han, X., & Wang, F. (2019, November). Blockchain-Enabled Smart Contracts: Architecture, Applications, and Future Trends. *IEEE Transactions on Systems, Man, and Cybernetics. Systems*, *49*(11), 2266–2277. doi:10.1109/TSMC.2019.2895123

Chapter 8
Development of an Efficient Electricity Distribution Model and Making the Model Robust Using Blockchain

Dishi Jain
Vellore Institute of Technology, India

L. Sharmila
Dhaanish Ahmed College of Engineering, India

ABSTRACT

There are villages and rural areas in modern India that still do not have electricity connection. And there are those places where there is a connection, but low voltage and frequent cut-offs are still a major problem. Electricity companies face a lot of issues regarding the efficient power supply to all their customers, and the customers have issues regarding their needs of power. Wastage of power during transmission is another big issue. This chapter aims to develop and implement algorithms to improve the above-mentioned situation. To make the pricing model more robust, blockchain is used as it ensures secure transactions. Blockchain can give users more efficiency and authority over their energy sources.

INTRODUCTION

Theoretical Background

The theoretical background and main considerations for the model used in the project is presented by Sianaki (Aydinalp et al., 2002; Faruqui et al., 2009; Fischer, 2008). As the meaning of the power for the public economy was expanding, an ever-increasing number of endeavors were put to track down the ideal network advancement techniques. Perceived, that there are various techniques applied by and by, which bring about possible and respectable arrangements. Nonetheless, it is clear that these strategies can be

DOI: 10.4018/978-1-7998-9274-8.ch008

improved. The advancement of an ever-increasing number of proficient techniques for the arranging of distribution models is continually huge. This can be clarified by the high ventures engaged with support and activity of the organization. In industrialized world almost 50% of the interests in power industry is spent in appropriation models.

Different studies show that force supply is unpredictable and even there are where there is no power. On the off chance that investigation of Haryana, because of expansion popular of power, individuals are changing to biomass fuel and other sustainable fuel (Joon et al., 2009). Smart grid is albeit a productive answer for power distribution (Vineetha & Babu, 2014). It is continued in six territories of India and government is attempting to grow it. India is utilizing chunk based evaluating strategy for power conveyance, subsequently new valuing technique is proposed.

The exploration on load shedding issues traces all the way back to 1972 and work date numerous examinations were acquainted by the exploration local area with address the issues. A nearer survey of existing methods shows that still the viability of load shedding plans is not yet benchmarked and lion's share of the current framework simply believes the procedures to be very indicative to one or the other recurrence or voltage (Wood & Newborough, 2003). With a development of savvy matrices, dominant part of the controlling highlights of force framework and organizations are represented by a computational model. Nonetheless, till date insufficient confirmations of potential computational model has been seen that professes to have better harmony between the load shedding plans and nature of force framework execution.

Subsequently, survey of some huge writings and features the exploration hole with the current techniques of load adjusting that is intended for helping the scientist to close after the determination interaction of existing framework as a source of perspective for future course of study (Mozina, 2007).

Motivation

India is the world's third biggest producer and also the third biggest consumer of electricity. The public electric network of India has capacity of installed of 382.15 GW as of 31st March'21. Yet we face low voltage and frequent cutoffs. Electricity is not there when required during peak hours. Future development of electric distribution systems must pursue a number of different goals. The power system should be financially proficient, it ought to give solid energy supply and ought not harm the climate (Nishio & Asano, 2006). Simultaneously, activity and improvement of the framework is impacted by an assortment of questionable and arbitrary elements. Thus, in this project I have kept the following points in mind to design the proposed system.

Objectives Of the Study

Most of the residential colonies and areas in India are scattered and don't follow a specific distribution. The second major problem related to electricity in India is unavailability of required amounts of electricity. The traditional pricing model followed for consumers is based upon the slots which depends on consumption. The proposed model can be used where the new cities are building like gift city or new township. For the conservation of energy in university buildings it is necessary to know when and how much electricity is being consumed (McNeil & Letschert, 2010; Swan & Ugursal, 2009). To analyze this utilizing the clustering strategy, it is important to choose the appropriate number of bunches.

Applications

- In development of new cities like gift city.
- Developing new township or colonies like Shobha dream acres.
- The model can also be used in campus of universities and colleges.
- The two-way transmission of electricity approach is a very futuristic way to use energy and renewable resources.
- The dynamic pricing method will avoid wastage of electricity.

Challenges

Future power networks ought to have a bunch of new highlights that will be important to fulfill the requirements of the lattice administrators and the buyers, targeting expanding the flexibility of the force frameworks thinking about the incorporation of new rising innovations. A lot of information is being produced at various fragments of the force area as it is changed by the continuous advanced unrest. This information has a possible incentive for power organizations, network administrators, and end clients, and can be abused utilizing huge information calculations for various purposes as force market interest gauging. Later on, a proficient and economical usage of the electrical energy in an end-client driven methodology is normal, guaranteeing the unique association of buyers alongside the attention to the necessities of the force lattice. For such reason, the advancement of progressively productive, self-ruling, and strong innovative arrangements is required, permitting to play out the interface between the various frameworks coupled to the force matrices, just as correspondence frameworks empowering end-client association in a unique manner (Klüser, 2009).

Pointing the difficulties and new standards of future force frameworks, it will be important to coordinate endeavors for working regions in the extent of force hardware, utilized for the improvement of innovative answers for power lattices, including progressed control calculations zeroing in on the enhancement of the whole framework. In this unique circumstance, new improvements of force gadgets are normal for the future force frameworks (Yamamoto et al., 2008).

BACKGROUND

The power conveyed by electric utilities is communicated from producing plants to mechanical destinations and the substations that disperse capacity to private and business clients; the utility business is in this way partitioned into these two regions: transmission and circulation. Transmission networks associate generators to substations through transmission organizations. The dispersion framework conveys power from substations to private and business clients. This segment covers the parts of the electric utility business managing dissemination (Barriers to the adoption of efficient electricity technologies, 1998; Wood & Newborough, 2007).

Distribution addresses around 35% to 50 percent of a utility's speculation. The mission of the utility is to give capacity to customers at a suitable voltage with a specific level of unwavering quality. Dissemination parts regularly include:

- Sub transmission circuits with voltage evaluations that reach somewhere in the range of 12.47 and 345 kV.
- Distribution substations that believer energy from transmission network levels to bring down essential framework voltages.
- Feeders, or essential circuits, that work somewhere in the range of 4.16 and 34.5 kV and supply burden to explicit geographic regions.
- Distribution transformer ordinarily evaluated from 10 to 2,500 kVA that change essential voltage to usage voltage.
- Secondary circuits that convey current from the transformer along the road.
- Service drops that convey current from transformers to client point of usage.

Ordinary essential overhead dispersion frameworks are worked as spiral circuits from the substation outward. In overhead frameworks, designs, for example, shafts and H-outlines support the essential and auxiliary conductors. Surface constructions, for example, cushions encase and ensure electric gadgets on the ground (Golmohamadi & Keypour, 2017; Newsham & Bowker, 2010).

Survey work on under-voltage plans was additionally distributed by Verayiah (Verayiah, 2014) and Mozina (Lakra & Kirar, 2015). Audit of burden shedding procedures has additionally been found in the work did Lakra and Kirar (Pferschy, 1999) where a decent and basic depiction of burden shedding plans can be found. Consequently, we complete further updation of the new works as for various types of technqiues that has not been examined or surveyed in the current framework. Consequently, the foundation study can be discovered to be more associated with distinguishing the methods of burden shedding as for voltage or recurrence, and very little on different procedures. The following area features about the difficult that has been recognized in the current framework.

TECHNICAL SPECIFICATIONS

K-means Clustering

K-Means is one of the algorithms utilized for clustering which will part the information into a few gatherings. K-Means clustering is one of the strategies for information non-progressive clusters/group that can bunch the information into a few groups' dependent on the similitude of the information. This component empowers information which have similar attributes are assembled into one cluster and information that have various qualities are gathered in different groups (Young, 2008). To decide the group name of any information, determined the distance between the information with each cluster/group center. There are a few different ways that can be utilized to perform distance estimations, for example, Euclidean distance, Manhattan distance, and Chebichey distance.

The K-Means strategy expects to limit the number of squared distances between all points and the centre of cluster. This method comprises of the accompanying advances.

1. Select k out of the given n designs as the underlying bunch communities. Appoint every one of the leftover n−k examples to one of the k bunches; an example is relegated to its nearest focus/group.
2. Compute the bunch focuses dependent on the current task of examples.
3. Assign every one of the n examples to its nearest focus/bunch.

4. If there is no adjustment of the task of examples to bunches during two progressive cycles, at that point stop; else, go to Stage 2.

It is clear in this depiction that the last grouping will rely upon the underlying bunch focuses picked and on the worth of K. The last is of the most concern since this requires some earlier information on the quantity of bunches present in the information, which, practically speaking, is profoundly far-fetched. Compare and contrast with what has been, or is currently being done as it relates to the article's specific topic and the main theme of the journal.

Knapsack Algorithm

To achieve efficient energy consumption and distribution in my proposed map, the end-clients are furnished with some energy regulator advancements. We as well as the consumers note the pattern of consumption and distribution and we get details about the cost of operation in that locality. This sort of valuable information is given to help them in settling on choices about modifying their utilization. In the case of 2-way distribution of electricity, such information can help users decide the amount of energy they need to consume and the amount they need to outsource. Knapsack can help us in finding the optimum layout of transmission wires for maximum profit and minimum wastage of power. This will help massively in peak hours of consumption (Sianaki et al., 2010). With this methodology, not exclusively will the expense of power not raise during top hours, yet additionally client inclinations, their fulfilment and minimal change to current way of life will be thought of. Also, during peak hours in a day, the power consumption of the household is normally high and the demand exceeds the amount of power, thus the substation has to resolve this problem. We distribute electricity to different houses using different substations in such a way that maximum no. of houses receive electricity and demands are matched. To implement this, we use the knapsack approach.

Outlier Detection

Outliers are information objects with attributes that are viewed as totally different from other information objects inside the dataset. Exception showed up as outrageous worth to a solitary variable or mix of factors. Information cleaning with eliminating exceptions is vital. Remembering exceptions for information mining calculation is at some point will create mistaken outcome in light of the fact that the harming qualities of the information handled. Performing anomaly discovery and eliminate exceptions will make the information more precise, consequently the information mining result will be more exact.

DESIGN APPROACH AND DETAILS

Distribution of Model into Levels of Electricity Distribution

Stage 1 involved Construction of the Graph that is defining the levels of the model and how they will be involved in the algorithm.

In the graph we are provided with the coordinates of the houses (the 4th layer). The locations of the transformers will be decided on the basis of the clusters formed basically we will find the clusters of houses and assign a transformer at the mean position of the coordinates of the locality. The clusters will form the third layer of the structure i.e., the localities. After the substations have been assigned the coordinates, we will connect the substation with every house present in the locality. Thus, we will obtain several connected components. Here the Substations are the second layer of our city structure. Now the first layer i.e., the Source will now be connected to the substations and the location of the Source will be the mean of the coordinates of the substations. Thus, the city will be now a fully connected graph with the root of the city as the Source (the first layer).

Framework of Proposed Model

Making algorithms for graph selection and preparation. This will include making clusters of houses based on average electricity consumption by each house. Average consumption will be calculated on the basis of historic data, survey and sampling. This data will include appliances in each household and their power consumption along with the time of consumption. Individual household electric power consumption Data Set by UCI. Making algorithm for load-shedding during peak consumption period. This will be done by prioritizing commercial, industrial and household consumptions. Electricity will be sent only to the clusters of houses with maximum earning potential. The load-shedding will be done on a rotatory basis to meet demands of each and every sector. Adding and modifying algorithms for enabling solar energy accumulation by households. This will allow us to include the fact that extra solar energy collection can be sent back to the grid.

Figure 1. Graph showing various levels of distribution model

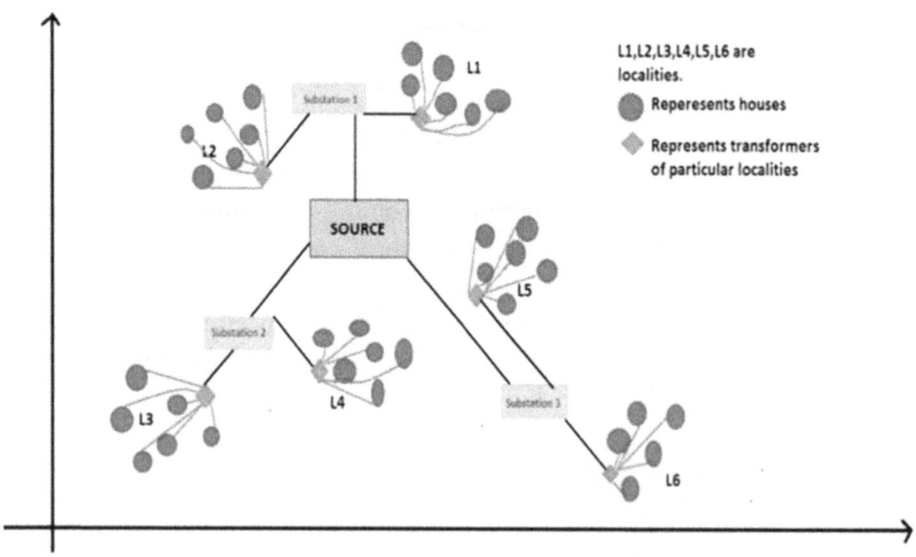

Figure 2. Typical hourly consumption of electricity of each household

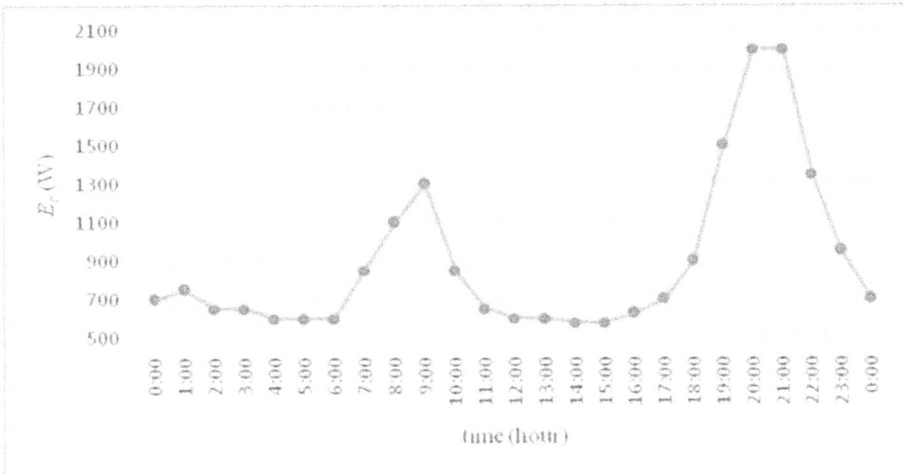

PROPOSED MODEL

Clustering of Houses

Problem

Most of the residential colonies and areas in India are scattered and don't follow a specific distribution, while western countries have a well-planned city layout and structure. Hence it becomes a problem in India to cluster houses in an area which follows a specific pattern of electricity consumption.

Goal

To make clusters of houses which are near to each other in a city by making localities.

Solution

K-Means is an algorithm that has been utilized to resolve the issue. We utilize topographical directions of a given arrangement of houses as sources of info. K-Means clustering is an iterative approach that attempts to segment the dataset into k pre-characterized non-covering subgroups (clusters) where every information point has a place with just one gathering. We classify these houses into gatherings.

Design

The calculation fills in as follows:

- First, we instate k points, called means, which is taken randomly.

- Then arrange every item to its nearest mean and update the mean's directions, which are the midpoints of the items sorted in that mean up until now.

Then rehash the cycle for a given number of emphases and toward the end, we have our groups or clusters.

Load Shedding

Problem

The second major problem related to electricity in India is unavailability of required amounts of electricity. While western grid and southern grid are surplus with electricity, northern and eastern grids are production deficits. This is solely due to large demand for electricity in households during peak hours of the day. This problem is dealt with by illogically cutting powers of localities (Sempértegui et al., 2002). Can we find mathematical ways to find localities to manage this?

Goal

Finding an algorithm which helps in solving the above-mentioned problem. Making the algorithm work for the clusters of houses so that load shedding can be done.

Solution

Our first approach is the Greedy algorithm to minimize the number of houses to be cut-off. Meters in houses calculate the current power consumption and with this data, greedy can be applied. A better approach would be to maximize profit by using Knapsack algorithm. Values of weight and benefit will be analogous to current consumption and profit, which is calculated by multiplying the consumption with respective prices according to the meter installed. Arbitrary data: ₹4 for 2kW meter, ₹4.5 for 5kW meter and ₹6 for 10kW meter. Considering the larger scale on which these algorithms will be applied, it is better for the company to cut off the entire cluster (all houses connected to a transformer). This is carried out by Knapsack using weight and benefit as the cumulative consumption and profit for a cluster. This process is repeated after fixed intervals so we get new sets of localities which must be provided with electricity.

Design

Figure 3. Greedy algorithm for load shedding

Input: Set S of n houses, such that house i has meter type value m_i with price p_i and consumption w_i . m_i has values from a set V. Total instantaneous available electricity is W.
Output: Number of houses that receive electricity N and profit as P. (arr[ ][ ] is taken as buffer)

```
N ← 0
P ← 0
for i ← 0 to n do
    arr[i][0] ← i + 1
    arr[i][1] ← w_i
sort arr[i][1]
for i ← 0 to n do
    if W ≤ 0 then
        return P, N
    N ← N + 1
    W ← W - arr[i][1]
    for j ← 0 to n do
        if arr[arr[i][0]][1] = m_j then
            P ← arr[i][1] * p_j
```

Figure 4. Knapsack algorithm for load shedding

Input: Set S of n houses, such that house i has positive benifit b_i which is profit and positive integer weight w_i which is consumption of the house; total available electricity for consumption is W.
Output: For w = 0,...., W, maximum benefit K[i][w] of a subset of S with total weight w.

```
for i ← 0 to n do
    for j ← 0 to W do
        K[i][j] ← 0
for i ← 0 to n do
    for j ← 0 to W do
        if i = 0 or j = 0 then
            K[i][j] ← 0
        else if w[i-1] ≤ j then
            K[i][j] ← max(b[i-1] + K[i-1][W-w[i-1]] , K[i-1][j])
        else then
            K[i][j] ← K[i-1][j]
```

Time and Space Analysis

Greedy Algorithm for load shedding
Time complexity = O(nlogn)
Space Complexity = O(n)
Knapsack algorithm for load shedding
Time Complexity (n*W) = $O(n^2)$
Space Complexity (n*W) = $O(n^2)$

Pricing Model

Problem

The traditional pricing model followed for consumers is based upon the slots which depends on consumption. For instance, 10-100 units' cost ₹x, then 100-150 units cost ₹y and so on. This slab-based pricing method does not account for wastage of electricity as well as very uneven pricing based on consumption demands. It is logical to decide price based on consumption demands as different topography has different consumption patterns (Ziel et al., 2015).

Goal

Develop 2 pricing models: Static and Dynamic. Base the pricing on the consumption pattern of a cluster rather than a fixed slab.

Solution

We develop two pricing models: Static Pricing and Dynamic Pricing. Static Pricing includes different slabs that have a fixed price per unit and is used for houses that opt for it. Dynamic Pricing depends on various factors like the cost of production, fuel costs, infrastructure costs, etc. The city is divided into clusters of static and dynamic pricing, each applied to clusters based on the consumer demands (Celebi & Fuller, 2007).

Design

Figure 5. Algorithm: Dynamic pricing for electricity consumption

Input: fuels = {coal, gas, diesel}, cost of fuel required for producing 1 unit and its percentage in production is C_i and P_i and where i ε fuels, hourly consumption of last 7*24 hours as A[168] and consumption of last hour as cons_last_hour and fixed price as fixed_price.
Output: Dynamic Price as dynamic_price for current hour

```
repeat
    for j ← 0 do
        for k ← 0 to k ← 168 do
            sum ← sum + A[k]
        average_consumption ← sum / 168
        for i in fuels do
            price_per_unit ← price_per_unit + C_i * P_i
        price_per_unit ← price_per_unit + fixed_price
        dynamic_price = (cons_last_hour / average_consumption) * price_per_unit
        print dynamic_price
    j ← j + 1
until program is stopped
```

Figure 6. Algorithm: Static pricing for electricity consumption

Input: S_i for four paritions of consumption slab where $0 < i \leq 3$, P_j for prices of respective slabs where $0 < j \leq 4$ and last month's consumtion as consumption.
Output: Static Price as static_price for last month

```
if consumtion ≤ S_1 do
    static_price = consumption * P_1
    return static_price
else if S_1 < consumtion ≤ S_2 do
    static_price = S_1 * P_1 + (consumption - S_1) * P_2
    return static_price
else if S_2 < consumtion ≤ S_3 do
    static_price = S_1 * P_1 + (S_2 - S_1) * P_2 + (consumption - S_2) * P_3
    return static_price
else if S_3 < consumtion do
    static_price = S_1 * P_1 + (S_2 - S_1) * P_2 + (S_3 - S_2) * P_3 + (consumption - S_3) * P_4
    return static_price
```

Time and Space Analysis

Static Pricing
Time complexity = O(n)
Space Complexity = O(n)
Dynamic Pricing
Time Complexity = O(n)
Space Complexity = O(n)

Two-Way Transmission Approach of Electricity

Problem

Under normal distribution method, there is one supplier and many consumers. That is, the relationship for supply to demand is one to many. This has several drawbacks as there is much load on one head as well as there is no promotion of renewable sources of energy on grass root level. In this modern era, the focus is on futuristic sources of energy, minimizing wastage and pollution that occurs during production of electricity.

Goal

Develop a two-way electricity distribution system where the consumer itself can produce electricity. Implement futuristic methods of production of electricity.

Solution

We design a two-way electricity distribution model where the consumer (by renewable methods like solar energy) can distribute in its cluster when the demand increases. This not only reduces the demand load on the supplier but is also beneficial for the environment. Also, this reduces the bill for the consumer as it can use the electricity developed in-house.

RESULTS AND DISCUSSION

Clustering of Houses

Table 1. Output of clustering of houses

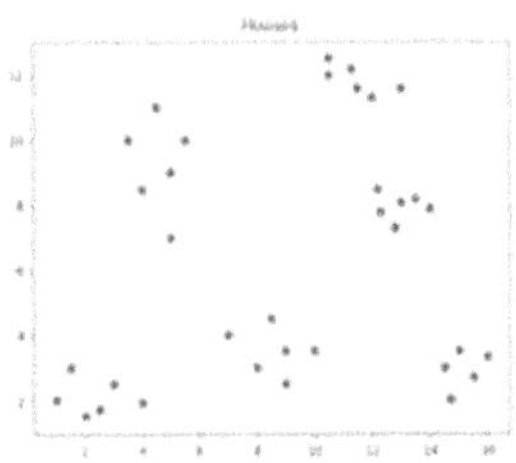 *forming clustering of houses*	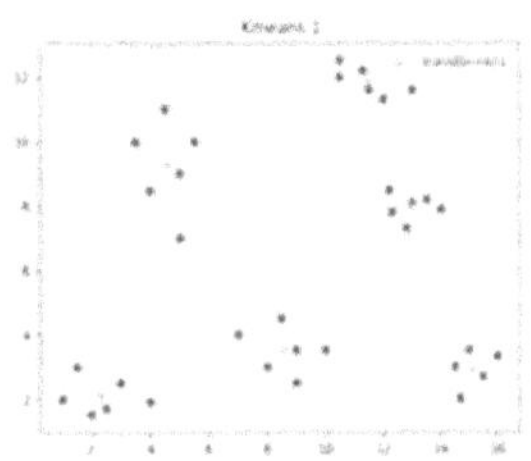 *Location to place transformers*
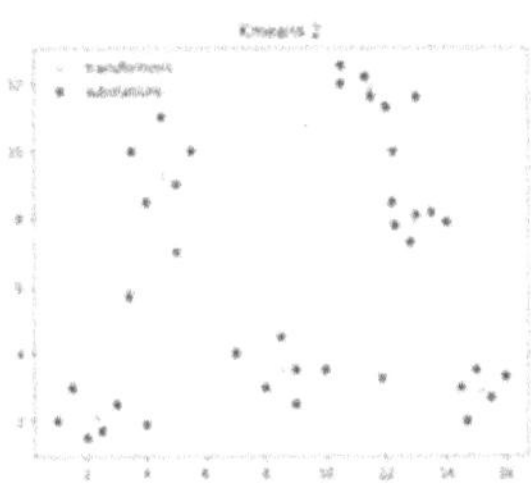 *Location to place substation*	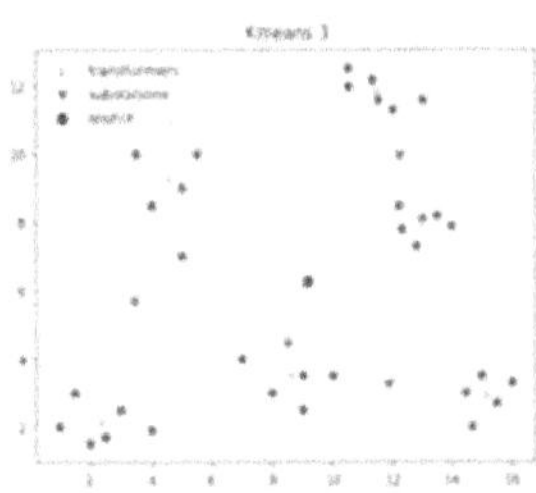 *Location to place source*

Load Shedding

Table 2. Output of load shedding

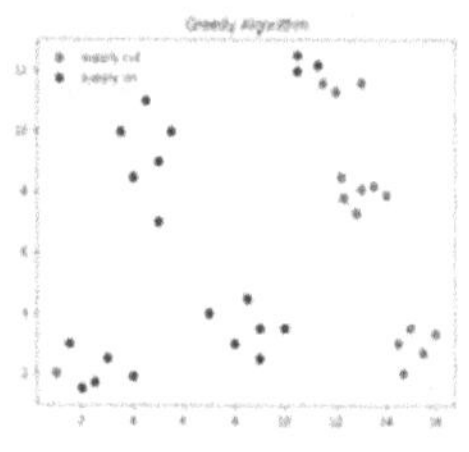 **greedy algorithm on houses**	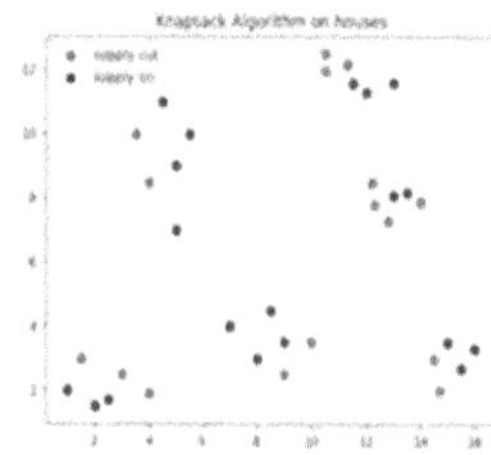 **Knapsack algorithm on houses**
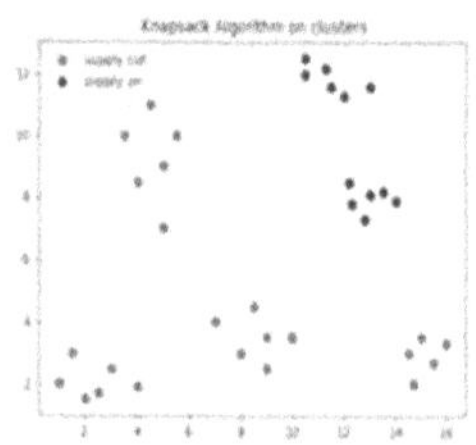 **Knapsack algorithm on clusters**	

Simulation

For Load Shedding We also created a simulation to check load shedding algorithms. It generates random values of average power consumption per hour in appropriate ranges for a day. The output is shown as follows.

Table 3. Simulation screenshots

According to Greedy Algorithm

Hour	Profit	Number of houses
00	2445.0	23
01	2531.5	19
02	2567.0	20
03	2492.0	21
04	2841.5	18
05	2420.5	17
06	2631.5	19
07	2713.0	19
08	2717.0	18
09	2535.0	17
10	2540.5	18
11	2484.5	17
12	2461.0	20
13	2386.5	21
14	2615.0	18
15	2582.0	20
16	2632.0	18
17	2589.5	20
18	2576.0	18
19	2550.5	19
20	2505.5	18
21	2533.5	18
22	2621.0	19
23	2602.5	22

Simulation 1

According to Knapsack by house Algorithm

Hour	Profit	Number of houses
00	2803.5	15
01	3000.0	8
02	2992.0	12
03	2992.0	10
04	2968.5	14
05	3000.0	7
06	3000.0	7
07	3000.0	8
08	3000.0	8
09	3000.0	8
10	3000.0	6
11	3000.0	8
12	3000.0	9
13	3000.0	7
14	3000.0	7
15	2998.5	10
16	2992.5	12
17	3000.0	9
18	2994.0	12
19	3000.0	10
20	3000.0	8
21	2955.0	11
22	3000.0	9
23	2992.0	11

Simulation 2

According to Knapsack by cluster Algorithm

Hour	Profit	Number of clusters
00	2513.0	2
01	2219.0	2
02	2474.0	2
03	2428.0	2
04	2186.0	2
05	2254.5	1
06	2459.5	2
07	1734.0	1
08	2133.0	1
09	1825.0	1
10	1786.0	1
11	1902.0	1
12	2671.5	2
13	2451.0	2
14	2569.0	2
15	2281.5	2
16	2176.5	2
17	2188.0	2
18	2137.5	2
19	2399.5	2
20	1816.0	1
21	1672.0	1
22	2443.0	2
23	2238.5	2

Simulation 3

As we can see, the number of houses is maximum in greedy algorithm, but profit is lower. But knapsack prioritizes benefit, so we get more profit but at the expense of more houses being cut.

Figure 7. Comparison of static and dynamic pricing

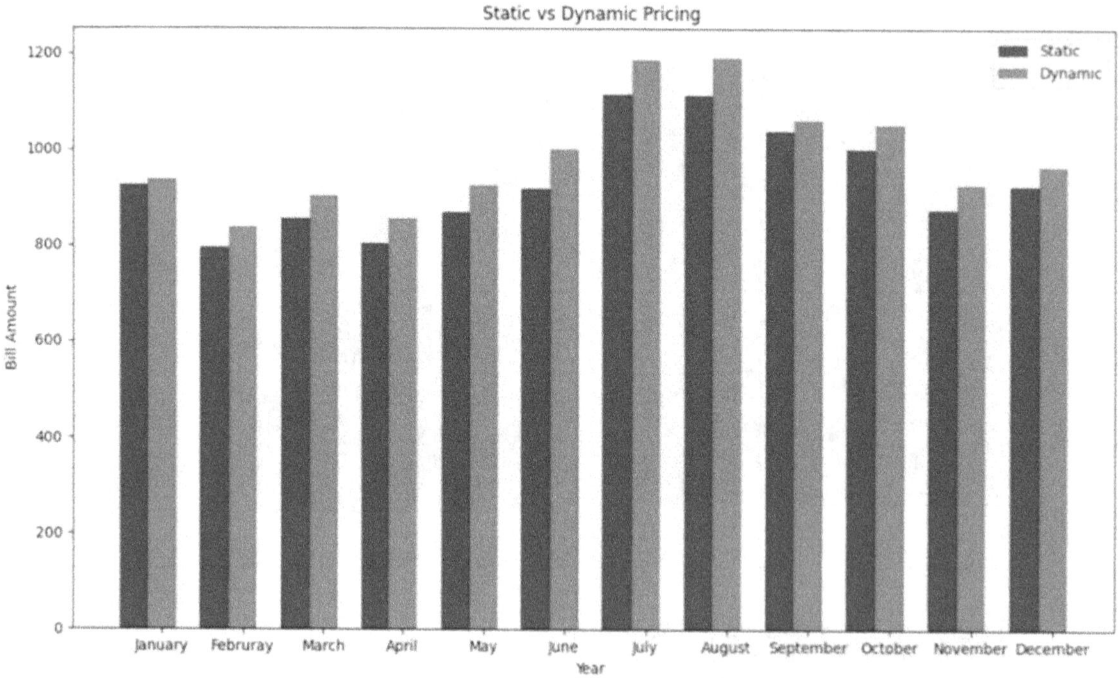

Static and Dynamic Pricing

The graph represents a comparison of monthly bills calculated by using the dynamic and static price of a particular house throughout the year. For calculating dynamic price for an hour, firstly we calculate a base price which depends upon the price of fuel (coal, diesel, natural gas, etc.) and their percentage contribution in total electricity production. Infrastructure costs are also taken into account while calculating base price. Previous hour demand and average consumption per hour calculated from the past 7 days are also required for calculating price per unit for a given hour. Static price is set equal to existing rate of electricity in the state of residence; in my case, Uttar Pradesh.

Two-Way Transmission Approach of Electricity

Figure 8. Two-way transmission of electricity on clusters

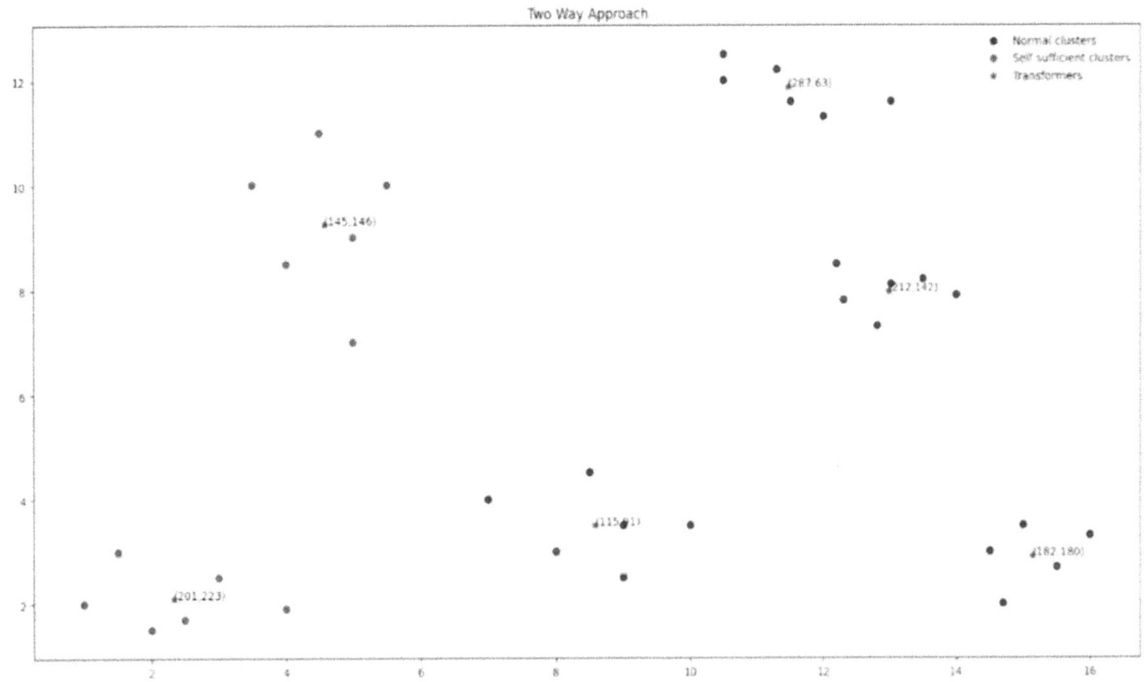

A two-way electricity distribution model where the consumer (by renewable methods like solar energy) can distribute in its cluster when the demand increases. This not only reduces the demand load on the supplier but is also beneficial for the environment. Also, this reduces the bill for the consumer as it can use the electricity developed in-house.

CONCLUSION

Distribution system planning is a main component to achieve the service in a fast-growing demand market, both from a technical view and from the economic costs management. In coming future, electric companies will require faster and cheaper planning tools to estimate various scenarios and their consequences for the remaining system and provide their customers with safe reliable and economic energy.

Present distribution models do not consider the demand growth on the studied period. The main objective is reducing the costs of installation of substations and feeders, also the implied costs related to maintenance and operation, while satisfying several other constraints related to the allowable voltage and values, reliability, service maintenance, and more. The k-means clustering applied not only reduce the cost but also maintains the efficient supply. For the load shedding issue, as we can see, the number of houses is maximum in greedy algorithm, but profit is lower. But knapsack focuses on advantage, so we get more benefit yet to the detriment of more houses being cut.

Apart from the distribution, the two-pricing model are developed and compared. The static pricing use slab-based pricing method which doesn't account for wastage of electricity so dynamic pricing is introduced. In dynamic pricing, various other factors like cost of coal, gas, diesel are also considered while calculating it. Previous hour demand and average consumption per hour calculated from the past 7 days are also required for calculating price per unit for a given hour. Static cost is set equivalent to existing pace of power in the condition of home; in this case, Uttar Pradesh. The proposed design of a two-way electricity distribution model where the consumer (by renewable methods like solar energy) can distribute in its cluster when the demand increases. This not only reduces the demand load on the supplier but is also beneficial for the environment. Also, this reduces the bill for the consumer as it can use the electricity developed in-house. This model can be used to build future township and new cities that are being developed like gift city.

ACKNOWLEDGMENT

We thank Dr. Swarnalatha P for providing constant guidance and support in completion of this paper.

REFERENCES

Aydinalp, M., Ismet Ugursal, V., & Fung, A. S. (2002). Modeling of the appliance, lighting, and space-cooling energy consumptions in the residential sector using neural networks. *Applied Energy*, *71*(2), 87–110. doi:10.1016/S0306-2619(01)00049-6

Barriers to the adoption of efficient electricity technologies: A case study of India. (1998). *Fuel and Energy Abstracts, 39*(3), 196.

Celebi, E., & Fuller, J. D. (2007). A Model for Efficient Consumer Pricing Schemes in Electricity Markets. *IEEE Transactions on Power Systems*, *22*(1), 60–67. doi:10.1109/TPWRS.2006.888956

FaruquiA.SergiciS.SharifA. (2009). *The Impact of Informational Feedback on Energy Consumption—A Survey of the Experimental Evidence*. doi:10.2139/ssrn.1407701

Fischer, C. (2008). Feedback on household electricity consumption: A tool for saving energy? *Energy Efficiency*, *1*(1), 79–104. doi:10.100712053-008-9009-7

Golmohamadi, H., & Keypour, R. (2017). Retail Energy Management in Electricity Markets: Structure, Challenges and Economic Aspects- a Review. *Technology and Economics of Smart Grids and Sustainable Energy, 2*(1).

Joon, V., Chandra, A., & Bhattacharya, M. (2009). Household energy consumption pattern and socio-cultural dimensions associated with it: A case study of rural Haryana, India. *Biomass and Bioenergy*, *33*(11), 1509–1512. doi:10.1016/j.biombioe.2009.07.016

Klüser, R. (2009). Current challenges for efficient electricity grids. *Poiesis & Praxis: International Journal of Ethics of Science and Technology Assessment*, *6*(3–4), 265–271. doi:10.100710202-009-0070-x

Lakra & Kirar. (2015). Load Shedding techniques for System with Cogeneration: A Review. *Electrical and Electronics Engineering: An International Journal, 4*(3).

McNeil, M. A., & Letschert, V. E. (2010). Modeling diffusion of electrical appliances in the residential sector. *Energy and Building*, *42*(6), 783–790. doi:10.1016/j.enbuild.2009.11.015

Mozina, C. (2007). Undervoltage Load Shedding. *IEEE Power Systems Conference: Advanced Metering, Protection, Control, Communication, and Distributed Resources*, 39-54.

Newsham, G. R., & Bowker, B. G. (2010). The effect of utility time-varying pricing and load control strategies on residential summer peak electricity use: A review. *Energy Policy*, *38*(7), 3289–3296. doi:10.1016/j.enpol.2010.01.027

Nishio, K-I., & Asano, H. (2006). *A residential end-use demand model for analyzing the energy conservation potential of new energy efficient technologies.* Academic Press.

Pferschy, U. (1999). Dynamic Programming Revisited: Improving Knapsack Algorithms. *Computing*, *63*(4), 419–430. doi:10.1007006070050042

Sempértegui, R., Bautista, J., Cubero, R., & Pereira, J. (2002). Models And Procedures for Electric Energy Distribution Planning. A Review. *IFAC Proceedings Volumes, 35*(1), 395-400. 10.3182/20020721-6-ES-1901.01217

Sianaki, O., Hussain, O., & Tabesh, A. (2010). A Knapsack problem approach for achieving efficient energy consumption in smart grid for endusers' life style. *2010 IEEE Conference on Innovative Technologies for an Efficient and Reliable Electricity Supply, CITRES 2010*, 159 - 164. 10.1109/CITRES.2010.5619873

Swan, L. G., & Ugursal, V. I. (2009). Modeling of end-use energy consumption in the residential sector: A review of modeling techniques. *Renewable & Sustainable Energy Reviews*, *13*(8), 1819–1835. doi:10.1016/j.rser.2008.09.033

Verayiah, R. (2014). *Review of Under-voltage Load Shedding Schemes in Power System Operation.* Przegląd Elektrotechniczny.

Vineetha, C. P., & Babu, C. A. (2014). Smart grid challenges, issues and solutions. *2014 International Conference on Intelligent Green Building and Smart Grid (IGBSG).* 10.1109/IGBSG.2014.6835208

Wood, G., & Newborough, M. (2003). Dynamic energy-consumption indicators for domestic appliances: Environment, behaviour and design. *Energy and Building*, *35*(8), 821–841. doi:10.1016/S0378-7788(02)00241-4

Wood, G., & Newborough, M. (2007). Energy-use information transfer for intelligent homes: Enabling energy conservation with central and local displays. *Energy and Building*, *39*(4), 495–503. doi:10.1016/j.enbuild.2006.06.009

Yamamoto, Y., Suzuki, A., Fuwa, Y., & Sato, T. (2008). Decision-making in electrical appliance use in the home. *Energy Policy*, *36*(5), 1679–1686. doi:10.1016/j.enpol.2008.01.022

Young, D. (2008). When do energy-efficient appliances generate energy savings? Some evidence from Canada. *Energy Policy*, *36*(1), 34–46. doi:10.1016/j.enpol.2007.09.011

Ziel, F., Steinert, R., & Husmann, S. (2015). Efficient modeling and forecasting of electricity spot prices. *Energy Economics*, *47*, 98–111. doi:10.1016/j.eneco.2014.10.012

Chapter 9
The Role of Blockchain Technology in Harnessing the Sustainability of Chinese Digital Finance

Poshan Yu
https://orcid.org/0000-0003-1069-3675
Soochow University, China & Krirk University, Thailand

Shengyuan Lu
Independent Researcher, China

Emanuela Hanes
Independent Researcher, Austria

Yijun Chen
Independent Researcher, China

ABSTRACT

In recent years, blockchain, with its features of decentralization, security, reliability, transparency of information, and low maintenance cost, has been applied in the fintech industry in more and more cases in China. The maturity of blockchain-enabled technology has improved. This is expected to rebuild the underlying internet technology or will trigger a new round of development and innovation in the Chinese fintech industry. This chapter aims to investigate the current application of blockchain technology in China's fintech industry by analyzing case studies. Through the comparison and analysis of the cases in the fintech industry, this chapter attempts to examine how and whether digital transformation and other innovative technologies can drive new business models and create valuable experiences for consumers and/or firms in China's fintech industry. A variety of challenges and opportunities is analyzed to draw a blueprint for Chinese blockchain development. Finally, this chapter also provides some recommendations to the government.

DOI: 10.4018/978-1-7998-9274-8.ch009

INTRODUCTION

Blockchain technology use in China has seen substantial institutional and policy change from 2013 to 2021. Under the unique economic environment with Chinese characteristics, the institutional setting designed by the Chinese government is used as a unique tool to guide the development of digital finance integrating with blockchain-enabled technology.

This study aims to interpret the interlinked relationships between the policy changes associated with the promotion of blockchain-enabled technology and the development of digital finance in China from an institutional perspective. Case studies of successful and unsuccessful applications in China are used to analyze why some firms and projects can keep running stably and expand rapidly while some applications ended in failure. By analyzing the formation of blockchain and digital finance policies, this chapter helps the readers better capture the future dynamics of China's digital finance development and understand the interrelated reactions of various stakeholders in the industry.

Blockchain technology has attracted much attention not only in people's work and lifestyle (Guo et al., 2021) and academia, but also in finance and industry (Beck & Müller-Bloch, 2017). The Ministry of Industry and Information Technology of the People's Republic of China (2018) posted a White Paper on China's Blockchain Industry stating that the various characteristics of blockchain technology provide trust mechanisms and have the potential to change the financial infrastructure. Different kinds of financial assets, such as equity, bonds, bills, warehouse receipts, fund shares, etc., can be integrated into the blockchain ledger and become digital assets on the chain, which can be stored, transferred, and traded on the blockchain.

The development of blockchain technology can be broken down into three phases: Blockchain 1.0, 2.0 & 3.0 (Swan, 2015). When it comes to the situation in China, the development nowadays is at the transitory stage from Blockchain 2.0 to Blockchain 3.0. The Chinese government and most enterprises are in the Blockchain 2.0 stage of rapid development and capacity expansion. Therefore, this paper mainly introduces the background of the Blockchain 2.0 stage of Chinese blockchain technology application in the financial sector and its economic impacts on Chinese business. This stage marks the relative application of blockchain technology for other extensive financial applications and the advanced use of more than just simple cash or digital payments and transactions (Albayati et al., 2020). This entails advantages and potential for the deeper application in financial smart contracts, capital markets, and so on.

As the industrial blockchain has been steadily promoted, and the value of the blockchain-enabled industry has been fully verified, one can describe it with the equation: Industrial blockchain 2.0 = industrial chain + asset chain + data chain + technology convergence +CBDC (central bank digital currency) application (Ning, 2021). It means that on the one hand, at present blockchain technology together with 5G, artificial intelligence (AI), cloud computing and other cutting-edge technologies are accelerating the deep integration of industrial development, thus becoming a key cornerstone of the digital economy from the perspective of the industrial chain. On the other hand, it shows that China is presently at the Blockchain 2.0 stage, mainly in relation to the financial sector in China.

The largest difference between Blockchain 1.0 and 2.0 should be smart contracts. Because of the feature of the blockchain technology, smart contracts are a safe and secure technology which can be adopted to stimulate innovation in many kinds of industries, such as supply-chain management, digital healthcare management, finance, and business (Abhinav et al., 2021).

For the current Chinese stage, Blockchain 2.0, the main focus is on the financial sector. As the combination of blockchain technology and the financial industry is bound to create more and more business models, service scenarios, business processes and financial products, it has deep impact on the development of financial markets, financial institutions, financial services, and financial formats. As the smart contract is a feature of blockchain technology (Varfolomeev et al., 2021), it can be applied to all transactions involving resources, movement of finance, material and people. This safe and reliable smart contract feature will be helpful for effective tracking, visibility, security enhancement and cost reduction in the overall business process (Abhinav et al., 2021).

Due to the disintermediation, secure encryption, the transparent and immutable nature of the distributed ledger with blockchain technology, the application of blockchain technology in China at the present stage is mainly optimizing the original mechanism from the perspective of cost saving and efficiency improvement. This coincides with the goal of transaction cost economics. The significant advantage of reducing transaction costs has been gradually reflected in financial services, supply chain management, and the Internet of Things, with financial services being the most significant. The application of blockchain technology in China brings cost reduction for market transactions and a material impact on corporate performance in a highly emotionally volatile market (Cioroianu et al., 2021).

Benefiting from the Chinese government's strong support via beneficial policies and laws as well as the massive capital investment from Chinese private equity investors, China has been one of the pacemakers in the global blockchain application field. So it is significant to do deep researches to find out the proved advantages that can be learned from to harness the sustainability of digital finance, since there is a high possibility for China to be the first to enter the stage of Blockchain 3.0 in the whole world.

Through the study of China's policy changes and the analysis of success and failure cases in the corresponding period of policy promulgation, this paper gives some specific and feasible suggestions for countries like China to develop policy oriented blockchain application. This paper studies previous literatures in the West and China and finds that there is almost no literature to analyze China's blockchain application from the perspective of policy change, which is an academic gap that this paper can fill. Few papers systematically analyzed the reason for the success of the mature and typical applications in China. The paper is to fill in the blanks in this field. The paper is to help the countries or companies who want to use Chinese experience for example understand faster of the characteristics, impact of institutional change, and specific blockchain applications in China.

This chapter introduces the characteristics of China's blockchain-enabled technology in the financial sector, then proceeds to the roles of government in assisting Chinese businesses in adopting blockchain technology via institutional setting. By giving an overview of the relevant policies and their impact on Blockchain technology, the institutional shaping by the Chinese government is shown. Through the comparison and analysis of the cases in the fintech industry, this chapter then examines how and whether digital transformation and other innovative technologies can drive new business models and create value experiences for consumers and/or firms in China's fintech industry. Then a variety of challenges and opportunities is analyzed to draw a blueprint for Chinese blockchain development. The chapter ends by providing some recommendations to the government for optimal Blockchain implementation.

CHARACTERISTICS OF CHINA'S BLOCKCHAIN-ENABLED TECHNOLOGY IN THE FINANCIAL SECTOR

With the Chinese socialist market economic system continuously developing and gradually reaching maturity, some transaction modes and methods of the traditional financial model need to be improved and optimized. The emergence and development of blockchain technology has a strong financial application background. Therefore, Chinese blockchain planning and constructing work closely focuses on many financial applications such as digital currency, smart contracts, financial transactions, and other financial areas with broad development prospects (Tang, 2019).

As shown in Figure 1 below, the distribution of blockchain services in China was highest in sectors such as Government affairs and public services, in finance, logistics and supply chain. The chart below also clearly shows the rapid development of fintech in financial institutes and is another strong evidence that the current focus of Blockchain application in China is on the financial sector.

Figure 1. Proportion of firms of blockchain application by industry segment in China in 2021
Data source: 2021 China industry blockchain ecological map

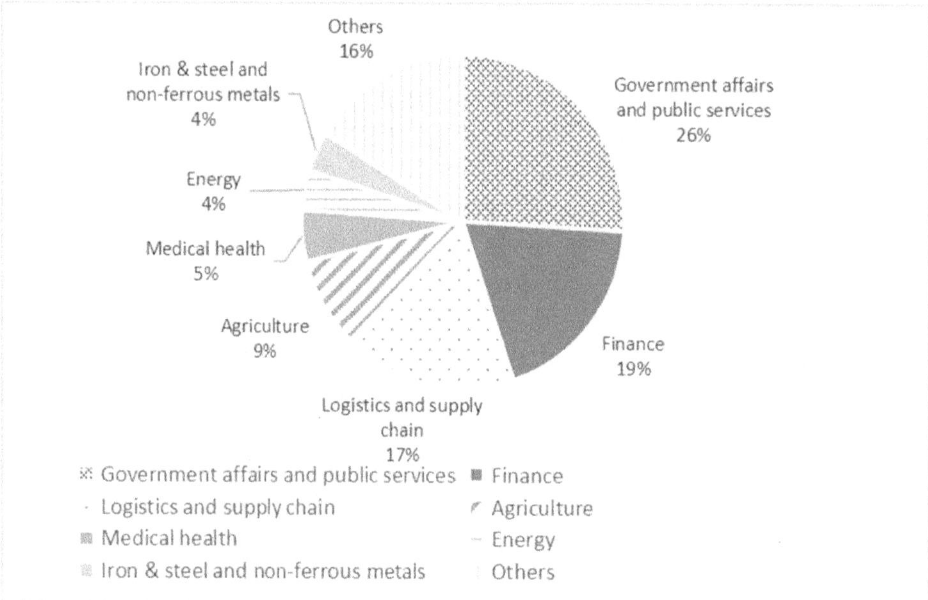

Blockchain is the result of a P2P network distributed structure. Network accounting, smart contracts and other technologies were developed on this basis. Therefore, blockchain technology has a certain technical and application foundation. The application forms of blockchain technology include public chains, private chains and alliance chains (Sanka et al., 2021). The important premise of blockchain technology application is a broad industry consensus and the overall layout (Rong & Yu, 2020). Commercial banks' compliance, legal, security and other departments still have strong doubts about the large-scale application of blockchain technology. At present, the application of this technology in the field of commercial banks is mainly in corporate asset and intermediate business.

First, blockchain technology in commercial banks is mainly applied for corporate business. The application for retail business faces more difficulties compared to corporate business. The use of blockchain technology can give more security and transparency for consumers when it comes to the application of retail business: By optimizing logistics costs, facilitating freight tracking, building customer trust and earning premium by tracking all supply chain data on the product and preventing counterfeiting by retailers providing customers with information about the origin of their products at every step of the supply chain. Nevertheless, Mu (2020) and other research members in Blockchain Research Group of Digital Currency Research Institute of the People's Bank of China said that Chinese current blockchain technologies are not suitable for high concurrency scenarios such as traditional retail payments on the account of many deficiencies in the initial stage of blockchain development in the finance sector in China. For instance, the cost of data synchronization and verification increase and system processing performance decreases with growing numbers of participating nodes.

In addition, in terms of functions, the decentralized feature of blockchain technology creates conflict with the centralization requirements of China's central bank. The Chinese government and law stipulate that payment services provided by the central bank cannot be separated from centralized account arrangements and must be built on a centralized system. The decentralized feature of blockchain technology conflicts with the promulgation of Measures for the Administration of Online Payment Business of Non-Bank Payment Institutions (2015) and the establishment of The Payment and Clearing Association of China Working Committee on Bank Card Based Payment, Online Payment Application and Prepaid Cards in Beijing (2011). Therefore, the application of retail business faces many difficulties, and the current focus is on the corporate business (Mu et al., 2020). It is not recommended to transform the traditional payment system into one based on blockchain at present.

Second, blockchain technology has a large application rate in the field of asset business. Applications in asset business include cross-border trade financing and supply chain financing (Zhang & Li, 2020). As is generally accepted in the industry, the cost for credit in the financial sector is relatively high. Specifically, the cost and difficulty of building mutual trust are high. Blockchain can not only solve the problem of trust in the financial system, but also reconstruct the order of financial business with its characteristics of reliability, real-time, fault tolerance and traceability (Liu, S.P., 2019). For instance, the securities industry, can achieve peer-to-peer securities issuance and trading with the help of blockchain and thus greatly simplify the process.

Figure 2 and *Figure 3* show the monthly blockchain transaction closing prices and volumes from 2016 to 2021. As the first graph illustrates, the closing price was 915.5577 *Yuan* in February 29, 2016 at the beginning of Blockchain listing on the Chinese stock market. Because of the rapid development of blockchain in the Chinese financial sector, the stock price was also increasing fast in the first quarter after it was issued. And when the investment fever died down, the closing price started falling for more than one year until the end of 2018. Then the closing price of Blockchain began to recover and with the introduction of relevant policies, it gradually reached stability and provided some small opportunities with its closing price of 887.5872 *Yuan* by August 29, 2021.

Figure 2. Blockchain transaction closing price (monthly) until August 29, 2021
Data source: Eastmoney Choice database

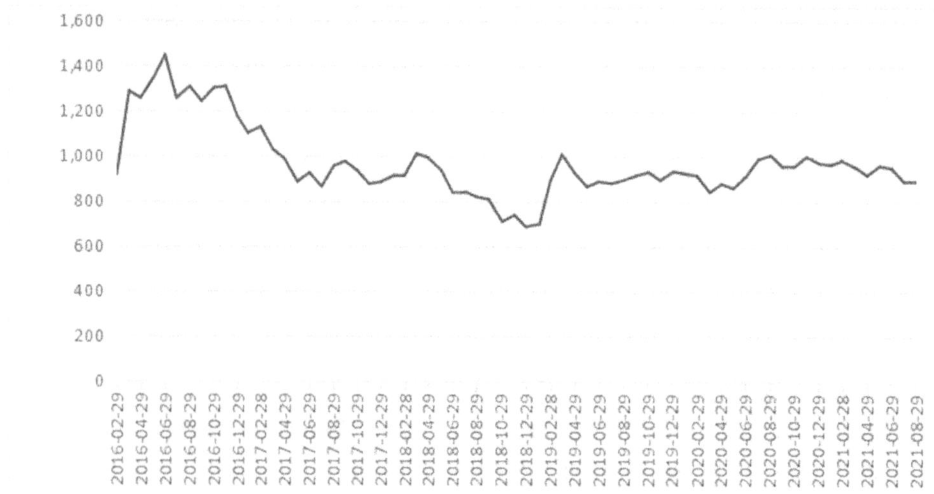

There is a clear rising trend of Blockchain as can be seen in Figure 2. From the monthly volume graph, the blockchain transaction volume fluctuated between 2016 and 2019, when the Chinese government and enterprises tried to receive and develop the relative applications and policies. After 2019, Chinese Blockchain transaction started to rise and the transaction volume reached up to 2,974,392 million *Yuan* in July of 2020. The explosive growth of blockchain technology applications in China not only stimulates the development of many entity industries, but also promotes the blockchain technology itself in the securities industry sector.

Figure 3. Blockchain transaction volume (monthly) until August 29, 2021
Data source: Eastmoney Choice database

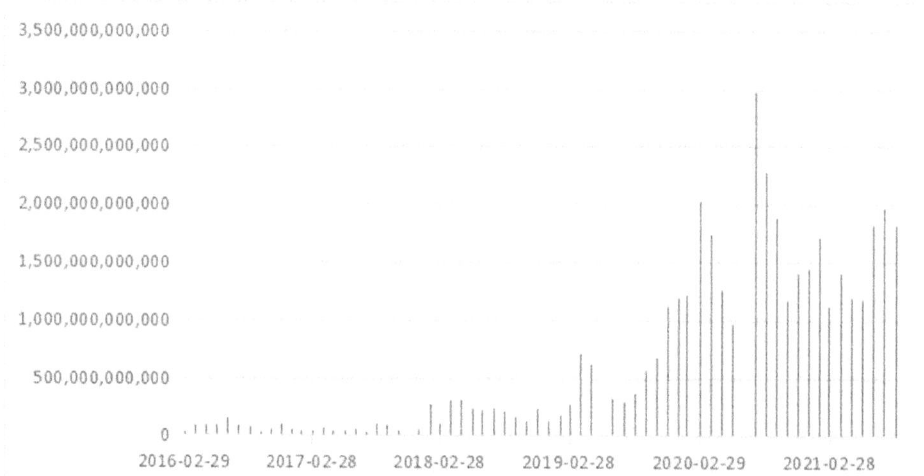

With the continuous development and progress as well as its integration with other financial technologies, the blockchain technology will gradually adapt to a variety of large-scale financial scenarios.

Third, blockchain technology helps commercial banks reduce costs and increase efficiency of intermediary business. Blockchain technology will help improve the efficiency of cross-border payment business. To be specific, the banking sector can effectively reduce the risk of cross-border payment and settlement and save payment costs through blockchain technology. The application of blockchain technology has improved the efficiency of financial transactions and settlements and reduced the cost of financial regulation.

The industrial development Blue Book "Regulatory Technology Blue Book: China Regulatory Technology Development Report (2020)" (Beijing Blockchain Technology Application Association, 2020) states that the essence of fintech is finance. The integration of technology and finance has not changed the essence of financial inter-temporal transactions and credit exchange. Risks existing in the financial industry will be reflected in the fintech field and new risk characteristics will evolve.

Since the financial crisis of 2008, with the technological revolution of financial regulation, and as regulatory requirements continue to rise, business models based on regulatory requirements needed adjustment and great advances have been made in financial technology innovation. Following the background of the financial crisis in 2008, the nature of financial markets, services, and institutions is changing owing to new entrants' disruptive and innovating technological practices (Gomber et al., 2017). Because of these, the fintech generated with the change of global drivers of the value chain continuously stimulates the business modes to adapt to the change of the economy in the future (Thurber, 2012).

As the financial services industry evolves and fintech constantly emerges and develops, the application of RegTech in China will have the potential for beneficial outcomes for consumers and public welfare as well as for regulatory and supervisory gains in the financial sector (Anagnostopoulos, 2018). RegTech is a subclass of the financial technology sector that focuses on technologies that may facilitate the delivery of regulatory requirements more effectively and effectively than existing capabilities. The rapid increase of fintech motivates the demand of RegTech (Ernst & Young, 2016) and at the same time it can help regulators better enforce prudential supervision of its financial institutions in order to strengthen the application of blockchain technology in financial regulation in China.

The logic of blockchain technology applied in the field of financial scene and application has been relatively clear. The relevant applications are being gradually and thoroughly explored, constantly adding scenarios and practice cases. With the support of policies, blockchain has been attracting attention and welcomed in a period of rapid development (Xiao, 2018).

In terms of application scenarios, multiple application scenarios are explored and closely combined with business requirements in the Chinese financial sector. The focus is on meeting potential market needs and addressing financial industry pain points. Blockchain application scenarios mainly involve supply chain finance, trade finance, insurance technology, cross-border payment, asset securitization and other sectors, with respective proportion of 32.6%, 11.2%, 11.2%, 7.9%, 6.7% and 30.4%.

Figure 4. Chinese financial applications, 2020
Data source: Research Report on Blockchain Financial Application and Development in China (2020)

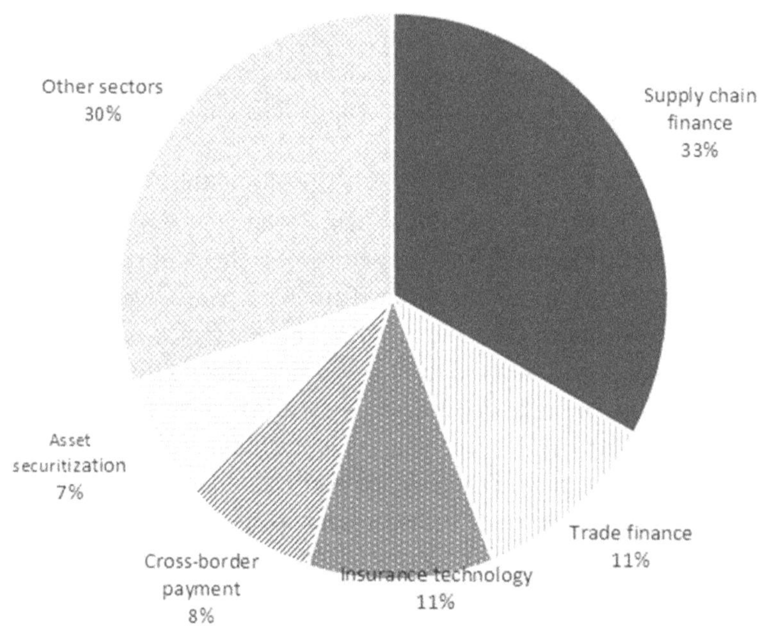

As shown in Fig. 4 above, the main application scenario of blockchain technology in Chinese financial field is supply chain finance. The application proportion of this scenario reaches 33%, which is the highest proportion. And no application scenario except supply chain finance accounts for more than 12%. This shows that the current supply chain application is well established and that other scenarios still need to be developed. This phenomenon may be due to the good performance of supply chain finance (SCF) in utilizing for capital optimization and cost reduction in many industries (Yan et al., 2016), so that such an application of blockchain technology is the hot spot of the financial field in China.

Apart from supply chain finance, only trade finance and insurance technology accounted for more than 10% of the application scenarios, and both of them accounted for 11.2% of the application scenarios. Applications accounting for more than 5% included cross-border payments (7.9%) and asset securitization (6.7%).

THE ROLES OF GOVERNMENT IN ASSISTING CHINESE BUSINESSES IN ADOPTING BLOCKCHAIN TECHNOLOGY VIA INSTITUTIONAL SETTING

Institutional settings influence economic development. As shown in Yu et al. (2018), "Institutions can be defined by commonly known rules used to structure recurrent interaction situations that are endowed with a sanctioning mechanism. Institutional theory looks at the deeper aspects of social structures with a high level of resilience (Scott, 2008). North (1990) defined institutions as 'the humanly devised constraints that shape interaction' and distinguished formal institutions created by entities such as the government and public services from informal institutions such as social convention or customs and implicit constraints. Economic activities are conducted under a certain institutional setting that define the interest of each member in the marketplace and how they can be achieved and protected."

Institutions are able to structure incentives in human exchange, whether political, social or economic (Scott 2008). Boettke (2016) pointed out the immaturity of the research of managing incentives and critiqued the contemporary economic focus as a continuation of much of the thinking laid out by previous mainline economists and Nobel Prize winners in his speech delivered in Southern Economic Association Meetings. "The existence of imperfect information and constraints on governments means there is a perpetual need for institutions to be crafted, tested and adapted in practice to ensure the effectiveness of institutions in aligning incentives, metering performance and monitoring behaviors. As stated by Boettke (2016), besides the rule of the general game in the society, there are consistent interactions amongst the markets, firms, regulations, politics and the whole society, which could be referred to as 'games within the games' played inside the nested institutions of such a complex society. Scott (2008) also stressed that institutions are subject to both incremental and discontinuous processes of movement, in which the economic system's performance could be modified." (Yu et al, 2018) Thus, analyzing the relevant policies and their impact on Blockchain technology, a better understanding of the institutional framework as well as the roles of government in assisting Chinese businesses in adopting blockchain technology via institutional setting can be realized. (Yu et al., 2018: 401).

With the transformation of the Internet of information to the Internet of value, Yang Lu (2021) indicates that blockchain technology has entered the stage of scale application in China, so legislative regulation and policy support of blockchain, in order to assist Chinese businesses in adopting blockchain technology has become a new focus of the government. The rapid development of China's blockchain industry is inseparable from the attention of the government and the support of relevant policies (Wang et al., 2020). In order to investigate the policy support behind the blockchain industry in China, Table 1 enumerates the relevant specific policies, laws and regulations promulgated by the nation and relevant departments from 2013 to 2021, gives detailed policy content and summarizes their impact.

China is becoming strong and has been striving to be a leader in technological innovation. Having been the first to promote 5G and artificial intelligence (AI), these technologies have by now been widely promoted on a global scale. Now, China has done the same with blockchain, declaring the blockchain technology as a national development priority and letting local governments implement their own blockchain policies and enforce blockchain standards.

In the second half of 2013, the bitcoin market rapidly heated up and the price rose rapidly, which set off the first round of virtual currency investment boom in the world and drew wide attention from society. At the same time, the financial risk that virtual currency may bring has caused the governments of many countries to attach great importance to it. Since 2013, the Chinese government has issued regulatory policies on virtual currency. For example, "the Notice on Preventing Risks of Bitcoin" was issued, indicating that the management of Bitcoin Internet sites should be strengthened to prevent money laundering risks that may be generated by Bitcoin.

Table 1. The summary of national blockchain policies in China 2013-2021

Issuing date	Issuing departments	Policies	Impact summaries (Sources: the discussed Policies)
December 2013	People's Bank of China, Ministry of Industry and Information Technology, China Banking Regulatory Commission (CBRC)	"Notice on Bitcoin Risk Prevention"	Clarified the nature of Bitcoin, required financial institutions and payment institutions not to carry out bitcoin-related businesses, and strengthened the management of Bitcoin Internet sites.
2014	People's Bank of China	Set up a special research group on legal digital currency	Plan to carry out the initial technical reserve and knowledge accumulation to demonstrate the feasibility of the central bank issuing legal digital currency.
2015	People's Bank of China	Some key issues in the field of digital currency were investigated	The central bank commissioned a series of research reports in the field of digital currency. For the first time, some important new ideas are put forward, and suggestions such as prohibiting ICO for policy supervision are proposed.
October 2016	Ministry of Industry and Information Technology	"China Blockchain Technology and Application Development White Paper (2016)"	Summarizes the development status and typical application scenarios of blockchain at home and abroad, and introduces the development roadmap of blockchain technology in China and the standardization direction and process of blockchain technology in the future
December 2016	State Council	"Notice of the State Council on the issuance of the '13th Five-Year' National Informatization Plan"	New technologies such as blockchain are driving the evolution of cyberspace from the interconnection of everyone to the interconnection of everything.
December 2016	Ministry of Industry and Information Technology	"Circular of the Ministry of Industry and Information Technology on the issuance of the Development Plan for Software and Information Technology Service Industry (2016-2020)"	Key points of guidelines / signals issued by the Ministry of industry and information technology to local governments and local companies: Conducting research and development in cutting-edge technologies; Preemptively laying out and developing intelligent software; Developing new service models and forms of business.
January 2017	General Office of the State Council	"Opinions of the General Office of the State Council on Innovating Management, Improving Services, Cultivating and Strengthening New Driving Forces of Economic Development, and Accelerating the Transition Between Old and New Driving Forces of Economic Development"	Blockchain related policies: Innovating the system and mechanism, break the management restrictions of institutions and disciplines, and build several industrial innovation centers and innovation networks in artificial intelligence, blockchain, energy Internet, intelligent manufacturing, big data application, genetic engineering, digital creativity and other cross-integration fields.
January 2017	Ministry of Commerce, National Development and Reform Commission, Ministry of Land and Resources, Ministry of Transport, State Post Bureau	"Notice on the issuance of the 13th Five-Year Plan for the Development and Development of Trade and Goods"	Promoting the use of advanced technologies and equipment such as automatic identification, electronic data interchange, cargo tracking, intelligent transportation, and the Internet of Things, exploring the application of blockchain technology in the field of business logistics, and vigorously developing smart logistics.
March 2017	Ministry of Industry and Information Technology	"Circular of the Ministry of Industry and Information Technology on the issuance of the Three-Year Action Plan for the Development of Cloud Computing (2017-2019)"	Key points of guidelines / signals issued by the Ministry of industry and information technology to local governments and local companies: Supporting innovation and entrepreneurship based on cloud computing. Supporting small and medium-sized enterprises and individual developers to carry out research, development and industrialization of blockchain and other new technologies and businesses based on cloud computing platform.
July 2017	State Council	"Circular of the State Council on the issuance of the New Generation of Artificial Intelligence Development Plan"	Key points of guidelines / signals issued by the Ministry of industry and information technology to local governments and local companies: Promoting the integration of blockchain technology and artificial intelligence, establishing a new social credit system, and minimizing the cost and risk of interpersonal communication.

Continued on following page

Table 1. Continued

Issuing date	Issuing departments	Policies	Impact summaries (Sources: the discussed Policies)
August 2017	Ministry of Commerce, Ministry of Finance	"Circular of the General Office of the Ministry of Commerce and the General Office of the Ministry of Finance on the Construction of Supply Chain System"	Key points of guidelines / signals issued by the Ministry of industry and information technology to local governments and local companies: Supporting the innovation and upgrading of the traceability system of core supply chain enterprises. Promoting the innovation and upgrading of the traceability system.
September 2017	People's Bank of China, Office of the Central Cyberspace Affairs Commission, Ministry of Industry and Information Technology, State Administration for Market Regulation, China Insurance Regulatory Commission, China's Securities Regulatory Commission, China's Bank Regulatory Commission	"Notice on Preventing the Risk of Token Issuance Financing"	Since the date of the announcement, various coin issuance and financing activities shall be stopped.
September 2017	China Insurance Regulatory Commission	"Circular of CIRC on Printing and Distributing the 'Construction Plan of the Second Generation Project of Compensation' "	Key points of guidelines / signals issued by the Ministry of industry and information technology to local governments and local companies: Tracking the development trend of financial technologies such as cloud computing, big data, artificial intelligence and blockchain, carrying out research on the application of regulatory technologies, and actively exploring prudential supervision which is a multidimensional, three-dimensional, open solvency risk analysis and monitoring system of the insurance industry under the conditions of new technologies.
October 2017	General Office of the State Council	"Guiding Opinions of the General Office of the State Council on Actively Promoting Innovation and Application of Supply Chain"	Key points of guidelines / signals issued by the Ministry of industry and information technology to local governments and local companies: Researching and utilizing emerging technologies such as blockchain and artificial intelligence to establish a credit evaluation mechanism based on supply chain.
December 2017	General Postal Office	"Guiding Opinions of State Post Bureau on Promoting the Construction of "One Belt And One Road" in Postal Service"	Key points of guidelines / signals issued by the Ministry of industry and information technology to local governments and local companies: Leveraging the role of national engineering laboratories and other scientific research institutions within the industry, exchanging experience with countries along the Belt and Road on the integrated development of the postal industry with the Internet, big data, cloud computing, artificial intelligence and blockchain, and jointly carrying out demonstrations of science and technology application.
February 2018	General Office of Ministry of Industry and Information Technology	Circular of the General Office of the Ministry of Industry and Information Technology on the Application of the Pilot and Demonstration Projects of Information Consumption	Pilot demonstrations including the "10 directions in the field of the 5 classes" (living, public service, industry, new product, platform) Supporting the development of modern logistics services for the whole process of information consumption. Supporting the innovative application of integrated logistics for multi-modal transport.
April 2018	Ministry of Education	"Notice of the Ministry of Education on the issuance of the Action Plan for Educational Informatization 2.0"	Actively exploring effective ways of recording, transferring, exchanging, and certifying the effects of intelligent learning based on new technologies such as blockchain and big data, forming ubiquitous and intelligent learning systems, and promoting the deep integration of information technology and intelligent technology into the whole process of education and teaching

Continued on following page

Table 1. Continued

Issuing date	Issuing departments	Policies	Impact summaries (Sources: the discussed Policies)
May 2018	Ministry of Industry and Information Technology	"China Blockchain Industry White Paper 2018"	Deeply analyzes the development status of China's blockchain technology industry, summarizes the development characteristics of China's blockchain industry, deeply expounds the application of blockchain in the financial field and the real economy, and looks forward to the development trend of the industry.
October 2018	Ministry of Industry and Information Technology	"Notice of Ministry of Industry and Information Technology of China"	Actively building and improving the blockchain standard system, accelerating the development, application and promotion of key standards, and gradually building a perfect standard system.
February 2019	Cyberspace Administration of China	"Blockchain Information Service Management Regulations"	It standardizes the development of China's blockchain industry, which means that China is officially ushered into the "supervision era" for blockchain information services.
March 2019	Cyberspace Administration of China	"On the notice of the first batch of domestic blockchain information service record number of the announcement"	The first batch of blockchain information service names and record numbers were publicly released.
May 2019	State Council	"Opinions on Deepening Reform and Strengthening Food Safety"	Promotes the application of blockchain and other technologies in the field of food safety supervision. New requirements and suggestions are put forward for the food safety application of blockchain technology.
August 2019	State Council	"Opinions on Supporting Shenzhen in Building a Pilot Demonstration Zone for Socialism with Chinese Characteristics"	Supports the development of digital currency research and innovative applications such as mobile payment in Shenzhen; Promotes market access connectivity and mutual recognition of financial (fund) products with Hong Kong and Macao; Efforts are made to advance the internationalization of the RMB and explore innovative cross-border financial regulation.
October 2019	Standing Committee of the National People's Congress	"Code Law"	As a comprehensive and fundamental law in the field of cryptography in China, it also has far-reaching and important regulatory significance for blockchain technology.
February 2020	People's Bank of China	"Technical security specification for financial distributed ledger"	The standard is the first block chain specification in domestic and even international financial industry. It can be seen that blockchain is increasingly becoming one of the core technologies in the digital age. Meanwhile, it will also play a foundational role in the further standardized, comprehensive, and truly in-depth application of block chain in the financial industry.
April 2020	National Internet Finance Association of China	"Risk tips for participating in speculation on overseas virtual currency trading platforms"	Clarifies that virtual currency transactions and ICOs are illegal financial activities, defines clean-up and rectification work.
July 2020	People's Bank of China	"Notice to promote the application of blockchain technical specifications" "Blockchain technology financial application evaluation rules"	The first block-chain related normative documents have been issued by China's top authority.
May 2021	National Internet Finance Association of China, China Banking Association, Payment & Clearing Association of China	"Notice of the prevention of virtual currency trading speculation risk"	Formally bans all non-sovereign countries in the world to give endorsement to virtual currencies in China. Announces Crack-Down upon trading speculations of Bitcoin and other virtual currencies in view of the production link of Bitcoin.

Source: Foresight Research Institute, Ministry of Industry and Information Technology of the People's Republic of China, the Chinese Government Legal Information Website

In 2016, The State Council issued the "13th Five-Year National Information Plan", for the first time adding the blockchain technology into the scope of new technology and frontiers. This was a formal mark of China beginning to promote the development of blockchain technology and application. Since then, the central and local governments have respectively issued relevant regulatory or supportive policies, creating a beneficial policy environment for the development of blockchain technology and industry. As it shows in the figure below, the numbers of blockchain policies posted by Chinese government and ministries increased from 1 item in 2015 to 21 items in 2019. There is a clear trend of gradual improvement and reliability for Chinese blockchain laws and policies.

Figure 5. Number of state-level blockchain policies
Data source: Eastmoney Choice database

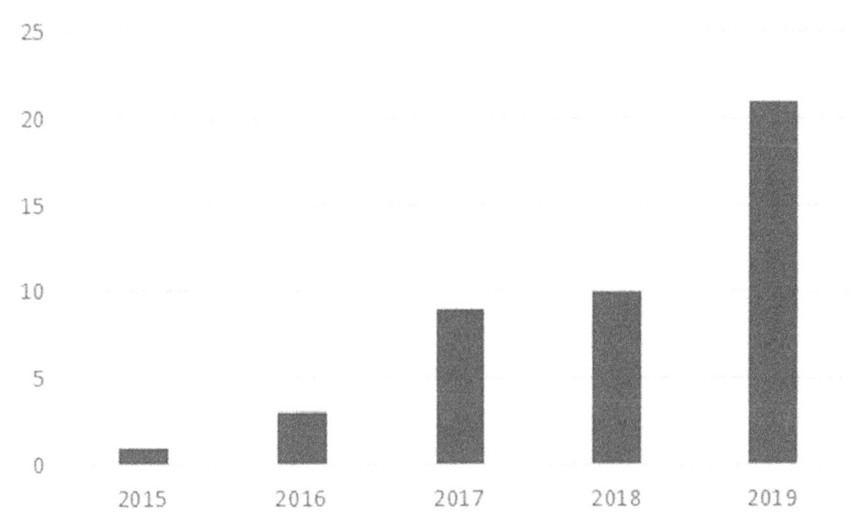

By analyzing the institutional changes of Chinese national blockchain-related policies from 2013 to 2021, it is clear that the development of blockchain applications in China is strongly associated with the attitude of the government. Two examples are the strict regulation of cryptocurrencies and banning the Initial Coin Offering (ICO) ("Notice on Bitcoin Risk Prevention", 2013) at the beginning stage of Chinese blockchain applications in 2013.With the "Notice on Preventing the Risk of Token Issuance Financing" (2017), any Chinese blockchain applications had been renormalized or prohibited from running.

From 2013 to 2017 we can discern the stage of building and improving the basic conditions for the development of blockchain applications and issuing relevant normalizing and supporting policies such as the "Set up a special research group on legal digital currency" (2014) researching central bank digital currencies, the "Circular of the Ministry of Industry and Information Technology on the issuance of the Three-Year Action Plan for the Development of Cloud Computing (2017-2019)"(2017) addressing basic technology like cloud computing and the "Circular of the General Office of the Ministry of Commerce and the General Office of the Ministry of Finance on the Construction of Supply Chain System", addressing supply chain issues.

The publication of the "China Blockchain Industry White Paper 2018" (2018) marks the institutional change into the stage of substantial support and appreciation by continuous introduction of favorable and standardized policies in the use of blockchain technology, for instance the issue of "Code Law" (2019),"Technical security specification for financial distributed ledger"(2020) and "Notice of the prevention of virtual currency trading speculation risk"(2021). With the development wave, several enterprises were able to develop rapidly into leaders in their fields, like our first case study, the Ant Group.

With the maturity and completeness of the regulatory policy of blockchain market and the further improvement of blockchain technology, the construction of a regulatory system for the Chinese industry has been further advanced, providing a strong guarantee for industrial block chain projects to deeply serve the real economy. Some illegal projects will be strictly regulated, the market is becoming standardized, and the environment for industrial development can be optimized.

From the perspective of supporting policy content, it can be divided into technical level, industrial level, industrial application, and other levels (Wang, 2021).

In terms of technology, on the one hand, the Chinese government encourages innovative research, the application of blockchain technology and emphasizes the need to speed up the formulation of blockchain technology standard system; On the other hand, it encourages the integration, research, and application of blockchain technology with big data, artificial intelligence, and other technologies in order to solve social credit, cost, efficiency and other problems (Xi, J.P., 2019).

From the perspective of industry, the government encourages the establishment of the blockchain industry guide fund. It also promotes the development of blockchain industry through special subsidies for talents, technology innovation subsidies, office housing subsidies and other measures. The Chinese government has issued a series of policies like the "Guidance on accelerating the application of blockchain technology and industrial development" (2021) to incubate blockchain projects and promote the development of the blockchain industry.

At the application level, the government mainly encourages and supports the deep integration of blockchain technology with finance, supply chain, cross-border e-commerce, supply chain finance, logistics, public welfare, agriculture, government, and other industries (Ministry of Industry and Information Technology of the People's Republic of China, 2018).

In general, the Chinese government encourages the exploration and research of blockchain technology standards and technology security applications to promote the combination of blockchain technology and practical application scenarios in order to serve the real economy and build a new digital economy; However, the potential risks of digital currency or ICO have not been addressed so far. The next section will further discuss the latest Chinese cases.

DISCUSSION OF CHINESE CASES (ANALYZING AND COMPARING SUCCESSFUL AND UNSUCCESSFUL CASES)

Blockchain, as a technology of digital finance, is rapidly entering social governance, enterprise production and the lives of people. The most important application areas of blockchain can be divided into three levels: the first level is the application of blockchain in financial business, such as clearing, exchange and reconciliation. The second level is how blockchain technology, as the underpinning of fintech development, can effectively empower other platforms and capabilities. The third level is the scene and ecology, implying all the elements related to the industrial ecology on the chain in order for the whole ecology to operate effectively (University Big Data and Artificial Intelligence Promotion Alliance, 2021).

When it comes to China, the application of blockchain in supply chain finance, cross-border payment, asset securitization, bills, credit investigation and risk management has begun to show a colorful bloom. By solving the problems of information asymmetry, intermediary centralization and "information islands", the application of blockchain technology in China can greatly improve the breadth, depth, and efficiency of traditional financial business.

As a solution to information and cost problems, some new business categories based on blockchain begin to appear in the core and the margins of traditional financial business. It can be said that blockchain is profoundly impacting the application of financial business in China. In this paper, several successful and unsuccessful cases of Chinese blockchain technology application will be introduced and discussed.

1. The Successful and Most Typical Case——Ant Group

As shown in the figure below, the bar chart compares the most-funded Chinese fintech and blockchain startups in 2020. With the funding of Ant Group having reached 22 billion U.S. dollars, this is the most-funded FinTech enterprise in China. Its funders included General Atlantic, Sequoia Capital, and others (CrunchBase & SCMP, 2020). And most funding is due to the Ant Group's great business conditions and development prospects in online financial service. Just like the rapid rise and success of Ant Group, the combination of blockchain technology and Chinese financial sector has given many investors great prospects for development. At the same time, the resulting additional investment will accelerate the development and further application Chinese fintech and blockchain technology, so as to enter the stage of Blockchain 3.0 earlier.

Figure 6. Most funded Chinese fintech and blockchain startups 2020 (in Million U.S. Dollars)
Data source: Statista, CrunchBase, SCMP

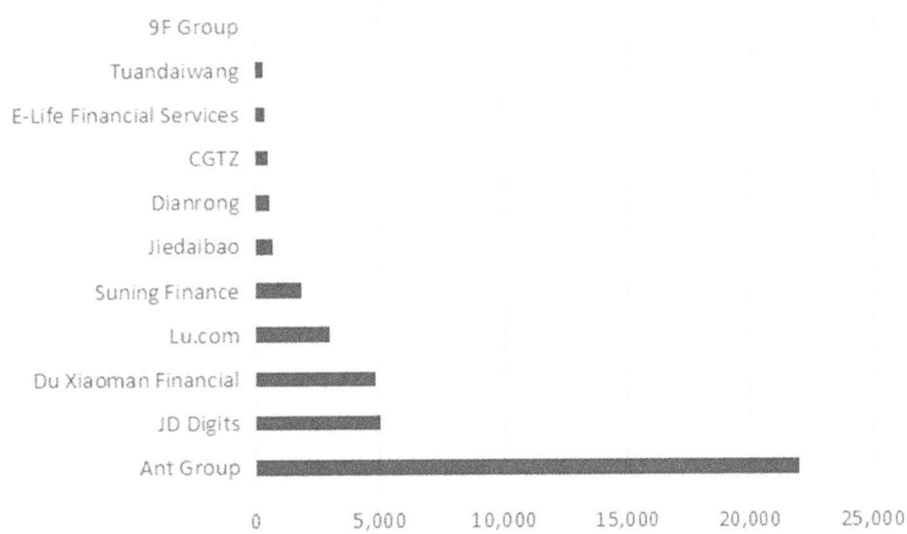

The Ant Group is the parent of mobile payment platform——Alipay, which is also the world's leading financial technology opening platform, committed to science and technology promotion, including financial services, modern service industry and global digitalization (Ant Group website, online). The reasons why the Ant Group became the largest fintech company in China are interesting lessons and a reference for not only many start-up Chinese companies based on blockchain applications in finance (CBRC General Office, 2021), but also every other company.

Firstly, the success of Ant Group is mainly owing to its product research and user acceptation. The Ant Group has done well in the application of blockchain technology in the financial sector such as digital payment, digital finance, and digital life. The digital mobile payment Alipay, is not only a means of digital payment for users, but also a digital financial service (Ant Group Co. Ltd, 2020). Digital payment has become the infrastructure of consumers' daily life and digital finances (Li et al., 2021). As *Figure 6 shows,* the number of monthly active users of Alipay APP have risen continuously from 499 million people in 2017 to 711 million users in 2020. The growing number of monthly users and the growing users' retention rate is the reason for the Ant Groups' growth and especially important for the development, survival and growth of fintech companies.

Figure 7. Monthly active user of Alipay APP (Million people)
Data source: Prospectus of Ant Technology Group Co. Ltd

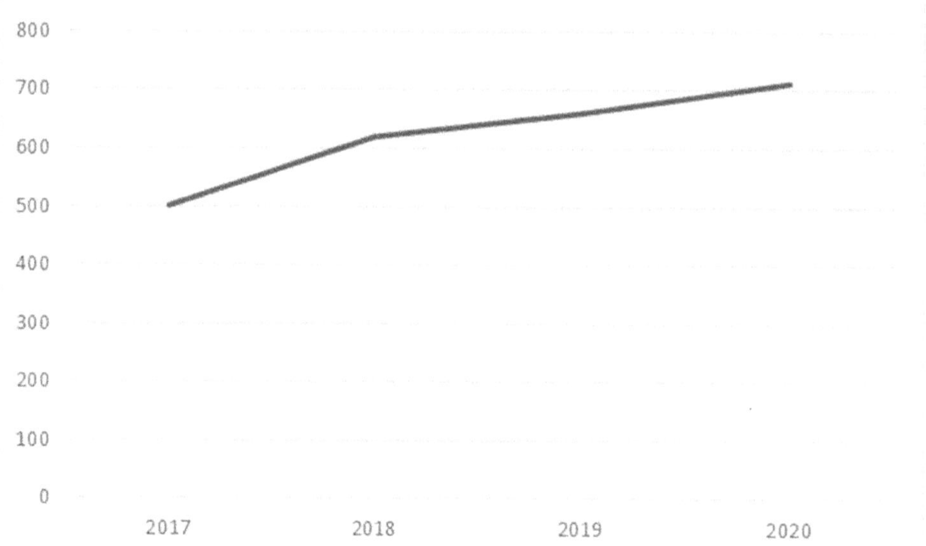

Meanwhile, digital financial services are highly innovative and attractive, and have high synergies with digital payment and digital lifestyle services such as online living payment platform and e-marriage certificate. Digital payment, lifestyles based on the Internet and various digital technology applications and digital financial services form a powerful synergy and a virtuous cycle, which is an important driving force for the growth of Ant Group.

The second important reason is the Ant Group's technological innovation and its active application of blockchain-authorized patents as well as the awareness of patent protection. Ant Group as a global technology leader constantly innovates its technology and applications, stays open to sharing with the whole industry especially in the fields of Blockchain, IoT technology, Database, AI and Regtech. For blockchain technology applications, Ant Chain is an internationally leading financial alliance blockchain platform which is independently developed by the Ant Group. With its high performance, high reliability and high security, Ant Chain can support super-scale applications of 1 billion accounts * and1 billion daily transactions (Ant Group website, online). For example, in June 2018, AlipayHK launched the world's first cross-border blockchain remittance service together with Philippine e-wallet GCash (*China Strategic Emerging Industry*, 2018), providing convenient, safe, transparent and low-cost remittance services for Hong Kong and Philippine residents. Based on deep research and the blockchain technology, Ant Group was able to increase the strength of its technology as well as open provide additional markets and cooperation.

When it comes the patents, it attaches great importance to protect its technological achievements. This is also why Ant Group as the largest fintech company in China can also play an important role in the word. In the global list of authorized blockchain patents of *Figure 8*, Alipay ranks first globally with 212 blockchain authorized patents, followed by IBM (136) and Coinplug (107).

Figure 8. Ranking of global patent holders in the number of blockchain patents granted 2020
Data source: 《2020 Global Patent Authorization report in blockchain field》

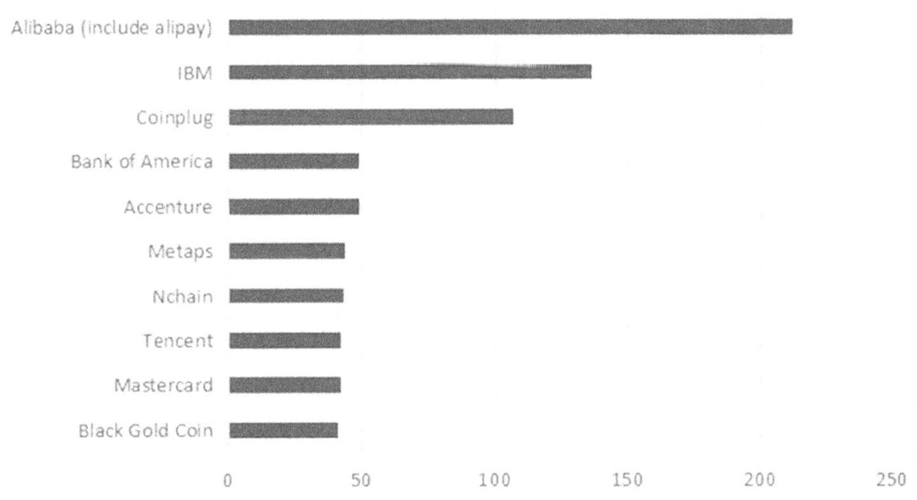

Figure 9. Chinese blockchain patent disclosure rankings 2019
Data source: Eastmoney Choice database

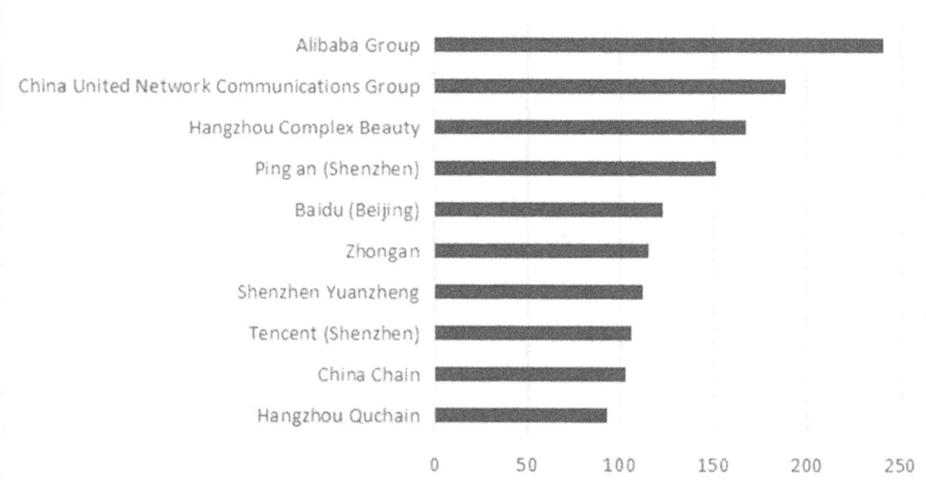

As the Chinese blockchain patent disclosure rankings of 2019 show in the *Figure 9*, the Ant Group was still the most active company of blockchain patent disclosure, with 241 patents, 52 more than the second company in blockchain sector of China.

The authorization and disclosure of the patent means that the blockchain patent has gone through a number of steps such as preliminary review, public review and substantive review (China Patent Protection Association, 2020), which can better reflect the "gold content" of the enterprise technology. Ant group's patent advantages are representative and typical and can thus be used to measure the technological progress of the industry and the R & D strength of the enterprise.

The other main advantage that makes the Ant Group the largest blockchain company in China is its product launch as well as the development of a company in line with the demand of Chinese national which seizes the market opportunity.

Chinese GDP is continuously growing, and China is gradually transforming to a consumption-oriented economy. The COVID-19 pandemic has further pushed digital transformation in China. In 2020, about 40.7 percent of the value added in the tertiary industry was contributed by the digital economy, such as remote working, online education, and online entertainment which accelerates the Chinese transition to a consumption-based economy as it is illustrated in the *Figure 10*. Meanwhile, as the F*igure 11* shows, the digital economy has gained a more important role in Chinese economic growth, contributing about 38.6 percent of the country's GDP that year.

Figure 10. Value added share of digital economy in China from 2016 to 2020, by industry
Data source:(CAICT 2021, p. 7)

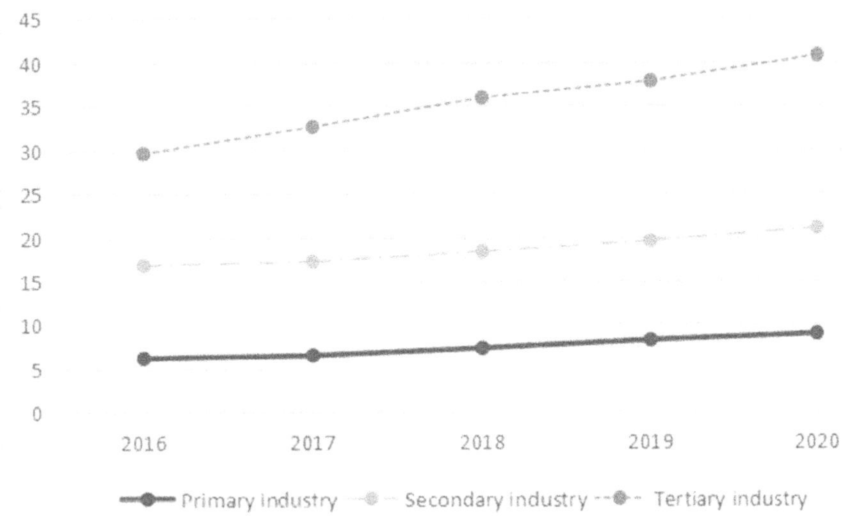

Figure 11. Share of the digital economy in the GDP of China from 2015 to 2020
Data source:(CAICT 2021, p. 6)

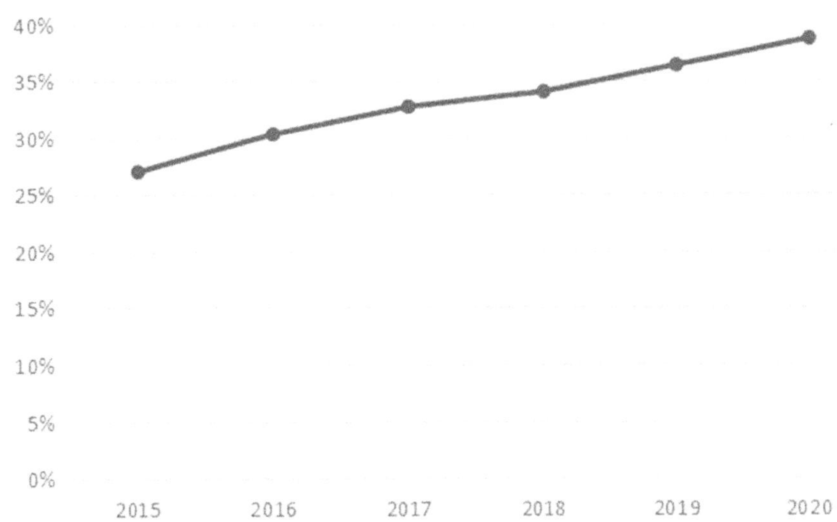

Moreover, China has entered a brand-new digital era and the digital economy is developing rapidly. In 2020, the economy reached a size of about 39.2 trillion Yuan, registering a nominal annual growth of 9.7 percent, much higher than the country's nominal GDP growth at 6.7 percent. Amid the COVID-19

pandemic, more and more companies have opted for digitization solutions for business operations. In many ways, China is the vanguard of the global fintech industry, with tech giants, such as Alibaba, bringing digital financial services to over one billion consumers.

Figure 12. Market size of the digital economy in China in selected years from 2005 to 2020 (in Trillion Yuan)
Data source: Statista

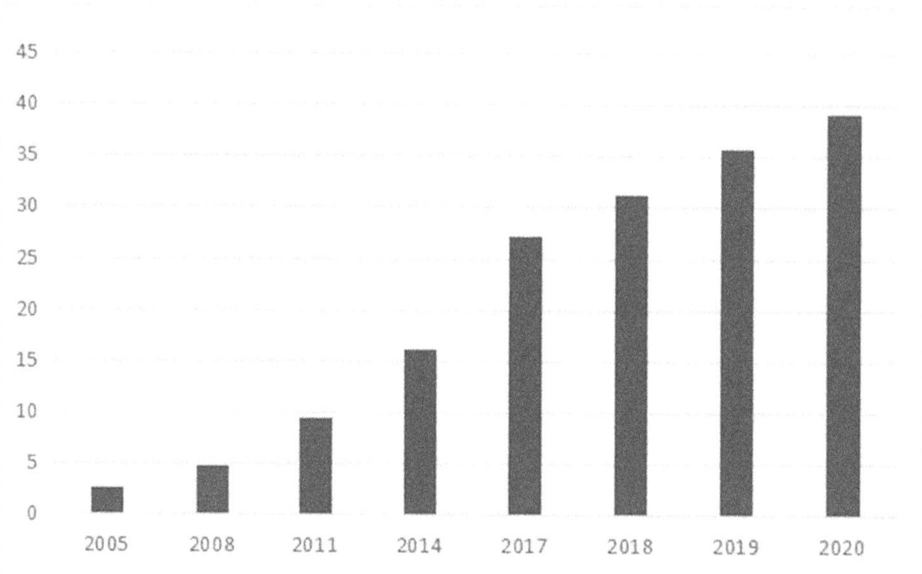

In response to China's national conditions and policies for the development of digital economy and fintech related technologies, the need of consumption transformation, the rapid growth of Ant Group conforms to the trend of development of the digital economy, for instance the applications of the digital mobile payment Alipay, the small and micro lending platform Ant Credit Pay, and the financial management technology platform Yu'E Bao.

Moreover, policy support and subsidies of Chinese government in order to vigorously develop the digital economy, large upfront capital investments, as well as a healthy and profitable business model are the indispensable advantages for Ant Group to become the largest fintech enterprise in China.

2. Successful Case——Chinese First ABS Supply Chain

The application of blockchain technology in asset securitization business has provided convenience for capital parties to understand the underlying assets, intermediaries to grasp the default risk of assets in real time, and regulators to effectively monitor financial leverage and prevent systemic risks in advance.

The "Deppon Securities and Zheshang Bank ChiRong No. 2 Asset Support Special Plan" ABS was successfully issued in Shenzhen Stock Exchange. The size of its issue was 336 million Yuan, and the underlying assets are trade financing. It is worth noting that this product was not only successfully applied the blockchain technology, but also was the first blockchain financial ABS product in supply chain (PuHuiMingYan Financial Research Institute, 2018).

The ABS is jointly created by Zhesang Bank and Deppon Securities. It is the first securitization project of the stowed-shelf type of trade financing receivables. Its basic assets cover four kinds of trade financing receivables: domestic letter of credit assets, bank payment guarantee assets, electronic bank acceptance bill assets, and electronic commercial acceptance bill assets. The asset-backed securities of each period under the shelf mode of this product are not classified as priority or inferior, and all grades have obtained the initial AAA rating of China Credit (Flush Finance, 2020).

3. Unsuccessful Case——Cloud Currency Network

With the appearance and rapid development of cryptocurrency, the ICO market experienced phenomenal growth in the last years (Bollaert et al., 2021). The Figure 13 shows that many national and regional governments focused on the ICO. The complete repel of the cloud currency network on August 31, this is one of the unsuccessful cases in Chinese blockchain technology related to cryptocurrency for the most important reason - the influence of National policy.

Figure 13. Countries with most research into state-sponsored interbank cryptocurrency 2020
Data source: BIS, PWC

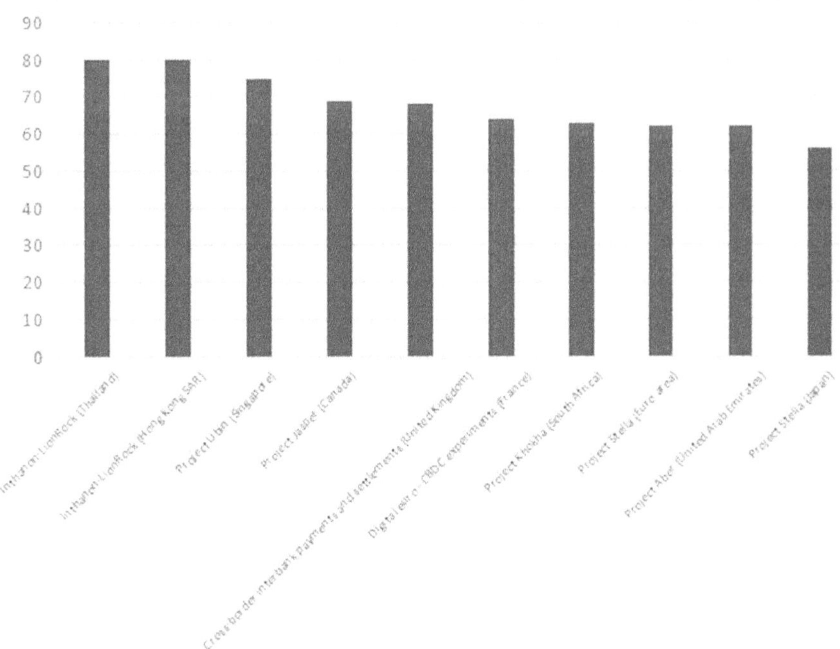

On July 16, 2018, the cloud currency network announced that under the guidance and supervision of the "Beijing Internet financial risk special rectification work leading group", it carried out the digital currency liquidation work on September 5, 2017 (TechWeb Report, 2018). After BTC China announced the closure of virtual currency transactions, the cloud currency network also announced that it will close all virtual currency transactions on September 20. In the announcement, the cloud currency network said that it will permanently close all kinds of trading functions at 00:00 on September 20, 2017 and has stopped new user registration and RMB recharge business in order to cater to China's Internet financial risks special rectification work (TechWeb Report, 2018).

Comparing the last two case of blockchain technology's financial application in China, there are many lessons to be learned mainly in the four fields: policy, capital, industry, sustainability. First, in terms of policy, which is the most significant factor, analysing the feasibility of blockchain application is necessary. In the unsuccessful case of the cloud currency network, we show that the feature of the application violates the attitude of the Chinese government towards cryptocurrency - the national policy forbids the trade of cryptocurrency and ICO in Mainland China. That was the main reason of the cloud currency network's complete repel. In the successful cases like the ABS product in China, the Chinese government supports the combination of blockchain and asset securitization. The success is also due to its large capital input on the scale of Hundred Million of *Yuan*, the new potentially increasing importance of the security industry and the strong sustainability of its reliable and cost-reducing features as well as filling up the gaps of trust and liquidity in the field of asset securitization.

OPPORTUNITIES AND CHALLENGES FOR CHINESE COMPANIES IN INTEGRATING BLOCKCHAIN-ENABLED TECHNOLOGY IN BUSINESS SECTOR

Opportunities

1. In the term of production and trade, blockchain can help develop sustainable supply chains (Kshetri, 2021) in order to support many manufacturing enterprises. It has also been introduced and employed in many traditional industries, such as supply chain finance, production management, logistics, and the Internet of Things, to optimize their operation processes (Jiang et al, 2021).

Meanwhile, China is looking for new export locations in Europe, the Middle East and Africa through the Silk Road Economic Belt. Having advanced blockchain technology will give China huge trading opportunities when it comes to dealing with emerging technology markets.

2. From the perspective of regulation in the Business Sector, Blockchain can address institutional deficiencies to implement sustainability, make the society stable and efficient. Moreover, Blockchain can reduce fake and low-quality ingredient products in supply chains (Sunny, J. et al., 2020).
3. In the term of new applications, the recent applications of blockchain technology in crowdfunding promise a possible solution to current obstacles and may facilitate socially oriented crowdfunding platforms (Nguyen et al., 2021).

Furthermore, the digital RMB (Chinese CBDC) could become a global currency. China is going full tilt into digital assets. As part of this, they are developing digital currency electronic payment platforms. It's planned to promote a central bank digital currency that the People's Bank of China has issued, with the anonymity in transactions and extensive crypto services in order to promote the new digital RMB as a global currency. The RMB in its current form lacks openness and liquidity in international foreign exchange markets (Tao Liu et al., 2019), so there is definitely room for improvement. With this new digital RMB, China will be able to make cross-border payments at lower cost and faster.

In addition, automated functions using smart contracts will provide easier liquidity management and transaction efficiency (Umer et al., 2021), and eventually allow the digital RMB to serve as an upgraded version of the current M0, or cash in circulation.

Challenges

1. First, technical hurdles remain. Financial businesses attach importance to privacy protection (Hao Wang et al., 2020). Under the current consensus mechanism, some sensitive information is transparent and visible, but lacks privacy. Blockchain technology involves many interdisciplinary and cross-field technologies such as cryptography, informatics, artificial intelligence, and computer. The technologies in these different fields are closely linked, and the research and development is extremely difficult with extreme development costs.

Moreover, the underlying technology is at risk of being attacked, and the speed of reaching consensus is relatively slow. The exponential adoption of blockchain puts higher demands on the underlying technology (Zhang & Li, 2020).

2. Second, the impact on the global economy is unclear (Xiao, 2018). Currently, there are insufficient assessments of the short and long-term economic impacts of blockchain technology. At present, it is still difficult for blockchain architecture to meet the requirements of financial system availability and business continuity. Trust mechanisms and data preservation methods still need to be accepted and recognized by mainstream financial institutions. In addition, the ecosystem is still relatively underdeveloped. A sound technical standard of the blockchain system has not yet been established globally, which leads to the lack of convincing practical data and affects the orderly development of the blockchain ecosystem.
3. Thirdly, the technology lacks universal rules, and the application scene is segmented. At present, there is no unified industry norm and regulatory norm for the application of blockchain technology (Zhang & Li, 2020). Regulatory policy in China remains murky. The government's policies and effective supervision restrict the sustainable development of blockchain to some extent. Due to the lack of effective supervision, the authenticity and integrity of the pre-chain data cannot be guaranteed. When the blockchain technology is used for the traceability of various assets, it is difficult to truly form a closed loop to reduce risks and reduce speculation (Xiao, 2018), and the risk of potential losses due to information distortion or distortion exist.
4. Fourthly, the application of commercial banks should consider the "double-sided" nature of blockchain technology (Zhang & Li, 2020). At present, the starting point of commercial banks' application of blockchain technology is to reduce costs, and the profit point is not clear. Major commercial banks in China have already made various trials such as asset custody system and cross-border clearing and settlement in field of blockchain, but most of them are still in the exploratory stage and have not yet entered the stage of large-scale commercial application.
5. Fifthly, in the blockchain, all smart contracts have been coded in advance and automatically executed. Normally, the basic transaction execution has no problem, but in the case of special complex

transactions, there may be some deviation from the user's understanding, which will bring losses to the user. At the same time, there is uncertainty in blockchain smart contract (Upadhyay et al., 2021), Turing smart contracts are too flexible, and once loopholes are exploited, this will cause irreparable losses. On the other hand, the tamper-proof feature of blockchain will also form some obstacles to business logic correction and contract upgrade.

CONCLUSION AND RECOMMENDATIONS

Since 2008, blockchain has a global experience development of 13 years. Chinese blockchain national policy has improved since the beginnings, the industry application standards are gradually established and the development is directed from technology leading into market penetration. Meanwhile, the integration of blockchain applications in finance, supply chain, government and other fields is being implemented, and the industrial scale will show rapid growth. However, the current development of Blockchain in China still faces many problems which need to be considered closely. For instance, core technology independent innovation capacity still needs to be strengthened, security and privacy issues should be continuously focused on, the talent gap needs to be improved, and ecological integration application scenarios also need to be further explored.

Since it is the central banking system which strengthens the central bank's macro-control function and improves the management system in Chinese financial sector, the government plays a primary role in the promotion of blockchain applications. Following areas and issues still need to be addressed by the regulating bodies to mitigate the risks of blockchain:

1. Strengthen supervision and continuously follow (Long et al., 2019). The Chinese government should focus on the impact and challenges that the application of blockchain technology in the financial sector may bring to the existing legal system and regulatory framework and then build up a sound and efficient financial supervision mechanism. Illegal acts and disturbances that violate the law of technological development and damage the financial order should be strictly punished.
2. The Chinese blockchain application industry should improve its technological innovation capacity. Chinese government should play a catalytic role in encouraging innovative research and application of blockchain technology by applying the "regulatory sandbox" mechanism to enable blockchain financial products and services to better participate in the development process of the micro-market. Meanwhile the Government should strive to build an independent and controllable blockchain development ecosystem with strong innovation capabilities (Xiao, 2018). Furthermore, the development of major commercial banks, insurance and securities institutions should be accelerated, and the potential risks and deficiencies of blockchain financial products and services in transactions should be effectively recorded in order to take measures to control them.
3. Increasing capital investment and setting up special funds to promote the research and development of blockchain basic theory and core technology is also a way to facilitate the development of blockchain. It is necessary to closely integrate industrial application scenarios, conquer key core technologies, accelerate the deep integration of blockchain with various economic and social fields and industries, then encourage the creation of blockchain integrated application and industrial development clusters (CCID Industry and Information Research Institute, 2021). All parties should actively carry out pilot and demonstration work on blockchain application under Chinese

"new infrastructure". The financial market should promote the deep integration of blockchain with artificial intelligence, big data, 5G and cloud computing.

Chinese national and regional governments should also actively carry out pilot demonstration work of new infrastructure applications based on blockchain and set up typical industry cases for demonstration effects. At the same time, improving the application level of blockchain technology in other real economy fields is also needed and using it to jointly promote the development of Chinese digital economy.

4. For introducing new policies and action plans by the government, the mode of public-private collaboration has shown positive effects. The attitude of the government should remain active in order to support the rapid development of blockchain technology since various advanced applications need to be endorsement by the People's Bank of China which also has a role of monitor and coordination. The next step for the government of China is to facilitate communications and negotiations with academia (the experts and professionals) and business (private enterprises of blockchain applications). By way of this three-parties communication, it can further accelerate the in-depth development of blockchain applications with relevant and precise policies support from the government, state-of-the-art academic researches and findings from experts and professionals, capital and specific production support from enterprises. This enables building a blockchain technology-based digital ecosystem which will contribute to sustainable development of the digital economy by including government-backed application, continuous innovation ability and adequate funding and support.

ACKNOWLEDGMENT

The authors extend sincere gratitude to:

- The Editor-in-Chief and International Editorial Advisory Board (IEAB) of this book who initially desk reviewed, arranged a rigorous double/triple blind review process and conducted a thorough, minute and critical final review before accepting the chapter for publication.
- All anonymous reviewers who provided very constructive feedbacks for thorough revision, improvement, extension and finetuning of the chapter.
- All colleagues, assistants and well-wishers who assisted the authors to complete this task.

REFERENCES

Albayati, H., Kim, S. K., & Rho, J. J. (2021). Accepting financial transactions using blockchain technology and cryptocurrency: A customer perspective approach. *Technology in Society*, *62*, 101320. doi:10.1016/j.techsoc.2020.101320

Anagnostopoulos, I. (2018). Fintech and Regtech: Impact on regulators and banks. *Journal of Economics and Business*, *100*, 7–25. doi:10.1016/j.jeconbus.2018.07.003

Ant Group Co. Ltd. (2020). *Prospectus of Ant Technology Group Co. Ltd.* Available online: https://www.antgroup.com

Beck, R., & Müller-Bloch, C. (2017). Blockchain as radical innovation: a framework for engaging with distributed ledgers. *Proceedings of the 50th Hawaii International Conference on System Sciences*, 5390–5399.

Beijing Blockchain Technology Application Association. (2020). *Regulatory Technology Blue Book: China Regulatory Technology Development Report (2020)*. Author.

Bing, T. (2019). *The application prospect of blockchain technology in the financial industry*. People Venture Blockchain.

BIS. PWC. (2020). *Central Bank Digital Currencies (CBDC) interbank or wholesale projects that are furthest in their development as of 2020, by country*. PwC CBDC Global Index. Retrieved from: https://www.statista.com/statistics/1229776/countries-with-most-mature-interbank-cbdg-worldwide/

Boettke, P. (2016). *Special Collections at the GMU Library – October is Nobel Celebration Month | Main | 2016 Southern Economic Association Meetings*. https://www.coordinationproblem.org/2016/10/and-the-nobel-goes-to-oliver-hart-and-bengt-holmstrom.html

Bollaert, H., de Silanes, F. L., & Schwienbacher, A. (2021). Fintech and access to finance. *Journal of Corporate Finance*, *68*, 101941. Advance online publication. doi:10.1016/j.jcorpfin.2021.101941

CAICT. (2021a). *Market size of the digital economy in China in selected years from 2005 to 2020*. White Paper on China's Digital Economy Development. Retrieved from: https://www.statista.com/statistics/1250080/china-digital-economy-size/

CAICT. (2021b). *Share of the digital economy in the GDP of China from 2015 to 2020*. White Paper on China's Digital Economy Development. Retrieved from: https://www.statista.com/statistics/1250092/china-digital-economy-gdp-share/

CAICT. (2021c). *Value added share of digital economy in China from 2016 to 2020, by industry*. White Paper on China's Digital Economy Development. Retrieved from: https://www.statista.com/statistics/1250100/china-digital-economy-penetration-rate-by-industry/

CBRC General Office. (2021). *Notice on further promoting the high-quality development of financial services for small and micro enterprises in 2021*. Retrieved from: http://www.gov.cn/zhengce/zhengceku/2021-04/26/content_5602140.htm

CCID Industry and Information Research Institute. (2021). *Blockchain development trends in China in 2021*. Retrieved from: https://baijiahao.baidu.com/s?id=1693991735268111467&wfr=spider&for=pc

China Insurance Regulatory Commission. (2017). *Circular of CIRC on Printing and Distributing the 'Construction Plan of the Second Generation Project of Compensation'*. Retrieved from: http://www.gov.cn/xinwen/2017-09/20/content_5226368.htm

China Patent Protection Association. (2020). *2020 Global Patent Authorization report in blockchain field*. Retrieved from: http://www.199it.com/archives/1077801.html

China Strategic Emerging Industry. (2018). The world's first cross-border blockchain remittance was born in Hong Kong. *China Strategic Emerging Industry*, *14*, 12–13.

Chinese IoT Blockchain Branch. (2021). *2021 China industry blockchain ecological map*. Retrieved from: https://mp.weixin.qq.com/s/zkwrOTSFPaM74pW5bUO_hg

Cioroianu, I., Corbet, S., & Larkin, C. (2021). The differential impact of corporate blockchain-development as conditioned by sentiment and financial desperation. *Journal of Corporate Finance*, *66*, 101814. doi:10.1016/j.jcorpfin.2020.101814

Cong, L. W., & He, Z. (2019). Blockchain disruption and smart contracts. *Review of Financial Studies*, *32*(5), 1754–1797. doi:10.1093/rfs/hhz007

CrunchBase, SCMP. (2020). *Total funding of the leading fintech and blockchain startups in China as of June 2020 (in million U.S. dollars)*. China internet report 2020.

Cyberspace Administration of China. (2019). *Blockchain Information Service Management Regulations*. Retrieved from: http://www.cac.gov.cn/2019-01/10/c_1123971164.htm

Cyberspace Administration of China. (2019). *On the notice of the first batch of domestic blockchain information service record number of the announcement*. Retrieved from: http://www.cac.gov.cn/2019-03/30/c_1124305122.htm?from=singlemessage&isappinstalled=0

Eastmoney Choice database. (2021). Retrieved from: http://choice.eastmoney.com/

Ernst & Young. (2016). UK FinTech census: The voice of FinTech, HM treasury. *EY and Innovate Finance*. Retrieved from: http://www.ey.com/Publication/vwLUAssets/EY-UK-FinTechCensus-2017/$FILE/EY-UK-FinTech-Census-2017.pdf

Finance, F. (2020). Fintech helps securities business transform and restructure. *China Securities Journal*. Retrieved from: https://baijiahao.baidu.com/s?id=1654762300223125200&wfr=spider&for=pc

General Office of Ministry of Industry and Information Technology. (2018). *"Circular of the General Office of the Ministry of Industry and Information Technology on the Application of the Pilot and Demonstration Projects of Information Consumption*. Retrieved from: http://www.gov.cn/zhengce/zhengceku/2021-03/04/content_5590162.htm

General Office of the State Council. (2017). *Guiding Opinions of the General Office of the State Council on Actively Promoting Innovation and Application of Supply Chain*. Retrieved from: http://www.gov.cn/zhengce/content/2017-10/13/content_5231524.htm

General Office of the State Council. (2017). *Opinions of the General Office of the State Council on Innovating Management, Improving Services, Cultivating and Strengthening New Driving Forces of Economic Development, and Accelerating the Transition Between Old and New Driving Forces of Economic Development*. Retrieved from: http://www.gov.cn/zhengce/content/2017-01/20/content_5161614.htm

General Postal Office. (2017). *Guiding Opinions of State Post Bureau on Promoting the Construction of "One Belt And One Road" in Postal Service*. Retrieved from: http://www.spb.gov.cn/zc/flfgjzc_1/201801/t20180112_1467509.html

Gomber, P., Koch, J. A., & Siering, M. (2017). Digital Finance and FinTech: Current research and future research directions. *Journal of Business Economics, 87*(5), 537–580. doi:10.100711573-017-0852-x

Guo, Y. M., Huang, Z. L., Guo, J., Guo, X.-R., Li, H., Liu, M.-Y., Ezzeddine, S., & Nkeli, M. J. (2021). A bibliometric analysis and visualization of blockchain. *Future Generation Computer Systems, 116*, 316–332. doi:10.1016/j.future.2020.10.023

Hao, W., Ma, S.L., Dai, H.N., Imran, M., & Wang, T.S. (2020). Blockchain-based data privacy management with Nudge theory in open banking. *Future Generation Computer Systems, 110.*

Jiang, S. R., Li, Y. Z., Wang, S. Y., & Lin, Z. (2021). Blockchain competition: The tradeoff between platform stability and efficiency. *European Journal of Operational Research.* Advance online publication. doi:10.1016/j.ejor.2021.05.031

Kshetri, N. (2021). Blockchain and sustainable supply chain management in developing countries. *International Journal of Information Management, 60*, 102376. doi:10.1016/j.ijinfomgt.2021.102376

Li, L., Freeman, G., & Wohn, D. Y. (2021). The Interplay of Financial Exchanges and Offline Interpersonal Relationships through Digital Peer-to-Peer Payments. *Telematics and Informatics, 63*, 63. doi:10.1016/j.tele.2021.101671

Liu, S. P. (2019). *Digital Economy New Engine (Blockchain Technology) Innovation Application Summit.* Retrieved from: https://www.cebnet.com.cn/20190419/102566686.html

Liu, T., Wang, X., & Woo, W. T. (2019). The road to currency internationalization: Global perspectives and Chinese experience. *Emerging Markets Review, 38*, 38. doi:10.1016/j.ememar.2018.11.003

Long, Y. A., Zhang, J., & Ai, R. (2019). Research on supply chain finance System Optimization based on Blockchain Technology. *Southwest Finance, 2019*(1), 74–81.

Lu, Y. (2021). Consistency, breakthrough and structure of the blockchain regulation in China. *Journal of Nanjing University of Posts and Telecommunications, 2*, 24–36.

Majeed, U. K., Khan, L. U., Yaqoob, I., Kazmi, S. M. A., Salah, K., & Hong, C. S. (2021). Blockchain for IoT-based smart cities: Recent advances, requirements, and future challenges. *Journal of Network and Computer Applications, 181*, 103007. doi:10.1016/j.jnca.2021.103007

Ministry of Commerce, the National Development and Reform Commission, the Ministry of Land and Resources, the Ministry of Transport & the State Post Bureau. (2017). *Notice on the issuance of the 13th Five-Year Plan for the Development and Development of Trade and Goods*. Retrieved from: http://www.mofcom.gov.cn/article/b/d/201702/20170202511705.shtml

Ministry of Commerce, Ministry of Finance. (2017). *Circular of the General Office of the Ministry of Commerce and the General Office of the Ministry of Finance on the Construction of Supply Chain System*. Retrieved from: http://www.mofcom.gov.cn/article/h/redht/201708/20170802627302.shtml

Ministry of Education. (2018). *Notice of the Ministry of Education on the issuance of the Action Plan for Educational Informatization 2.0.* Retrieved from: http://www.moe.gov.cn/srcsite/A16/s3342/201804/t20180425_334188.html

Ministry of Industry and Information Technology. (2016a). *China Blockchain Technology and Application Development White Paper (2016).* Retrieved from: http://www.shujuju.cn/lecture/detail/2134

Ministry of Industry and Information Technology. (2016b). *Circular of the Ministry of Industry and Information Technology on the issuance of the Development Plan for Software and Information Technology Service Industry (2016-2020).* Retrieved from: http://www.cac.gov.cn/2017-01/17/c_1120330752.htm

Ministry of Industry and Information Technology. (2017). *Circular of the Ministry of Industry and Information Technology on the issuance of the Three-Year Action Plan for the Development of Cloud Computing (2017-2019).* Retrieved from: https://www.miit.gov.cn/jgsj/xxjsfzs/zlgh/art/2020/art_fb1e14b54f234fc7b4f52c062b9d3d08.html

Ministry of Industry and Information Technology. (2018a). *Notice of Ministry of Industry and Information Technology of China.* Retrieved from: https://www.miit.gov.cn/jgsj/xxjsfzs/gzdt/art/2020/art_ebe80855c8234ba7bd1b7cf7ffcc5813.html

Ministry of Industry and Information Technology. (2018b). *China Blockchain Technology and Application Development White Paper (2018).* Retrieved from: https://www.miit.gov.cn/n1146290/n1146402/n1146445/c6180238/part/6180297.pdf

Ministry of Industry and Information Technology of the People's Republic of China. (2018). *White Paper on China's Blockchain Industry 2018.* Retrieved from: https://www.miit.gov.cn/n1146290/n1146402/n1146445/c6180238/part/6180297.pdf

Ministry of Industry and Information Technology, Office of the Central Cyberspace Affairs Commission. (2021). Guidance on accelerating the application of blockchain technology and industrial development. Joint Letter Of Ministry of Industry and Information Technology.

Mu, Geng, Lv, Qian, & Qing. (2020). Blockchain research Group of Digital Currency Research Institute, People's Bank of China: Existing blockchains are not suitable for high concurrency scenarios such as traditional retail payments. *Chinese Finance, 4.*

National Internet Finance Association of China. (2020a). *Risk tips for participating in speculation on overseas virtual currency trading platforms.* Retrieved from: https://baijiahao.baidu.com/s?id=1662828395598126938&wfr=spider&for=pc

National Internet Finance Association of China. (2020b). *Research Report on Blockchain Financial Application and Development in China.* Retrieved from: https://wenku.baidu.com/view/0193c7d6b6360b4c2e3f5727a5e9856a5712267e.html

National Internet Finance Association of China, China Banking Association, Payment & Clearing Association of China. (2021). *Notice of the prevention of virtual currency trading speculation risk.* Retrieved from: https://baijiahao.baidu.com/s?id=1700098234762596743&wfr=spider&for=pc

Nguyen, L. T. Q., Hoang, T. G., Do, L. H., Ngo, X. T., Nguyen, P. H. T., Nguyen, G. D. L., & Nguyend, G. N. T. (2021). The role of blockchain technology-based social crowdfunding in advancing social value creation. *Technological Forecasting and Social Change*, *170*, 120898. doi:10.1016/j.techfore.2021.120898

Ning, Y. J. (2021). *Speech on "Industrial Blockchain 2.0:2021 New Opportunities for Digital Economy"*. Retrieved from: https://new.qq.com/omn/20210128/20210128A08O3A00.html

North, D. (1990). *Institutions, Institutional Change and Economic Performance*. Cambridge University. doi:10.1017/CBO9780511808678

Pal, A., Tiwaria, C. K., & Haldar, N. (2021). Blockchain for business management: Applications, challenges and potentials. *The Journal of High Technology Management Research*, *32*(2), 100414. doi:10.1016/j.hitech.2021.100414

People's Bank of China, Ministry of Industry and Information Technology & China Banking Regulatory Commission (CBRC). (2013). *Notice on Bitcoin Risk Prevention.* Retrieved from: http://www.gov.cn/gzdt/2013-12/05/content_2542751.htm

People's Bank of China. (2014). *Set up a special research group on legal digital currency*. Annual Report of the People's Bank of China 2014. Retrieved from: http://www.pbc.gov.cn/eportal/fileDir/image_public/UserFiles/chubanwu/upload/File/中国人民银行年报2014-最终定稿.pdf

People's Bank of China. (2015). *Some key issues in the field of digital currency were investigated.* Annual Report of the People's Bank of China 2015. Retrieved from: http://www.pbc.gov.cn/chubanwu/114566/115296/3087023/3086996/2016062115305560130.pdf

People's Bank of China, Office of the Central Cyberspace Affairs Commission, Ministry of Industry and Information Technology, State Administration for Market Regulation, China Insurance Regulatory Commission, China's Securities Regulatory Commission & China's Bank Regulatory Commission. (2017). *Notice on Preventing the Risk of Token Issuance Financing*. Retrieved from: http://www.pbc.gov.cn/goutongjiaoliu/113456/113469/3374222/index.html

People's Bank of China. (2020a). *Blockchain technology financial application evaluation rules.* Retrieved from: https://www.mpaypass.com.cn/Download/202007/30153205.html

People's Bank of China. (2020b). *Notice to promote the application of blockchain technical specifications.* Retrieved from: https://www.jinse.com/news/blockchain/772215.html

People's Bank of China. (2020c). *Technical security specification for financial distributed ledger*. Retrieved from: https://www.sohu.com/a/375656914_120057347

People's Bank of China. (2021). *White paper of "The development of Chinese digital RMB"*. Retrieved from: http://www.gov.cn/xinwen/2021-07/16/content_5625569.htm

PuHuiMingYan financial research institute. (2018). *Application scenario of block chain in supply chain finance and the first block chain ABS case.* Retrieved from: https://mp.weixin.qq.com/s/2TbAxw7G_MD4OmEkfZ2npQ

Qiang, W., & Min, S., & Rong, R.L. (2020). Is China the world's blockchain leader? Evidence, evolution and outlook of China's blockchain research. *Journal of Cleaner Production*, 264.

Rong, Z.X. & Yu, L.M. (2020). The application of blockchain technology in the banking industry. *China Finance, 10*.

Sanka, A. I., Irfan, M., Huang, I., & Cheung, R. C. C. (2021). A survey of breakthrough in blockchain technology: Adoptions, applications, challenges and future research. *Computer Communications*, *169*, 179–201. doi:10.1016/j.comcom.2020.12.028

Scott, W. R. (2008). *Institutions and organizations: Ideas and interests* (3rd ed.). SAGE Publications.

Securities, M. (2017). *Driven by the dual track of blockchain policy and practice, the Fintech industry is heating up*. Minsheng Securities Institute.

Standing Committee of the National People's Congress. (2019). *Code Law*. Retrieved from: http://www.npc.gov.cn/npc/c30834/201910/6f7be7dd5ae5459a8de8baf36296bc74.shtml

State Council. (2016). *Notice of the State Council on the issuance of the '13th Five-Year' National Informatization Plan*. Retrieved from: http://www.gov.cn/zhengce/content/2016-12/27/content_5153411.htm

State Council. (2017). *Circular of the State Council on the issuance of the New Generation of Artificial Intelligence Development Plan*. Retrieved from: http://www.gov.cn/zhengce/content/2017-07/20/content_5211996.htm

State Council. (2019a). *Opinions on Deepening Reform and Strengthening Food Safety*. Retrieved from: http://www.gov.cn/zhengce/2019-05/20/content_5393212.htm

State Council. (2019b). *Opinions on Supporting Shenzhen in Building a Pilot Demonstration Zone for Socialism with Chinese Characteristics*. Retrieved from: http://www.gov.cn/zhengce/2019-08/18/content_5422183.htm

Sunny, J., Undralla, N., & Madhusudanan Pillai, V. (2020). Supply chain transparency through blockchain-based traceability: An overview with demonstration. *Computers & Industrial Engineering*, *150*, 150. doi:10.1016/j.cie.2020.106895

Swan, M. (2015). *Blockchain: Blueprint for a New Economy*. O'Reilly Media, Inc.

TechWeb Report. (2018). *Cloud coin network: the basic completion of the clean-up work plan to close the website on August 31*. Retrieved from: TechWeb.com.cn.

The People's Bank of China. (2011). *The Establishment meeting of Bank Card Base Payment Committee of Clearing Association of China, Network Payment Application Committee and Prepaid Card Committee of Payment*. Retrieved from: https://xueshu.baidu.com/usercenter/paper/show?paperid=7d2b16f58ef921bcfec6f8cdbbb55a60&site=xueshu_se

The People's Bank of China. (2015). *Measures for the administration of online Payment business of non-bank payment institutions*. Retrieved from: http://www.gov.cn/zhengce/2016-03/18/content_5055171.htm

Thurber, K. J. (2012). Do not invent buggy whips: Create! reinvent! position! disrupt! Digital Systems Press.

University Big Data and Artificial Intelligence Promotion Alliance. (2021). *Top 10 Cases of Blockchain technology application in China in 2020.* Retrieved from: https://mp.weixin.qq.com/s/Unt3QI6a7xy-dZdvxv4q2g

Upadhyay, A., Mukhuty, S., Kumar, V., & Kazancoglu, Y. (2021). Blockchain technology and the circular economy: Implications for sustainability and social responsibility. *Journal of Cleaner Production, 293*, 126130. doi:10.1016/j.jclepro.2021.126130

Varfolomeev, A., Alfarhani, L. H., & Oleiwi, Z. C. (2021). Secure-reliable smart contract applications based blockchain technology in smart cities environment. *Procedia Computer Science*, *186*, 669–676. doi:10.1016/j.procs.2021.04.188

Wang4185652. (2021). *Analyze the impact of the development of block chain on the financial field.* Retrieved from: https://www.doc88.com/p-17216081798929.html

Xi, J. P. (2019). *The 18th group study session of the Political Bureau of the CPC Central Committee.* Retrieved from: http://www.gov.cn/xinwen/2019-10/25/content_5444957.htm

Xiao, X. (2018). *Five features of the current application of blockchain in the financial field.* Internet Finance Association of China. Retrieved from: http://www.360doc.com/content/18/0623/22/44361760_764758542.shtml

Yan, N., Sun, B., Zhang, H., & Liu, C. (2016). A partial credit guarantee contract in a capital constrained supply chain: Financing equilibrium and coordinating strategy. *International Journal of Production Economics*, *173*, 122–133. doi:10.1016/j.ijpe.2015.12.005

Yu, P., Chen, Z., & Sun, J. (2018). *Innovative Financing: an Empirical Study on Public– Private Partnership Securitisation in China.* Australian Economic Papers. John Wiley & Sons Australia, Ltd.

Yu, P., Gong, R., & Sampat, M. (2022). Blockchain Technology in China's Digital Economy: Balancing Regulation and Innovation. In P. Tehrani (Ed.), *Regulatory Aspects of Artificial Intelligence on Blockchain* (pp. 132–157). IGI Global. doi:10.4018/978-1-7998-7927-5.ch007

Zhang, X.R. & Li, M.Y. (2020). The application of blockchain technology in the banking industry. *China Finance, 10*.

Chapter 10
Blockchain-Based Secured Voting Using Multi-Level Authentication

Megala G.
Vellore Institute of Technology, India

Prabu Sevugan
Pondicherry University, India

Venkatesan R.
SRM TRP Engineering College, India

ABSTRACT

Although electronic voting has been used for many decades, electoral organizations are still adopting it. Despite technologies that preserve the secrecy, authenticity, and privacy of data, vote-rigging and privacy breaches remain a risk in e-voting systems. Blockchain is a distributed technology that underpins bitcoin. It appears to be a fit for digital election systems, as it provides a chance to keep data secret and prohibit data manipulation. But blockchain alone is not enough to address other concerns that come up during elections such as how to make sure that the person voting has the right to do so while keeping their identity disconnected from their vote. The user can safely vote from the three-levels of authentication: OTP, QR codeshare, and blockchain. Here, the system contains a secured timestamp and the hashcode which is used to identify attacks on the distributed server. These protected services provide secure vote casting and remedies for security issues. The legal limitations are also analyzed in the blockchain distributed ledger.

DOI: 10.4018/978-1-7998-9274-8.ch010

INTRODUCTION

Block chain is a distributed transactional processing or distributed ledger and information management platform, originally designed for the Bit coin cryptocurrency. There has been a boom in interest in block chain technology since the notion was initially introduced in 2008. The essential qualities of Block chain is that it allow privacy, anonymity, and authenticity / integrity of information without the need of a trusted third-party authority for each transaction dependency, which opens up new study fields, particularly in terms of technological hurdles and constraints.

The three primary crucial elements to create a blockchain trust are (i) identity, (ii) possession and (iii) proof or verification.

1. Identity - The introduction of digital signatures in block chains solves the identification problem. Each participant is assigned a pair digital key (private key, public key). A "private key," which functions similarly to an account number, and a "public key," which functions similarly to a password that they may use to simply prove their identity and conduct allowed transactions.
2. Possession (Who holds or possess What) - Block chains use a technique called "cryptographic hashing" to tackle the possession challenge. A crypto hash is just an unit of information which has been turned into a smaller block of information by running it over a mathematical function. Each block in a block chain comprises a hashed version of the preceding block's content. Any changes you make to earlier bits of data will be reflected across the chain, making it easier for the system to spot and reject data manipulation attempts. This makes it possible to produce "immutable" information, sometimes termed as tamper-proof entries, using block chains.
3. Proof or verification - Ultimately, block chains address the authentication issue by providing a group of individuals to publicly check the truthfulness of a transaction without the requirement for a trusted middleman. This is known as "consensus mechanism" in block chain jargon. A significant benefit of block chains is their capacity to validate transactions with fewer intermediates, which might yield to cheaper costs for many of the applications mentioned in this article.

Block chain is gradually gaining traction as a technology. Though various peer-to-peer apps for file transfers, streaming video, and other services have been existed for a long time, the premise that these networks could use their own safety and infrastructure has been in use since 2008. Block chain's success in the decade since its birth has been largely essential to the achievement of the mechanism that inspired it, bit coin. Many people are discovering that the fundamental value of block chain is its potential to improve existing systems. From the beginning, astute onlookers realised the technology's promise, as bit coin provided a more stable and trustworthy payment processing and financial service over previous options. Block chain has modernized a variety of sectors in recent years, notably online data storage, intelligent contracts, and fundraising events. The two main consensus processes used by cryptocurrencies to authenticate new transactions, append it towards the blockchain, and generate fresh tokens are proof of work and proof of stake. Mining is used to attain those aims through proof of work, which was initially pioneered by Bitcoin. Staking is used in Proof of Stake, which is used by Ethereum blockchain, and others to achieve the same goals.

The security team has seen electronic voting machines as problematic, inadequate, especially due to tangible security issues. Anyone with access privileges to that kind of machine has the ability to ruin it, impacting all votes cast on that device. This is where block chain technology comes in. A block chain is a public ledger that is distributed, unchanging, and indisputable. The primary elements of this innovative technology are as follows: (i) The ledger can be found in a variety of places: There is no single point of failure in the decentralised ledger's upkeep. (ii) The ability to append new transactions to the ledger is disseminated. (iii) Each and every suggested "new block" to the ledger should refer to the prior version of the ledger, forming an immutable chain from which the block chain originally comes and prohibiting manipulation with earlier entries' integrity.

The following are some of the blockchain difficulties that anyone considering adopting it should be aware of.

1. There is a cost to blockchain in terms of the ecology.
2. A dangerous environment is created by a lack of regulation.
3. Because of its intricacy, end users have a tough time appreciating the benefits.
4. Blockchains could be complex and inconvenient to use.
5. The establishment has a financial stake in blockchain's failure.

A thorough mapping analysis is conducted with the purpose of gathering all relevant research on Block chain technology for this study. Primary goal is to gain a technical understanding of current research subjects, difficulties, and prospects in Block Chain technology. From scientific databases, we collected 41 primary papers. The findings suggest that over 80% of the articles focus on the Bit currency system, with fewer than 20% dealing with other Block chain applications such as smart agreements and authorising. The significant amount of research is focused on uncovering and addressing Block chain's security and privacy flaws, however many of the proposed remedies lack tangible evidence of their efficacy. Many additional aspects of Block Chain scalability, such as throughput and latency, have gone unexplored.

RELATED WORKS

Khan et al. (2020) implemented both the permissioned blockchain as well as permissionless one for secure voting system and its performances are investigated such as the generation rate of block, its size, the speed of transaction based on the population. The transaction between the miner (voter) and validator depends on the proof of work with average constant rate of block generation with reduced time interval of 15 seconds. Though this system was scalable with increasing number of nodes to the blockchain but the latency of the network increased due to bulk load of transactions. Khazaei and Rezaei-Aliabadi (2018) designed secure decentralized e-voting protocol with the proof of security. This protocol was designed based on security properties such as confidentiality, robustness, freeness and performance properties such as time complexity, throughput and number of rounds. Four main phases such as initialization of voters, vote casting phase, recovery phase and tallying phase are involved in decentralized protocol implementation. There exists security weakness of this protocol.

Larriba et al. (2020) developed voting system for multiple parties based on blind signature which allows to sign a masked communication. It involves three different functions such as signing, computation, and redundancy check. Li et al.,(2019) proposed electronic voting at multi role with multi level access in cloud. The author has used Chinese remainder theorem (CRT) for each role and sub access constructs are associated. Here the sharing of multiple secret is performed without the involvement of trusted third party service and is being pragmatic to distributed consent scheme. Benítez-Martínez et al.,

(2021) combined the blockchain and cloud services to design e-participation system by using tokenization concept in the blockchain. The author deployed a framework of neural distributed ledger (NDL) by interconnecting networks of blockchain. Each participant of the model is given a token to perform the transaction or the vote in this network.

Rathee et al. (2021) presented an apparent and safe e-voting scheme with the help of IOT devices involving blockchain technology. These devices may be intruded by the attackers and hence there is a security concern. The proposed scheme has prevented it by detecting the attacks such as denial of service (DoS), Distributed DoS (DDoS) and authentication delay. In this framework, each user can access the blockchain data and track their vote and hence there is a chance of illicit recordkeeping. Zaghloul et al. (2021) proposed anonymous e-balloting scheme in smartphones in order to preserve the privacy of vote. A decentralized blockchain is used to store the votes and it also prevents it from tampering. The security is enhanced by using non-interactive zero knowledge proofs (NIZKP) by the election body. Table 1 lists some of the blockchain platforms.

Table 1. Blockchain platforms

Platform	Description	Benefits
Ethereum	Enables true decentralization with support for smart contracts.	Provides high secure mechanism. High transaction processing cost and slow processing time.
IBM Blockchain	Private, decentralized blockchain network	Provides user friendly interface.
Hyperledger Fabric	Closed blockchain deployment and supports open smart contracts	Improves data privacy by isolating transactions in channels
R3 Codra	Uses consensus mechanism to cryptographically secure transactions.	Process transactions fast and cheap.
EOSIO	Uses complicated consensus mechanism based on Proof of Stake (PoS)	Provides better performance than Ethereum

Panja and Roy (2021) also used NIKZP to secure the provable voting scheme. Flexible and recurrent difficulties plague these designs. There are also some other elements to consider, including the existence of member nodes. Just after polling process, if the candidacy node is unavailable due toward being disconnected or continuing operations, the next accessible member should indeed be elected. To address this, the wu et al. and Sun et al. 2019 offers an alternative cluster head election model known as voluntary-tallying selection of cluster head, which meets the basic needs of ballet confidentiality, participant anonymity, and the consensual feature. The system has been defined, and security concepts for this democratic election framework have been developed, resulting in a safe and practical design.

ELECTRONIC VOTING

In today's world, the notion of implementing digital voting techniques to make the democratic political process simpler, quicker, and clearer is appealing. Making the voting process inexpensive and rapid generalises it in the perspective of participants, lowers a psychological hurdle between the voter and the public official, and puts a lot of pressure upon that elected representative. It also allows for an even more transparent type of democracy by letting people to voice their preferences on specific measures and initiatives. Experts have investigated leveraging blockchain potential to accomplish authenticity, anonymity, and non-repudiation, which seem to be crucial for a casting a ballot application. Digital or e-voting has become one of the developing adoptions of blockchain technology in which researchers have suggested leveraging blockchain ability to realize integrity, privacy and confidentiality which are critical for a voting application. The goal of present blockchain performance and scalability exploration is to attain similar performance to that of legacy systems.

To create a more stable and trustworthy voting mechanism, Blockchain is critical in ensuring that legitimate votes are cast and counted without being tampered together in any way. The Krishnamurthy et al. (2020) recommended layout would be focused on improved security via Blockchain in an E-voting service employing IoT devices, in which the person will sign up and adequate verification will indeed be accomplished using voter identity and perhaps other fingerprint means on a smart device. A miner will authenticate every transaction occurring through this profile, which in this instance would be recording votes. Liu and Wang (2017) also proposed a distributed electronic voting protocol via permissioned blockchain with access control structure using blind signature.

PROPOSED BLOCKCHAIN BASED VOTE SYSTEM

Electronic voting system using block chain service is proposed on identifying the technical and legal limitations of the existing systems. Blockchain is a data structure where each blocks written and can only be append to the chain structure. Each block consists of hash function of its previous block, assuring immutability and these blocks cannot be modified. Each user / candidate registers and obtain the OTP and individual QR code leads to secured voting. Permission based blockchain or consortium based chain is proposed using proof of authority consensus (POA) algorithm. POA validates the approved accounts transactions. Distributed ledgers relate to the chain system providing ease of access. Fig 1 shows the proposed architecture of three level authentication of electronic voting system.

Figure 1. Proposed secured voting with multilevel authentication

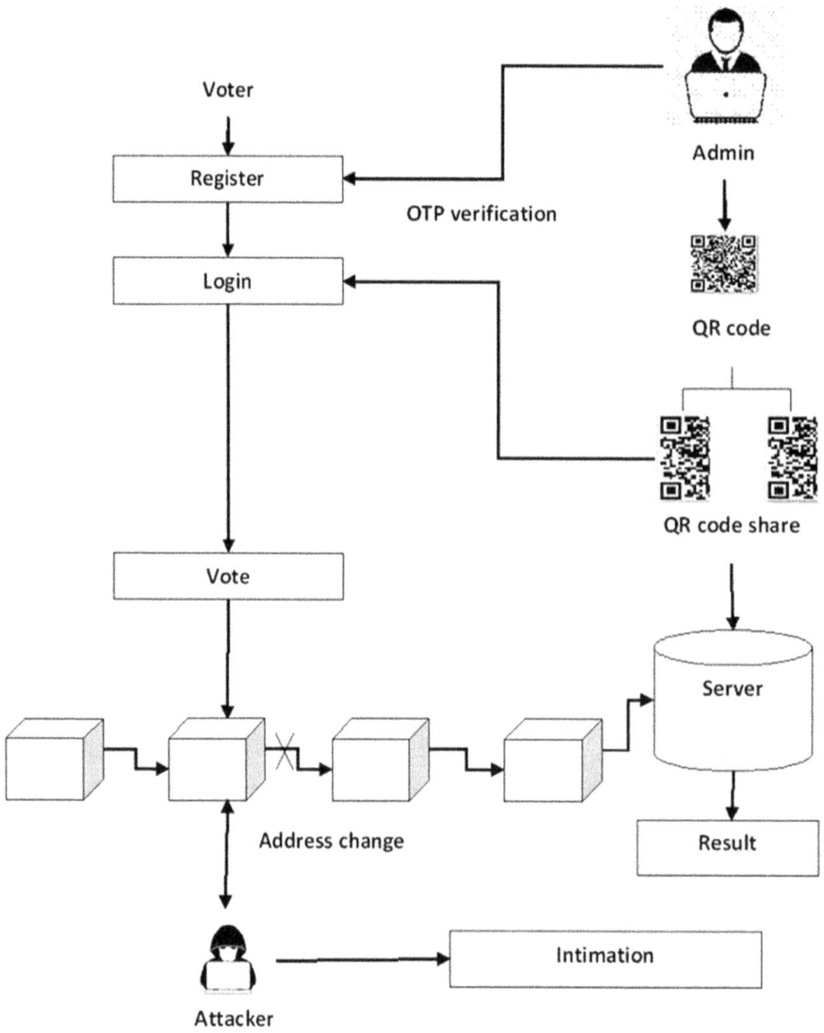

Proof-of-work Simulated miners oversee securing and verifying blockchains. Proof of work provides a lot of benefits, especially for a simple but extremely valuable cryptocurrency like Bitcoin. Proof of work might impose scalability constraints that might eventually have to be addressed. Proof-of-stake blockchains use a network of consensus mechanism who validate and update new transactions.

Registration

The registration unit serves as a basis for inputting data on users and candidates in a fashion which simplifies data capture and exactness by connecting MySQL database entries to datasets, and conveniently binds back to specific user and the candidate records to connect registers to the user details and the candidate details. It collects data elements to enable better handling for the voter's data and the candidate data. The data will be stored in the block chain based storage system. Data is prepared for storage by constructing data fragments or chunks, encrypting the fragments, creating a distinct hash for every fragment, and producing duplicate replicas of each fragment in a block chain-based storage facility. The replicated shards are then distributed across the decentralized nodes in the block chain infrastructure. Block chains are stored in computers within the system, also named as nodes.

Authentication

OTP Authentication systems generate highly secure one-time credentials, ensuring that only those who have been properly approved have access to critical data and applications. On a digital device, an OTP or dynamic passcode, is a credential that is effective for only one login session / activity. OTPs eliminate a number of flaws inherent in traditional passcode founded authentication; a few deployments often include two-factor authentication by making sure that the one-time password needs access to both something a person possess (including a modest keychain gadget with an OTP estimator constructed in, or a rfid card or particular smartphone) and knowledge of PIN access code. The biggest advantage of OTPs is that they are not vulnerable to active attacks like fixed passwords are.

User Login

The user logins the session using aadhar number and the created password. The arbitration server acts as intermediatory between user and the blockchain based voting system. Using authentication server, the user is verified while voting. The user then receives the unique QR code through the registered email. This code is further used for vting purpose on successful authentication.

Block Chain Voting

The blockchain technology is the one that handles the actual vote. Depending on the load on each node, each user vote is transferred to one of the system's nodes. Relying upon the smart contracts at each node, the node then uploads the transaction to the blockchain. Smart contracts are the policies that each nodes undertake in order to not only validate but also add votes to the system. To validate the votes, each node follows the smart contract. The blockchain is a closed environment that is not open to the general public. The system will have a node server in each state to guarantee that network traffic is spread throughout the scheme. In the old system, there were two steps to the voting process: registering voters and voting. The suggested approach will include an extra stage, allowing users to authenticate their votes. This is a vital phase in the process, since it allows the user to receive evidence of their vote. Other processes involve the organiser counting the votes and redoing if there are any inconsistencies. The user interacts with the Authentication Server via a webpage to begin the process of registering to vote. In a database, the AS keeps track of voter information. The user fills out his or her Personal Identifying Information (PII) and captures relevant documents to send into the system.

An image of the user is also captured for verification purposes. The user is authorised to set up an account if the information is confirmed and correct. To log in, the user supplies a username and password of their option. This data is kept apart from the users' personal information. During voting, this protects privacy and anonymity. In addition, an entry is placed beside the user's database entry indicating whether or not the person has entitled to vote. Users are not permitted to make accounts if their information cannot be confirmed.

The voting procedure is a multi-step procedure. It entails using the OTP to validate your identity and then voting using the QR code. On the day of the election, each candidate is issued a blockchain account through which they can receive votes. Abstention votes can also be sent to an "Abstinence account."

Notification

The vote count is saved in each block or node of the block chain server, which is interconnected to the inter server networks. Any data changes in any block will be noted with a time stamp, and indeed the admin will also be notified. The major process is that now the total count cannot be obtained since any change in the data would modify the server's hash code. If the hash code on the server changes, the data cannot be linked together to calculate the overall number of votes cast. Whereas if value changes, "data modified" would be alerted.

CONCLUSION

This chapter introduces a novel blockchain based e-voting system that employs smart contracts to provide a secure and cost-effective election while maintaining voter anonymity. The system framework has been described, as well as a security analysis. It is demonstrated that block chain technology provides a new opportunity for democratic governments to move away from the hard copy election mode towards a more cost reduced and time effective election system, while also enhancing the security measures in place today's system and providing new transparency opportunities. It is feasible to transfer dozens of operations per second into an Ethereum private blockchain, employing every component of the smart contract to lighten the burden on the blockchain.

REFERENCES

Benítez-Martínez, F. L., Hurtado-Torres, M. V., & Romero-Frías, E. (2021). A neural blockchain for a tokenizable e-Participation model. *Neurocomputing*, *423*, 703–712. doi:10.1016/j.neucom.2020.03.116

Dimitriou, T. (2020). Efficient, coercion-free and universally verifiable blockchain-based voting. *Computer Networks*, *174*, 107234. doi:10.1016/j.comnet.2020.107234

Khan, K. M., Arshad, J., & Khan, M. M. (2018). Secure digital voting system based on blockchain technology. *International Journal of Electronic Government Research*, *14*(1), 53–62. doi:10.4018/IJEGR.2018010103

Khan, K. M., Arshad, J., & Khan, M. M. (2020). Investigating performance constraints for blockchain based secure e-voting system. *Future Generation Computer Systems*, *105*, 13–26. doi:10.1016/j.future.2019.11.005

Khazaei, S., & Rezaei-Aliabadi, M. (2018). A rigorous security analysis of a decentralized electronic voting protocol in the universal composability framework. *Journal of Information Security and Applications, 43*, 99-109.

Krishnamurthy, R., Rathee, G., & Jaglan, N. (2020). An enhanced security mechanism through blockchain for E-polling/counting process using IoT devices. *Wireless Networks, 26*(4), 2391–2402. doi:10.100711276-019-02112-5

Larriba, A. M., Sempere, J. M., & López, D. (2020). A two authorities electronic vote scheme. *Computers & Security, 97*, 101940. doi:10.1016/j.cose.2020.101940

Li, J., Wang, X., Huang, Z., Wang, L., & Xiang, Y. (2019). Multi-level multi-secret sharing scheme for decentralized e-voting in cloud computing. *Journal of Parallel and Distributed Computing, 130*, 91–97. doi:10.1016/j.jpdc.2019.04.003

Liu, Y., & Wang, Q. (2017). An E-voting Protocol Based on Blockchain. *IACR Cryptol. ePrint Arch., 2017*, 1043.

Panja, S., & Roy, B. (2021). A secure end-to-end verifiable e-voting system using blockchain and cloud server. *Journal of Information Security and Applications, 59*, 102815. doi:10.1016/j.jisa.2021.102815

Rathee, G., Iqbal, R., Waqar, O., & Bashir, A. K. (2021). On the Design and Implementation of a Blockchain Enabled E-Voting Application Within IoT-Oriented Smart Cities. *IEEE Access: Practical Innovations, Open Solutions, 9*, 34165–34176. doi:10.1109/ACCESS.2021.3061411

Sun, X., Wang, Q., Kulicki, P., & Sopek, M. (2019). A simple voting protocol on quantum blockchain. *International Journal of Theoretical Physics, 58*(1), 275–281. doi:10.100710773-018-3929-6

Wu, T., Yang, G., Zhu, L., & Wu, Y. (2021). Privacy-preserving voluntary-tallying leader election for internet of things. *Information Sciences, 574*, 461–472. doi:10.1016/j.ins.2021.06.028

Yi, H. (2019). Securing e-voting based on blockchain in P2P network. *EURASIP Journal on Wireless Communications and Networking, 2019*(1), 1–9. doi:10.118613638-019-1473-6

Zaghloul, E., Li, T., & Ren, J. (2021). d-BAME: Distributed Blockchain-based Anonymous Mobile Electronic Voting. *IEEE Internet of Things Journal, 8*(22), 16585–16597. doi:10.1109/JIOT.2021.3074877

Chapter 11
Advanced Smart Educational Systems Using Blockchain Technology

Mannava Yesu Babu
Vardhaman College of Engineering, India

Balaji Vuppala
Vardhaman College of Engineering, India

Sharmila L.
Dhaanish Ahmed College of Engineering, India

S. Denis Ashok
Vellore Institute of Technology, India

Amedapu Srinivas
Sreenidhi Institute of Science and Technology, India

ABSTRACT

Blockchain is one of the new modern technologies of the tech world. Bitcoin was first introduced with this technology and is being extremely tested for several cryptocurrencies continuously. The research scholars and businesses have been concentrating on its pertinence in furthering many zones such as education, business, healthcare, logistics, and some more areas. This chapter shows the benefits of blockchain technology in the educational area. The advantage of blockchain technology in the educational sector is support in improving the convenience and value of educational areas. There are some zones in education that are essential: continuous observations of students and student correlated data and transcripts, decrease in physical and administration, and more. Further, the monitoring and tracing capability of the blockchain and some solutions to the problems have been proposed.

DOI: 10.4018/978-1-7998-9274-8.ch011

INTRODUCTION

Blockchain is a challenging advanced technical domain for consecutive records tracking of communications. This tracking was started over a period earlier in 2010, with the development of a commonly recognized Bitcoin as cryptocurrency. Then, normally commonly investigated and trialed to its exclusive landscapes containing fixity, removal of intermediate party, transparency, etc. Originally, blockchain was introduced for economic dealings but with the channel of period, scholars start to understand its possible usages in other zones such as commercial, health care, logistics and IOT etc. Hereafter blockchain is forecast to transform many zones of concentration outstanding to its extensive assortment of usages. Besides many submission areas previously declared, an inordinate possible usage of the technology can be observed in education area. The environment is motivated to quality and later extemporizing the ordinary of teaching laterally with making it merely reachable to everybody anywhere in the world. Blockchain systems are distributed, dispersed, attach only and clear systems that preserve the based-on time transactions records cryptographically protected from any interfering(Alammary et al., 2019). To accomplish the determination, training, knowledge, student data monitoring, credential issuing, transcription organization etc., are presence digitalized with the support of technology. It is forecast that production the accurate and optimal usage of technology like increased genuineness, Machine learning, cloud computing, artificial intelligence, hologram etc. Vuppala, B., & Swarnalatha, P. (2020), can renovation the educational sector I) enhancing capability for learning, ii) Tracing study notes easy to access using on cloud storage iii) student character analysis designs for modified programs, iv) decreasing the fee of study and several further support this technology. While teaching segment has previously succeeded in current days and is on a method of extra enhancement, conversely here are stagnant influenced zones are teaching that essential extra care Satoshi Nakamoto (2020). This regions comprise nonviolent and protected organization schemes for academic records and student data, secure and low-cost information storage structure, decrease in physical and form-filling etc. This is wherever the easy to access skill of blockchain technology is required. In this chapter emphases on the usage of blockchain technology in teaching area by examining the present government policies of the skill in the zone and discovering the upcoming teaching using this skill. The major elementary goal of this education is to appreciate how the records are monitoring and tracking structures of using with the blockchain system may deliver support to educate the knowledge, education and growth in Teaching areas improvements.

One of the maximum hopeful technical tendencies in information technology manufacturing these present time is the Blockchain technologies. In humblest standings, blockchain it may be observed as links of zones anywhere the zones discuss to numerical information deposited on a record of information storage db that is discussed as a links of chain. Blockchain may be distinct as a dispersed, unchallengeable, system to system and attach individual dispersed record that is protected with support of cryptoanalysis(GaoW et al., 2018). The development of Blockchain removed the essential of intermediate third party for business confirmation and data tracing. The main with this blockchain expertise, contributors in the system take the specialist of authenticate the dealings and finished common agreement. After an operations are activated, blockchain technology disseminated whole peer to peer link for confirmation, with the guarantees that not at all duplicitous contract receipts abode also encourages pellucidity. Here clients not essential to confidence any third party nevertheless slightly its comprehensive independence over his information. The primary technology of Blockchains, broadly recognized as Bitcoin, was familiarized around a period ago in Oct. 2008. Subsequent it, many more cryptocurrencies have been developed. Primarily presented for the economic trades of the crypto currency, in their blockchain progressively banquet to additional segments are fine(Chen, W et al., 2018). Completed the years, this technology has been abundantly investigated and investigated to carry its advantages to other request zones such as medical businesses, stream link organization structures, business, universities, medical, etc. The main in segment, the elementary overview to the blockchain technology, it showing and its enhancing the knowledge and improving in the teaching sector are deliberated Julian Hall(2019).

A block chain is an easy links of blocks anywhere individually block contains of numerical stage engraved data. Lump is an elementary module of the block chains. In this is a records encompassing system connected information may it not be changed. Zone in block chain is similar in the sheet of a transaction reserves anywhere in the information is updated. Every instance of time the data is complete, i.e., zone is accomplished, it transfers the device to the novel block in their link(Malik, et al., 2018). It essential be distinguished that for individually zone to be extra to the blockchain, all the communications in the zone essential be proved by the members of the system. For collectively and every record are added to this zone, the client essential demand the operation(Adnan, et al., 2015)

Subsequently this blockchain systems to dispersed and disseminated, the main appeal is transmission to system-to-system link for confirmation of the operation. An guarantees that completely the dealings are effective operation has been confirmed, the zone block is link to the next link. The replica of this novel link is over kept at each and every active system for transaction monitoring purposes. Subsequently the replica is existing at every system, it is enormously tough to interfere with the fillings of the block hereafter manufacture it an unchallengeable account.

For every country to development, educational sectors are an important part and is consequently frequently the key emphasis for maximum established nations Raschka, S. (2018). This is since the development in every segment depending on the near of teaching system. Though, in new centuries, the universal has observed an outstanding development in novel technology and its more advantages to education system, though, evolving humanoid sources finished education immobile requirements to be attentive. This section defines many questions in teaching that can essential more contemplation of the scholars

Blockchain Technology Features

The generally identified this are important features of blockchain are:

1. Smart agreement:

It is a system procedure projected to alphanumerically support, authenticate, or execute the concert of an agreement. It documents the operation to be accomplished without 3rd party. Furthermore, these records are monitoring permanently. This self-running program executes on distributed systems.

2. Distributed:

The information storage on record is distributed which is common widely complete the systems grouped in a mess network.

3. Monitoring:

Later the data is kept in protected blocks, Blockchain encourages Monitoring of an incident on the system. This information always protected with support of hash function with support of cryptographic simplex and complete link of blocks is accomplished by mining tarns.

4. Immutability:

In block chain, all information kept in the records and through every adjustment by peripheral systems; here worth of connected crypto graphical functions using secured hash key functions variations with beside lumps and information tinkering can be intermittent.

5. Currency:

With the block chain technology Cryptocurrency is implemented with support of a of using alphanumeric or simulated currency, which authorizes as through secured and prod worthy operation. Then, dissimilar removal procedures is utilized with is to develop this crypto ; finance and accounts are monitoring the utilization.

6. Consensus Techniques:

It is a technique of common agreement between all system connected in the networks. Thus, intermediaries are not compulsory to any further extent. Some methods and the novel techniques are the Authority of work, POS. The subsequent figure evaluations the significant possessions of using this blockchain technology

Blockchain Technology

It is complex to discovery new technologies similar blockchain to connecting educational organizations easily. The different advantages of blockchain can be realized as information loading, determinative assessment, smart contracts, information protection, worldwide database more. Many innovative and novel actions have remained occupied to accumulation information for scholars and abilities by instructive organizations.:

1. Single organization

Normal small Schools are using a software which offers project-based instruction as a selecting a course from college level syllabus. It has utilized blockchain to accumulation and allocate its distributed documentations to avoiding fake degree certification. To generate, credentials and saves the documents in the database, encoding and two-factor verification is applied.

2. Group of multiple institutions

As multiple group of educational institutions, the prerequisite for common certificates of documents and accomplishment archives become genuine of the degree certificates. For instance, EPFL, Boston, Delft, ANU and UBC are the group of universities that created a codeshare-like agreement on documentation. It can be uses to form worldwide suggestion among dissimilar groups of educational clusters to segment resources at low costs.

3. blockchain database

A countrywide connect to database at all regions as schools, colleges, institute, universities, examination boards, employers, relatives and others is importantly desirable. A complete supervisory frame is mandatory to sit on all and controller all for them to effortlessly run the record in system. All these problems are solution given by the block chain technology.

4. Universal Evaluation

The current valuation organization is an unwieldy process of endorsements and confirmation of records and accomplishments. The key distress is involvement of dissimilar bodies and domination of information. Block chain technology is distributed technique of storage of authorizations and accomplishments which platforms to valuation notches recording and distribution worldwide. All its requirements are that education organizations to utilize this service. It will provide universal service.

5. Blockchain and symbols

Undeveloped symbols offer interfere recorded structure to save and process the identifications and agreement with dependable authorizations on mass information. Thus, honest information can be obtainable anytime and wherever at inexpensive.

6. commercial training

The present education and aptitude development technologies are developed bungling to storage of vast quantity of information to provide training to their staffs and monitor their performances. A extra undeveloped and protected organization is mandatory to hold archives inside and outside as well.

Figure 1. Student details tracking

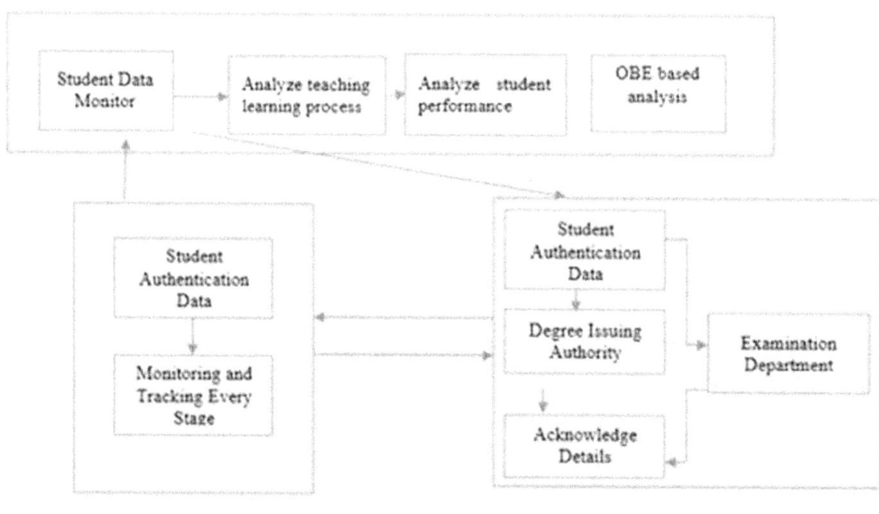

The major and prime sector in education is identifying the scam of false educational gradation and authorizations. Rendering to a connected paper false marks are presence traded in country Yimie Yong (2019). One more paper article prerogative in each 30 possible employment applicants are a false gradation holder in our country. The quantity of extra sources also emphasizes on these issues main in our nations currently. All the collected data it apparent that this is a thoughtful apprehension in teaching improvements that requirements to be began with greatest importance. In normal rehearsal, a sum of administrations needs the applicant to become their theoretical identifications confirmed earlier they can apply for further studies or professional jobs Wen, Y., & Looi, C. (2019). This organizes not guarantee the removal of the scam but decreases it to an excessive degree. Though, tracing and confirming applicants' identifications can be expensive and slow for university and industries similar. This types it a critical problematic in the teaching organization

Student Incapability to Select Best Job Arena

Additional tricky with teaching organizations is the incapability of the students to select an appropriate job path for them. Incapability to choose the greatest arena of education consume a longstanding consequence on societal, financial and intellectual happiness of a student. Selecting an incorrect optimal is not individually devise an adversative consequence on this discrete then it will also destructively influence of the goodness and quality of delivery the job work made by the specific person. efficiency is best value effort as not support the let humanity grow. Later, it is actual significant of a scholar to be bright to kind a precise optimal for their upcoming education(Gao, W et al., 2018).

Copying exams are one of the main and major important major difficulties in the universe of university currently. Copying substantial may simply be detached on the cyberspace(Zyskind, et al., 2015). Here mainly tops to original contravention to the types of intelligent truths to a discrete disposed to risk. To defend the honesty and truths of a person's effort i.e., article writing, novel innovative idea, etc. is major significant.

The outcome-based learning block chain contains 2 cross links: the pupils' capability link and subject linking. Here scholars' aptitude chain on behalf of completion necessities index earlier completion, feast education everywhere students' aptitude rendering to the directory, and might alter information into aptitude, pending scholars' advancement resolve the capability works(Junejo, et al., 2020). The prospectus equal links denotes that in the development of assessing scholars' consequences, the assessment of scholars' capability is used as a confirmation of development accomplishment degree. Since of those consequences harmfully disturb circumstances, the unremitting development of the syllabus has been shaped, the object of development is prospectus, the accomplishment gradation is appraised by students, and consequently, the excellence of the sequences qualified by each instructor can be recognized

Here the completion of a graduation, an instructor takes a complete checked of the pupil built with the grades of examination, in their session observation of student ability and the normal exercise, It is appraised with help on the dissimilar qualification necessities key and subject objects file mandatory by the sequence of course knowledge(Wang,, et al., 2016). When the progression originates to conclusion, once the regular of scholars' final marks beat edge, that is accomplish. The accomplishment gradation assessment of individually progression is produced knowledge in training area, and computerized gathering the appropriate information to the capability that scholars gained may be they get after the automatic estimation program.

Figure 2. OBE based Structure

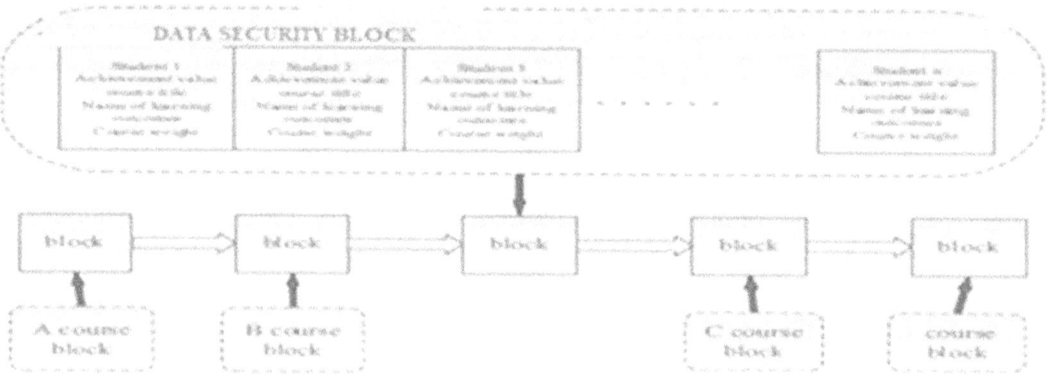

After implementation the information gathering of the graduation completion valuation, for everyone those not required to be restored over, built on the quantifiable and qualitative grouping of marks, procedure and indication to type out individually education consequence accomplishment worth of the program change provision. Then, the accomplishment worth collected with the course id, outcome-based learning name of the graduation, the weight of the progression and extra data will be designed as a best. And it resolves be uploaded into a chunk. Lastly, hash function, digital signature, timestamp and Meckler tree other methodologies will be utilized to add the data record to the blockchain

The education consequence builds on education block chain that projected in this Chapter, approves the quantifiable and qualitative method mechanism for every course, accomplishment gradation assessment of every course is not topic to the tutor's individual enthusiastic, but complete a compromise instrument, which brand it substantial(Liu, et al., 2017). First, gross the examination grades and other actual archives of student education development measurable gauges as the sustenance of accomplishment gradation appraisal, and take it as a measurable normal, which related with the quality valuation from scholars and employers to estimate accomplishment of students' capability quantitatively.

Secondly, the attainment gradation of the development is verified and inverted by the advancement necessities gradation and exercise purposes accomplishment gradation, and uninterruptedly better-quality on this base to ensure the consensus of the organization. In addition, the individual in responsibility of the estimation is the qualified documentation team but the teacher of this development. What's extra, the estimation classification of specialized occupation has been proved by authorities and establishments, and it is reliable(Cheng, et al., 2017). On the different, the occupation unsuccessful in the documentation is deficiency of compromise machine. This technique is favourable to the active communiqué amongst dissimilar universities, among campuses and business, and even in dissimilar nations and areas, what kinds the campus happenings additional open and comprehensive, and the predictable education consequence accomplished, and smooth an inaugural and distribution forthcoming manufacturing teaching scheme reassembled.

This chapter evaluates the obligatory sequence, that is, "data confidentiality", subsequently the teaching block chain is worldwide, this analysing to apply the other courses as well.

Basically, instructors' goal at resolving complex engineering difficulties, and sequentially converse the proposal of program purposes, schoolroom goals and training events, the training intergalactic is utilized to association the online and offline lessons to carry out the OBE training, using computerized accomplishment estimation software to accomplish the scrutiny of program aim to accomplishment gradation, which effectually improving students' performances capability to resolve difficult engineering problems. Table 1 shows the succeeding is the "data confidentiality" assessment in every project that from 10 identified students.

Table 1. Evaluating marks based on OBE

Number	Class performance	Spatial performance	test	Home Work Grade	Total Score
1	90	80	94	90	90
2	85	95	85	95	75
3	92	90	90	75	70
4	96	85	95	70	65
5	86	90	75	90	80
6	90	95	70	80	90
7	75	75	90	90	90
8	70	70	80	90	85
9	65	90	90	75	92
10	80	80	90	70	96

EDUCATION MOVING TO THE BLOCKCHAIN OPPORTUNITIES

Blockchain-Based Gradation Confirmation

To avoiding the danger of fake gradation certification, a blockchain based confirmation scheme can be implemented. This technology will encompass of a barcode or QR cryptogram it can be scanned by the investors to regain the material affecting with gradation. In this data may contain all student particulars of the specific laterally the completing issues of the academic record. Also avoiding the gradation scheme, it resolves marks issues the aggravation and price of gradation confirmation by advanced instruction establishments. Furthermore, the scheme determination make definite that hypothetical authorizations are they applicant is carefully and strongly kept. The completion of gradation container not necessity to fear about trailing the formerly confirmed certificate.

The organization might similarly to be utilized for additional short developments and learning documentations developed by a discrete. where situations from the standpoint of dissimilar investors i.e., students, employees, workers, parents etc.

Even though the statistic that block chain has exposed it probable for a teaching situation, here more and several contests that prerequisite to be measured after engaging block chain technology on teaching arena. we have briefed those problems into more standard groups. Though safety is the key concept of the block chain technology, in the danger different harmful malicious occurrences cannot be detached. It is very tough to offer safety and maintain confidentiality at the simultaneous time, and this problem develops more critical when an individual's job is at danger. To certify confidentiality, several blockchain outlines uses private keys and public keys. Though, blockchain cannot assurance transactional confidentiality due to the information of every key are everyone to noticeable and, in this method, consumer's communications can be connected to expose employer's data. Satisfactory storing and safety of private keys of every member is also a safety problem that would be cordially handled. Data outflow is a security anxiety and can happen due to common informs of data that would be measured. The measured speediness blockchain communications, known as the block chain's scalable problematic, is one of the questions that can be challenged in education organizations when utilizing the block chain technology.

Pupil Data and Performance Tracking

Monitoring and storage pupil academic information on a block chain are a more different advantage. From schools and college onwards they support the information analysis on estimates the zones where a particular person has been accomplishment and the zones the particular person undergraduate might not possess complete from scratch. It will helps apprentices and custodians that may know the probable of pupil and help them make apposite pronouncements for upcoming. Subsequently most the apprentices will opt for the profession tracks most suitable with their knowledge levels, so, the sum of positive specialists in all arena will surge. The growth of the efficiency in every areas and support the culture grow. Furthermore, the investigation may also allow the instructors, innovators in cluster the pupils in programs/ for actions conferring their capabilities. Here alliance of researchers on the source of the capacities may guarantee a improved and considerable further capable education atmosphere.

Block chain technology may be utilized to organizer the distribution of copy righted quantifiable crosswise the network. The main purpose of this Method is the protected database of the data verified with a link. As more, information inside the link may not be reformed physically subsequently the progressive encoding procedures shield it. This will the theoretical constituents reachable but protected and unchanged. Each procedure of satisfied is verified into the link, and the holder can simply regulator the access. The consumption can be traced connected and tenure is effortlessly verified

CHALLENGES OF BLOCKCHAIN TECHNOLOGY WITH INTEGRATION IN EDUCATION

Block chain has a sum possible uses of convert in educational structure for improved. Appropriate and simply available teaching is the support for some country to bourgeon. In this repute, it container be securely supposed that block chain systems can progress the entrée to education and affecting procedures for individually discrete as already conversed in the chapter. Though, the knowledge of block chain has the predisposition to develop to teaching structure, though, earlier the skill is positioned in the instruction area, there are numerous tasks to be measured. It may encounters embrace, it may not incomplete the subsequent:

1. Student data Alteration

Blockchain is unchallengeable by countryside, i.e., if the information has been deposited on the link, it cannot be improved. If blockchain is applied in teaching segment, the feature of immutableness can performance as two-edged sword. It eliminates the opportunity of adjusting educational record for authentic explanations for approximately students.

2. Confidentiality and Safety

Later Blockchain is a community record, the connections in blockchain are reachable to all of its accomplices. This sources privacy topics for the entities as their individual information and all speculative data is accessible for others to interpretation and entree. Furthermore, loss of isolated key incomes the client will be incompetent to privilege his/her speculative identifications and other connected data. This is a serious problem to be measured.

3. Scalability and Difficulty

Maximum block chain take really great storage expenses, subsequently the perfect block chain requirements to may be warehoused on every system in the link, tremendously huge drive procedure, outstanding to the calculation control essential to procedure the cryptanalysis. Storage pupil information wealth resolve acquire difficult expenses then resolve also growth the scalability difficult of the block chain systems(Gao, W et al., 2018). Moreover, a main

methodological blockade is computational density. For instance, the characteristic Impervious of Work compromise machinery wildernesses dynamism and has a unfortunate presentation in relations of quantity of connections each subsequent, which the price an additional expenditure, and hamper its request in universities and new instructive organizations.

RESEARCH OUTCOMES OF EDUCATION IN BLOCK CHAIN PLATFORM

The novel information documented in teaching block chain platform per month the of the student reports of analyses ultimate time period of the final conclusion of the semester, In Table 2 shows there was a consider the highest and huge in verified information, the information accounts that are reorganized in finalized performance analyse these technology are exposed in below, it embraces the maximum period, session, pupils' quantity, sequences, accomplishment needles, accomplishment gradations and extra data. The education consequences in this technology of block chain.

Table 2. Different courses block chain records

COURSE NAME	Index Points							
COURSE NAME	1-1	1-2	2-1	2-2	3-1	3-2	4-1	4-2
INFORMATION SECURITY	0	1	1	0	1	1	0	1
DATA SCIENCE	1	0	0	1	0	0	1	0
ARTIFITIL INTELIGENCE	0	1	0	1	0	1	0	1
MACHINE LEARNING	1	0	1	0	1	0	1	0
CLOUD COMPUTING	1	1	0	1	0	1	0	1
DIGITAL SYSTEMS	1	0	1	0	1	0	1	0
IOT	0	1	1	1	1	1	0	1
VLSI	0	1	0	1	0	1	1	0

Table 3 shows a mini-graduation of a convinced pupil those are completes the degree prerequisite of degree certification. In their mainly analyse the student performance associated with general graduation, the mini-graduation of capability and outcome based education developed by the stored in the teaching block chain system consists keys of degree necessity resulting of from notches 4 years of engineering. In this structure of education procedure and approval may be confirmed, and therefore the obvious of the modification among pupils.

This is a major difficult problem for block chain technology, which its implements with bitcoin methodology may be implemented with individual five to nine connections per second. Educations display that this problem can complete the delay of the block chain growth in teaching. Several researchers are focusing being done to fix this problem and many academics are focusing have suggested some suggestions and ideas such as: However, scalability complicated problems issues mainly need to be addressed effectively previously blockchain can be accepted on a large scale. Blockchain is an emerging technology that must be integrated into the bequest structure. However, the cost of this introduction and execution can be actual high. In addition to these application costs, the transaction or computation costs of many

Table 3. Ability engineering student

Requirement of Degree	Index Value	Target Value	Result Value	result
Engineering knowledge	1-1	2	1.45	Not Reached
Problem analysis	1-3	2	3.2	Reached
Design/development of solutions	2-4	2	3	Reached
Conduct investigations of complex problems	1-3	2	2	Reached
Modern tool usage	3-4	2	1.8	Not Reached
The engineer and society	4-8	2	4	Reached
Environment and sustainability	3-9	2	3	Reached
Ethics	4-6	2	2	Reached
. Individual and team work	7-8	2	2.5	Reached
Communication	4-6	2	4	Reached
Project management and finance	5-9	2	2.5	Reached
Life-long learning	4-6	2	3	Reached

blockchain technologies are also high. As mentioned above, the cost of managing and storing such large student data will also increase as the block size growths as the number of users increases. Without managing these growth and functioning costs, University politicians or other decision makers should consider what the blockchain can trust for how long. There is similarly a privation of transparency in significant the limitations of complete authorized persons. Some companies are not required to utilize the blockchain technology for every business types. A core task is conventional deprived of substantial the possible advantages of this technology in old-style structures. In their major key to response who sets the restrictions, based on the institution really required a updating technology, and which way the more blockchain technology processes this institute should take on. Early government or higher education system are collaboration with their depending all private sector on monitoring agreement policy may be set in bound for blockchain acceptance to implements in the teaching segment. One more main difficult problem through these technology at consistent in moment is the poor usability and the utilization is not effective in the education system of its harvests, particularly the former. There is main a some gap and deficiency of clearness about the vocabulary how to utilize and the supposed no one knows irresponsibility of the skill. In addition, there is so several multipart situations difficult to understand and that the employer can need the save for safety explanations, that as: B. Prilimary key, public key, and recapture spores. Blockchain has a completely novel technology that is challenging for educational users to appreciate. There is a need to improve usability through simple blockchain product interfaces so that non-technical people can simply appreciate and utilize the structure. Therefore, more enhancement for utilizing this technology work is required in the specific area of utilizabulity difficult in the arena. Worthy design and simple technology helping with blockchain.

CONCLUSION

Teaching is single of the significant zones of motivation in furthermost established nations as development in other segment hinges on the equal of teaching. Irrespective of the noticeable development in technology, there is silent a essential to motivation further on emerging humanoid sources complete education, and blockchain varieties this easier to accomplish. Comparable to the extra segments that presently exploit the blockchain of the attractive in the facilities and procedures, Here easy forecast in what way it will effort the teaching segment. Although more possible of a huge usage in teaching segment, this may profitable to producing stimulating to identify the usage.

Then, education outcome is straightforward of common acknowledgement and acknowledgment transmission of advanced education universally and the footstone of internationalization of complex teaching Learning outcome might come from dissimilar informative establishments, working preparation understanding, on-line education and additional education procedures. Hence, using this blockchain technology monitoring the student performance continuously also focusing the outcome based education student performance is analysed with focus in the student where the performance is week. the preceding locked education actions converts the protected type now, it is appropriate to size communal institution structure not incomplete by every time and changes are developing requirements of period in the classical of novel finances it is categorized by segment finances

ACKNOWLEDGMENT

We are thankful to Dr Swarnalatha P, Associate Professor, Scholl of computer science and engineering, Vellore Institute of Technology, for supporting us to complete a chapter and for giving valuable suggestions

REFERENCES

Alammary, A., Alhazmi, S., Almasri, M., & Gillani, S. (2019). Blockchain-Based Applications in Education: A Systematic Review. *Applied Sciences (Basel, Switzerland)*, *9*(12), 2400. doi:10.3390/app9122400

Chen, W., Xu, Z., Shi, S., Zhao, Y., & Zhao, J. (2018). A Survey of Blockchain Applications in Different Domains. *Proceedings of the 2018 International Conference on Blockchain Technology and Application - ICBTA 2018*. 10.1145/3301403.3301407

Chu & Yong. (2019, November 28). *Fake degrees: A worrying trend in Malaysia*. Retrieved from https://www.thestar.com.my/news/nation/2019/05/06/a-worrying-trend-in-msia/

Fake degrees in Malaysia offered online from S$4,000. (n.d.). Retrieved from https://www.todayonline.com/world/fake-degrees-malaysia-offered-online-s4000

Gao, W., Hatcher, W. G., & Yu, W. (2018). A Survey of Blockchain: Techniques, Applications, and Challenges. *2018 27th International Conference on Computer Communication and Networks (ICCCN)*.

Hall. (n.d.). *Online cheats: The rise of 'fake' degrees in Malaysia*. Retrieved from https://thepienews.com/analysis/documentary-highlights-malaysian-students-with-fake-degrees/

Junejo, A. Z., Memon, M. M., Junejo, M. A., Talpur, S., & Memon, R. M. (2020). Blockchains Technology Analysis: Applications, Current Trends and Future Directions—An Overview. *Intelligent Computing and Innovation on Data Science Lecture Notes in Networks and Systems*, 411-419.

Li, J., Jin, M., Wang, L., Qin, B., & Wang, K. (2016). MDS clinical diagnostic criteria for Parkinson's disease in China. *Journal of Neurology*, *264*(3), 476–481. doi:10.100700415-016-8370-2 PMID:28025665

Li, Y., Li, T., & Liu, H. (2017). Recent advances in feature selection and its applications. *Knowledge and Information Systems*, *53*(3), 551–577. doi:10.100710115-017-1059-8

Malik, G., Tayal, D. K., & Vij, S. (2018). An Analysis of the Role of Artificial Intelligence in Education and Teaching. *Advances in Intelligent Systems and Computing Recent Findings in Intelligent Computing Techniques*, 407-417.

Nakamoto. (2020). *Bitcoin: A Peer-to-Peer Electronic Cash System*. Retrieved from https://bitcoin.org/bitcoin.pdf

Olaniyi, E. O., Oyedotun, O. K., & Adnan, K. (2015). Heart Diseases Diagnosis Using Neural Networks Arbitration. *International Journal of Intelligent Systems and Applications*, *7*(12), 75–82. doi:10.5815/ijisa.2015.12.08

Palaniappan, S., & Awang, R. (2008). Intelligent heart disease prediction system using data mining techniques. *2008 IEEE/ACS International Conference on Computer Systems and Applications*. 10.1109/AICCSA.2008.4493524

Raschka, S. (2018, December 3). *Model Evaluation, Model Selection, and Algorithm Selection in Machine Learning*. Retrieved from https://arxiv.org/abs/1811.12808v2

Siyal, A. A., Junejo, A. Z., Zawish, M., Ahmed, K., Khalil, A., & Soursou, G. (2019). Applications of Blockchain Technology in Medicine and Healthcare: Challenges and Future Perspectives. *Cryptography*, *3*(1), 3. doi:10.3390/cryptography3010003

Vuppala, B., & Swarnalatha, P. (2020). Quantitative evaluation method of cloud resources based on work scheduling. *Journal of Ambient Intelligence and Humanized Computing*, *12*(7), 6969–6973. doi:10.100712652-020-02342-y

Wen, Y., & Looi, C. (2019). Review of Augmented Reality in Education: Situated Learning with Digital and Non-digital Resources. *Smart Computing and Intelligence Learning in a Digital World*, 179-193.

Zhu, L., Shen, J., Xie, L., & Cheng, Z. (2017). Unsupervised Topic Hypergraph Hashing for Efficient Mobile Image Retrieval. *IEEE Transactions on Cybernetics*, *47*(11), 3941–3954. doi:10.1109/TCYB.2016.2591068 PMID:28113794

Zyskind, G., Nathan, O., & Pentland, A. S. (2015). Decentralizing Privacy: Using Blockchain to Protect Personal Data. *2015 IEEE Security and Privacy Workshops*.

Chapter 12
Visual Blockchain for Intelligent Surveillance in a Smart City

Kasun Moolika Gedara
Auckland University of Technology, New Zealand

Minh Nguyen
Auckland University of Technology, New Zealand

Wei Qi Yan
Auckland University of Technology, New Zealand

ABSTRACT

Visual blockchain takes the idea of how to output accurate intelligent surveillance data by using computational methods in a smart city. Intelligent surveillance provides location-based visualization from different cameras which are deployed in the smart city. It may result in limited tamper-resistance of the gathered data from cameras. The decentralized nature of blockchain is to integrate video frames as data output by surveillance systems over the smart city by the way of selecting the most appropriate cryptographical algorithms using blockchain prototype to create sustainable computation method of visual blockchain and make reliable solutions for the existing solutions. In this chapter, the contributions are to enrich the blockchain security shield by using the most appropriate cryptographic algorithm to achieve each unique task and create a web-based image prototype through the visual blockchain to minimize the gaps in intelligent surveillance.

INTRODUCTION

Modern society has been dramatically updated in every aspect of human beings. People are incentive to imagine these sophisticated and complicated technologies, which are crucial to the development of this community, bring great convenience for ordinary day-to-day lifestyle. However, with the modern technology, considering its negative and positive aspects is much crucial before the implementation.

DOI: 10.4018/978-1-7998-9274-8.ch012

Therefore, with the presence, its security gets a unique place automatically. In smart cities, the real world depends on sensitive information and its most secure repository.

Blockchain was introduced in 2009 and operated as a distributed network in the large-scale industries (Nakamoto, 2009), such as global trade, insurance, banking, distributed energy, and healthcare, etc. Blockchain is adaptively connected to smart transportation systems, food supplier management, government, identity (Singhal, Dhameja, & Panda, 2018). It has been proven that blockchain has been transplanted as a state-of-the-art technology. Bitcoin (Duong *et al.*, 2018) is a popular currency implemented for blockchain, which makes bitcoins difficult to be phony in the transactions. This bitcoin was operated since 2009 (Taylor, *et al.* 2020), which has been extended to other high-tech areas such as intelligent surveillance. On the other hand, blockchain has been developed by the smart connector to end the third party between any transactions. With blockchains, transactions or any process could be conducted without involvement of the third party. Moreover, the blockchain is able to resolve the data integrity problem that is related to medical record verification, gas and electricity monitoring system in smart cities.

The idea of "smart city" is an advanced objective of many cities around the world, in response to the expanding complexity of urban areas and the potential of associated intelligent technologies (Pramod & Sankaran, 2019). Whilst considering the surveillance system, it is also important and should be a subject of investigation in modern community at present. However, the special goal of smart cities is to ensure the safety and security of the residents by reducing crimes and accident rates (Khan, Byun, & Park, 2020).

The technology that has been widely applied to address those issues is video surveillance (Gipp, Kosti, & Breitinger, 2016). In this case, the mission is to select suitable cryptographic algorithms to create the blockchain and secure the data communication without tampering (Tian, *et al.* 2021) as well as resist any types of attacks. The kind of video surveillance designs are able to resist crimes, anomalous event detection, privacy policy breaching and many. Therefore, the recorded video footages are most important for a pretty rich assortment of purposes. But, malicious attackers, knowledgeable hackers, or other unauthorized third parties can manipulate the cameras and video repositories illegally and create unreal directions to make them useless in the events of crimes (Khan *et al.*, 2020). These types of attacks are happened potentially against the loyalty of stored data captured in intelligent surveillance.

Moreover, blockchains take the place in the tamper resistance of the saved data, the nature of decentralized chain was employed to data verification and data integrity (Fattahi, Makanju, & Fard, 2020). Hashing is one of the reliable ways that take advantage of creating the trust between each and every block in the chain.

Furthermore, cryptographic Hash function is an algorithm that maps arbitrary data to a fixed size string. Usually, Hash functions are required to meet the security requirements of one-witness and collision-resistance. There is no same Hashing value of data in this world. In order to ensure at least 80-bit security, the output length of Hash functions should be at least 160 bits. Hash function in blockchains is SHA256 that is one of the algorithms from a family of cryptographic Hash functions named SHA (Secure Hash Algorithms). Approximately, Hash functions in blockchains achieve proof-of-work (PoW), address generation, block generation (as a part of Merkle-tree paradigm), understand in signatures, pseudorandom number generation, and bridge components (e.g., Fiat-Shamir mechanism, FSM) etc.

The decentralized nature of blockchains takes use of invention of the newest trend other than cryptocurrency. At present, the research trends in intelligent surveillance and blockchain are related:

(1) A smart city collects surveillance videos and enhances the video integrity. The method related to transferring the visual data through blockchains has been proposed (Gipp, Kosti, & Breitinger,

2016) to ensure the integrity of accident videos which was recorded from a dashboard of vehicles. In the event of a vehicle accident, digital videos are collected by using built-in accelerometers, relevant videos are cryptographically hashed and recorded on distributed storage with blockchain at present.

(2) IoT devices are connected through blockchains to monitor air and water quality, transport services, worldwide goods delivery and tracking, food distribution services. The collected data will gather much accurate data timely. The methods are called as a mobility as a service (MaaS) (Anwer, Saad, &Ashfaque, 2020).

In order to ensure the integrity of the recorded videos, using the unique feature of the distributed and tamper proofing characteristics in the blockchain has been employed. In the blockchain, timestamping features are applied to verify and transfer unaltered data to a distributed repository. Similarly, the captured data from a closed-circuit television (CCTV) cameras in smart cites are further explored (Anwer, Saad, &Ashfaque, 2020).

The blockchain-based system is able to guarantee that the recorded data are stored without altered or tampered. It aids to avoid manipulating the data integrity from original videos. Because a distributed ledger of blockchain records the data with metadata (Deepak *et al.*, 2020) from CCTV system, mechanism will assist the law enforcement and clients to secure data from surveillance. As shown in Fig. 1, videos and photos are employed to create blockchains so as to check the functionality of the implemented computational method.

Figure 1. Design of the recommendation system

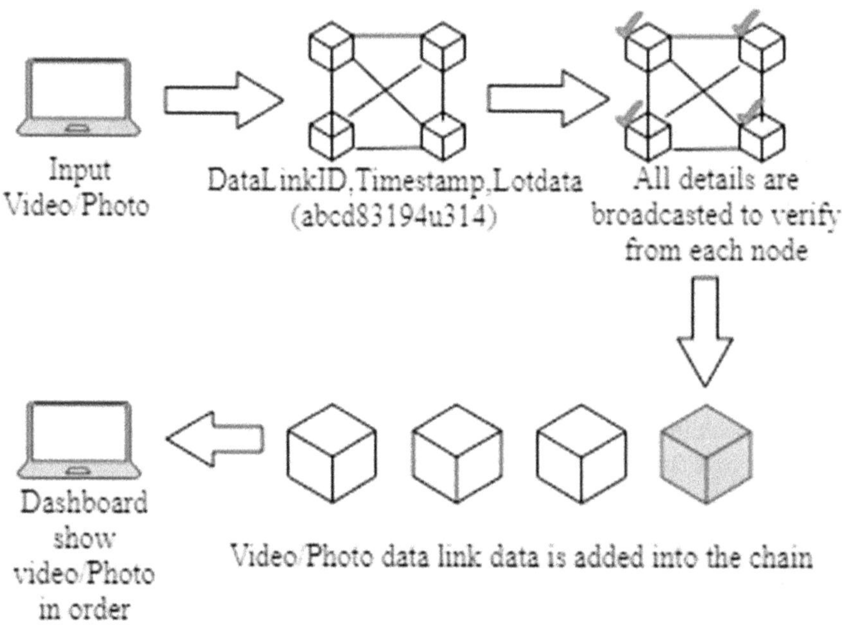

In this book chapter, following related work, the proposed methods will be iterated. The result analysis will be explicated which leads to the final conclusion of this book chapter.

RELATED WORK

Nowadays, blockchain-based development takes the pivotal place in today's world, but computer vision, intelligent surveillance, artificial intelligence, and machine learning have not combined with blockchain yet. The multidisciplinary knowledge could be combined together as the solutions effectively and efficiently.

BleddSPS is a public safety system (Xu, *et al.* 2020) with decentralized secure architecture, which supports immutability, auditability and traceability to ensure smart city safety. SD-IoD is the software-defined Internet architecture (Liao, *et al.* 2021) with smart contractor and blockchain to secure real-time monitoring systems by using drones. Maintaining the correct order of the recording data is able to ensure reliability and integrity. On the other hand, blockchain for smart cities (Hakak et al., 2020) requires obvious characteristics, which are robust, incorruptible and secure, consensus, transparency, and validation of information. The data processing time and capacity as well as resistance take part to evaluate the computational methods.

According to the analysis of past research work, blockchains connect most smart cities to continue their operations without a single point of failure (Treiblmaier *et al.*, 2020). In this book chapter, the computational methods of visual blockchains are taken into consideration, a new prototype is created for visual data storage by using blockchains, all visual data is linked together by using blockchain with decentralizing or flatting method, thus it ensures that the surveillance data as evidence chains from surveillance in digital and smart cities will not be tampered by using the computational methods of blockchains.

The Design

The main objective of this research project is to construct computational methods of visual blockchain for smart cities. Consequently, an efficient way is found to connect blockchain and select cryptographic functions so as to achieve this goal. Plentiful attacks are happened against the blockchain platform (Li, *et al.*, 2020). Therefore, the main consideration is to make the solution by combining the most appropriate cryptographical function and ensure the robust fortification than anticipated.

Visual blockchain (Dziembowski, *et al.*, 2015) is a new concept for intensifying the cybersecurity resistance in intelligent surveillance. At present, blockchain is mostly widespread among the industrial, academic, and military sections (Gallo, Pongnumkul, & Nguyen, 2018). Because an effective and efficient solution is highly dependable on the functionalities, intensive solution is needed for blockchain. As the core of blockchains, cryptography and game theory (Singhal, Dhameja, & Panda, 2018) address these three main components of a blockchain and deliver a better result for blockchain-based solutions. Moreover, the existing problems of cryptographical functions include data leaking, DDoS attacks, man in the middle attacks by blockchain (Khan & Salah, 2017). Utilizing the chosen cryptographic functions solves these problems and the other main problem of attacks against blockchains.

The goal of this book chapter is to secure the captured visual data in intelligent surveillance so as to record and link all video frames by using visual blockchain in a peer-to-peer way (Novotny, *et al.*, 2018) and increase the availability of visual data without centralized control. With regard to the surveillance in a smart city, computational methods were employed for sorting (Hu, 2019; Hu & Yan, 2020), merging, searching, and indexing the data and mitigating complicated designs by assisting the understanding the data in experimentation (Anwer, Saad & Ashfaque, 2020), such as the position and source of spectroscopic scenes. Intelligent surveillance is a kind of pervasive and invasive security methods which are accommodated to automatically analyse digital images, videos, audios, or other types of monitoring data with limited human interventions due to real-time requirements (Majdoubi, Bakkali, & Sadki, 2020). Fig.2 shows that how the proposed system works with visual blockchains and how each relevant layer in the blockchain be involved in this process.

Figure 2. The design of visual blockchain

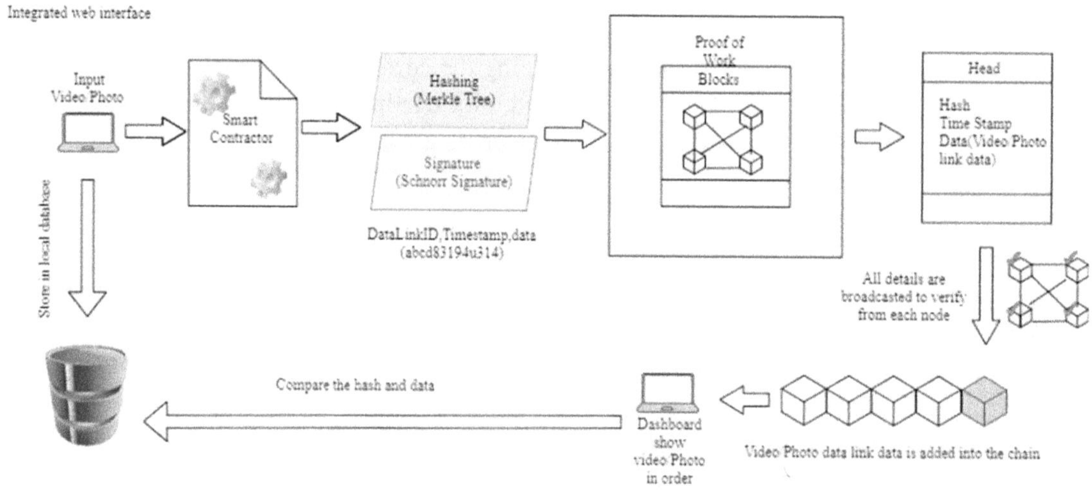

The Implementation

In this book chapter, the first goal is to select the cryptographic algorithms so as to develop the computational method for visual blockchain. Thus, the private blockchain is employed for testing most appropriate cryptographic algorithms.

Previously, a method (Hao, *et al.* 2020) was proposed to find the most appropriate cryptographic algorithm. Initially, the lightweight, security, and low energy consumption were achieved. This method was based on the communications between vehicles and control centres; therefore, the lightweight functions were considered for implementation.

On the other hand, the security problem was considered in implementation (Treiblmaier *et al.*, 2020), because of the attacks like the man-in-the-middle attacks, sniffing the transferring data between blocks is avoided (Kalbo *et al.*, 2020). Moreover, nowadays, the main focus of implementing blockchain is on low energy-consumption solution (Khrais, 2020). Thus, the best function is found for the algorithms which are relevant to the requirements.

At the first stage, the most appropriate method has been created, it is important to verify what is the correct combination to create the strong and secure computational method. Therefore, it is needed to carefully analyse the amalgamation between the selected functions. The blockchain prototype was encoded in the JavaScript to run each function and evaluated by using the parameters by selecting the most appropriate functions.

The way to implement the computational methods is figured out in this book chapter. The blockchain-based system (Xu, Weber, & Staples, 2019) needs to make decision about which part of the data and the computation should be placed on the chain or kept with the chain. Moreover, based on this explanation, the work of this book chapter is to achieve the most sophisticated implementation for the blockchain computational method, the proposed blockchain method is shown in Fig.2.

The implementation is based on the permission with less private blockchain, the implementation is conducted by using programming language Python, Golang, Java-scripts for preprograming (Panarello, *et al.*, 2018). Fig.3. shows how to select the cryptographic algorithms to develop the visual blockchain method.

Most of surveillance videos recorded with up to 30 *fps* (frame per second), in this chapter, the videos have up to 60 *fps* to get more content in the experiments. The sample videos of the Auckland city are applied to create the dataset for the experiments of this book chapter.

In the process of implementing blockchains, one of the main requirements is to ensure the data integrity (Raman & Varshney, 2018) and tamper resistance (Hao *et al.*, 2020). Furthermore, the similar methods such as Merkle trees are investigated, it unfolds that Hash list, H-Tree, and SM-Tree methods are most suitable for resolving the problem. First of all, the comparative analysis between these method is conducted to get the idea which one is pragmatic and appropriate to reach the required level of security.

Cryptographic methods and algorithms are able to be employed for blockchains, these computational methods extract the difference of video frames and solve the loophole problems. Therefore, the intention to reutilize these functions is able to achieve the high level of security for blockchain.

Moreover, by using Python programming to generate the keys and signatures, comparative understating is given for generating results and theoretical explanation, which is the most appropriate one to develop the blockchain-based applications by using the computational methods.

Signatures still keep core trust between two parties. By the way, elliptic curve cryptography takes the part of signatures in the past decades. Schnorr signature came with solving problems of the ECC, because it takes use of multiple signature instances to verify and prove the identity. The method (Neven *et al.*, 2009) shows that Hash function for Schnorr signature, also Bitcoin, takes use of the ECC for transaction throughput its limitation with the lest signature verification speed (Elbahrawy *et al.*, 2020). The digital signature scheme is a triple,

$$DS = (Kg, Sign, Vfy) \tag{1}$$

where Kg generates a public key Pk and corresponding secret key Sk for security parameter k; Sign(Sk, m) generates a signature σ based on message $m \in \{0,1\}$, and $Vfy(Pk, m, \sigma)$ outputs 1 if σ is a valid signature for m under Pk and 0 otherwise.

Figure 3. The evaluations of proposed methods

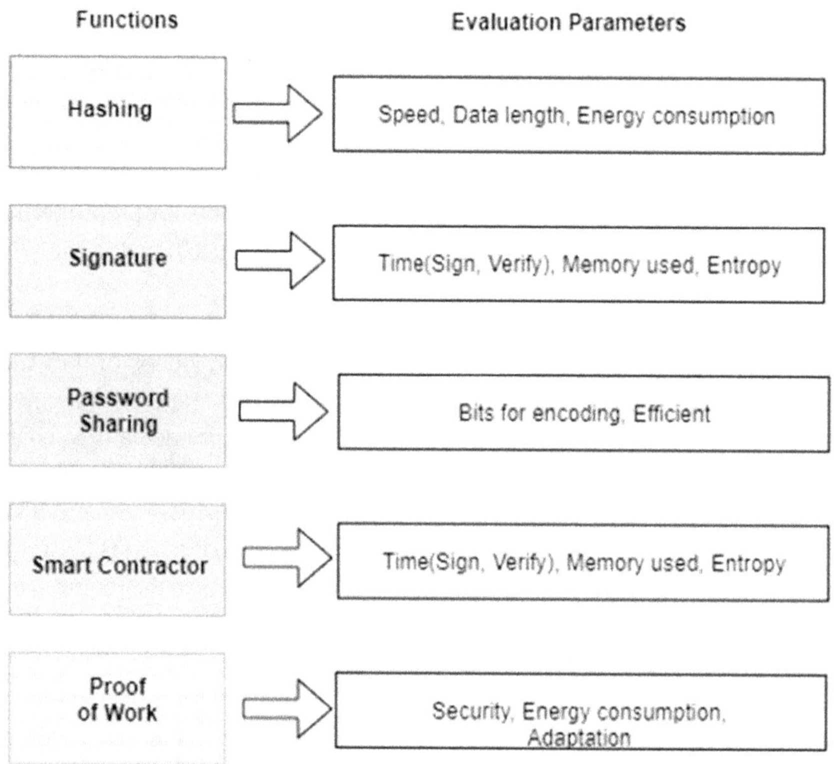

Figure 4. The selected cryptographic functions

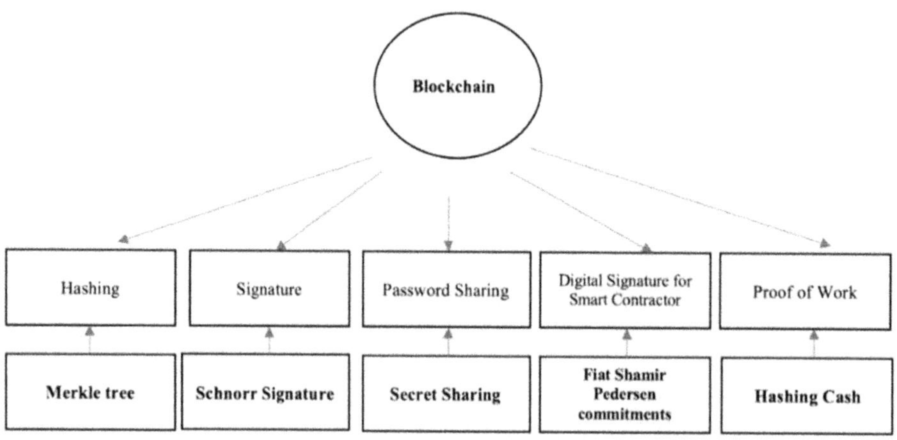

In this method, correctness is required,

$$Vfy\left(Pk, m, Sign\left(Sk, m\right)\right) = 1. \tag{2}$$

The requirements of Schnorr signature and Hashing functions are considered in this chapter. On the other hand (Maxwell, et al., 2019), a new multiple sign Schnorr signature-based method is proposed, this multiple sign protocol is one of the best solutions to enhance the security of the BApp. Schnorr signature has multiple ways to be implemented for various requirements such as SECP256K1, NIST P-256, MuSig, etc.

Compared with the existing methods without key aggregations (Maxwell, *et al.* 2019), the results show what the main requirement is for Schnorr signature which utilizes multisign method. In summary, the requirement of secure signature mechanism is fulfilled by using Schnorr signature multi-signature method accordingly.

In Shamir's Secret Sharing (SSS) for blockchain smart contractor (Harris, 2019), because secret sharing method is distributed way to share the secret for smart contractor in the blockchain. On the other hand, secret sharing with Chinese Remainder Theorem (CRT) (Shyu & Chen, 2008), RSA secret sharing, and SIKE secret sharing are the similar sharing methods implemented by various inventors.

There are different schemes that are similar to Fiat-Shamir's methods which were employed for sharing the secret between two parties without actually revealing the actual value, ZKP method has similar implementations, competitive analysis between these similar methods is conducted to figure out the best one for blockchain implementation. In addition (Gabay, Akkaya, & Cebe 2020), we take advantage of the ZKP method to create the privacy-preserving authentication scheme so as to link videos related to on-road vehicles by using blockchains, the research work aims to secure the user data among the electronic vehicles for charging at service stations. This design has been combined with zkSNARKs (Bitansky, *et al.* 2012) method to prove the knowledge in a single message transfer between one to another location.

In addition, with this method, the features of malleable are acquired. The problem was solved by using the method (Groth & Maller, 2017) with enhanced security of proof generating. From the results, the methods were identified in ZKP. In the ZKP, various implementation plans were conducted. But each of these method has different loophole. Thus, the way has been found to enhance the security of chosen methods. The ZKP is implemented by using Fiat-Shamir and connected with Pedersen commitment scheme. According to the findings of this book chapter, the improved ZKP method is secure more than others. With Fiat-Shamir ZKP method, the security of blockchain is enhanced, the analysis shows that the known value without revealing original value has been proved.

The similar method (Gabay et al., 2020) was employed for connecting electric vehicle charging station by using blockchain (Fill & Haerer, 2018). In future, this method will be employed to achieve the result by using experiments to prove this method immutable and tamper resistant. In summary, the Fiat -Shamir ZKP method is employed for this implementation.

Pedersen commitment is the necessary primitive in cryptography, in this method, the message is sent to a receiver but the receiver could not access to the message until given time period, this is the commitment between the sender and the receiver (Pedersen, 1992). In the method (Yu, 2020), the security of commitment scheme is enhanced by using Shcnorr signature to make the blockchain resistance against various attacks, Hash cash is a Proof-of-Work (PoW) algorithm (Back, 1997, 2002) in the cryptocurrencies and blockchain, the PoW is based on the consensus machines of blockchain.

Based on the outcome of this analysis, the selected structure of blockchain implementation is provided in this book chapter. Because the combination is the most sustainable one to achieve the required feature to satisfy this implementation. Thus, the focus of this book chapter is on the implementation of computational methods of visual blockchain.

In the well-designed experiments, the main part is to test the computational methods developed by using cryptographic algorithms. Therefore, it is needed to ensure that blockchain functionalities are implemented correctly with generating output into the tables by using video frames. The interface contains Hashes of each video frame uploaded to the web site that ensures each frame is able to create or update the entire chain with generating Hashes.

After sorted out the issues related to video frames, the developed interface is applied to check the functionality of the blockchain implementation. Then, the table is visualized with the video frames in the web site after this uploading. The goal of this book chapter is to generate the Hash value to video frames and identify the video frames, ultimately to resist the attacks. Multiple datasets were created with sample videos to run the experimental analysis for a number of images that can be upload into the system.

Figure 5. Video frames storing methods

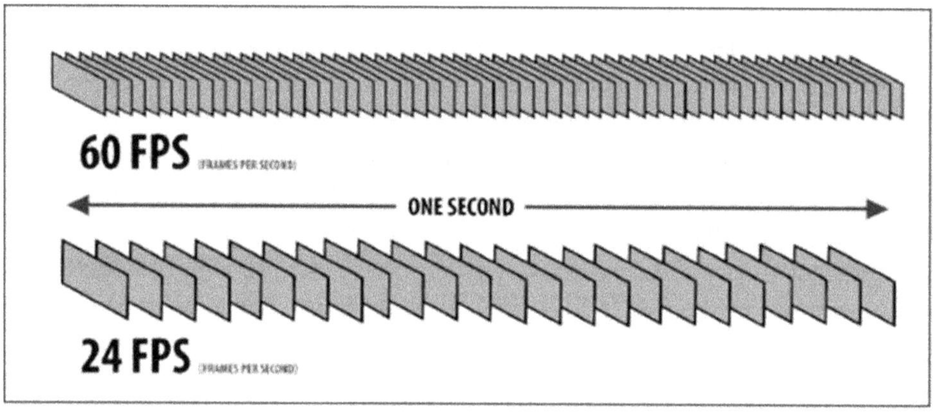

The surveillance data storing problem is resolved by utilizing blockchain-based file storing system. As a result, the verified data couldn't be changed, unauthorized access or modified after storing. Nowadays, most of the solutions came up with the cloud storing faculties. But indeed, they couldn't assure the security of store data other than backup for in case of lost. Therefore, we took a part to secure end of the tunnel by using IPFS (i.e., Interplanetary File System) a peer-to-peer network. Based on the process of the implementation, video frames go through the computational process and are stored in the IPFS network. Regarding surveillance data, we are able to observe video frame rate (*Fr*), video positions (*VPs*), number of video frames (*Vf*), and the storing address (*Sa*). The computational methods generate the Hash for the surveillance video frames with *HashCm* = Hash(*Fr*, *VPs*, *Vf*) which will ensure the integrity of the video data.

RESULT ANALYSIS AND DISCUSSION

In this project, the results of this project are obtained as shown in Fig.6. In order to continue the analysis, video frames are extracted from the road videos. This image set is uploaded into web interface for testing our system. In conventional surveillance, the video data is only uploaded to a web location. Some videos were encrypted but limited to video storing path or metadata only.

Figure 6. Web application frontend

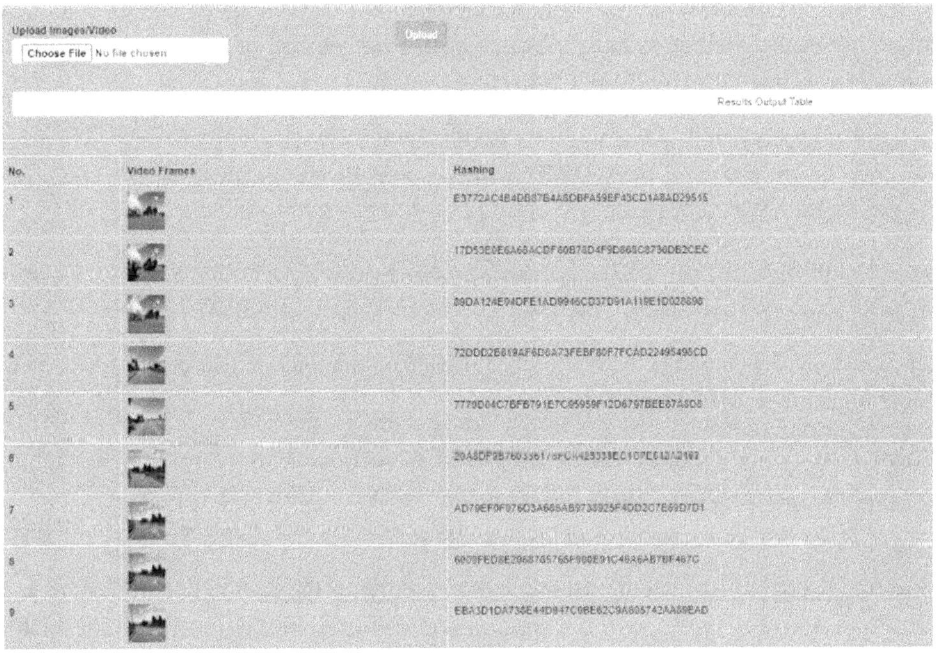

CONCLUSION

In this project, the main contribution is to identify the connection of video frames from intelligent surveillance to blockchains and enter that data into the decentralized repository for video surveillance. This is a different approach from other published work because the Hash and signature of visual blockchains are extracted based on cryptographical functions to enhance the security of surveillance data.

In this book chapter, a blockchain-based solution is presented to secure and increase the integrity of surveillance data and achieve the goals of higher loyalty, trust outcome, and uncontrolled disclosures in smart cities. With the combination of computational methods and visual blockchain, the focus is on the security of surveillance data, providing a solution to mitigate the tampering and accessing for unauthorised third-party. This project assists us closely to monitor law enforcement, insurance companies and traffic management systems, which makes necessary adjustments to achieve the video surveillance in the smart city with higher security and suitability.

REFERENCES

Anwer, M., Saad, A., & Ashfaque, A. (2020) Security of IoT using block chain: A review. In *Proceedings of International Conference on Information Science and Communication Technology* (pp. 1–5). 10.1109/ICISCT49550.2020.9079943

Bitansky, N., Canetti, R., Chiesa, A., & Tromer, E. (2012) From extractable collision resistance to succinct non-interactive arguments of knowledge, and back again. In *Proceedings of Innovations in Theoretical Computer Science Conference* (pp 326–349). 10.1145/2090236.2090263

Deepak, K., Badiger, A., Akshay, J., Awomi, A., Deepak, G., & Kumar, N. (2020) Blockchain-based management of video surveillance systems: A survey. In *Proceedings of Int. Conf. Adv. Comput. Commun. Syst* (pp. 1256–1258) 10.1109/ICACCS48705.2020.9074197

Duong, T., Chepurnoy, A., Fan, L., & Zhou, H. (2018) TwinsCoin: A cryptocurrency via proof-of-work and proof-of-stake. In *Proceedings of the ACM Workshop on Blockchains, Cryptocurrencies, and Contracts, Co-located with ASIA CCS* (pp 1–13). 10.1145/3205230.3205233

Dziembowski, S., Faust, S., Kolmogorov, V., & Pietrzak, K. (2015). Proofs of space. Lecture Notes in Computer Science, 9216(616160), 585–605. doi:10.1007/978-3-662-48000-7_29

Elbahrawy, J., Lovejoy, J., Ouyang, A., & Perez, J. (2020). Analysis of Bitcoin improvement proposal 340 — Schnorr signatures. MIT Report, 1–15.

Fattahi, S., Makanju, A., & Fard, A. (2020) SIMBA: An efficient simulator for blockchain applications. In *Proceedings of IEEE/IFIP Int. Conf. Dependable Syst. Networks* (pp. 51–52). 10.1109/DSN-S50200.2020.00028

Fill, H., & Haerer, F. (2018) Knowledge Blockchains: Applying Blockchain technologies to enterprise modelling. In *Proceedings of the Hawaii International Conference on System Sciences* (pp 4045–4054). 10.24251/HICSS.2018.509

Frisby, D. (2014) Who is Satoshi Nakamoto? In Bitcoin: The Future of Money (pp. 85–149). Academic Press.

Gabay, D., Akkaya, K., & Cebe, M. (2020). Privacy-preserving authentication scheme for connected electric vehicles using blockchain and zero knowledge proofs. *IEEE Transactions on Vehicular Technology*, *69*(6), 5760–5772. doi:10.1109/TVT.2020.2977361

Gallo, P., Pongnumkul, S., & Nguyen, U. (2018) BlockSee: Blockchain for IoT video surveillance in smart cities. In *Proceedings of IEEE Int. Conf. Environ. Electr. Eng., IEEE Ind. Commer. Power Syst. Eur.* 10.1109/EEEIC.2018.8493895

Gipp, B., Kosti, J., & Breitinger, C. (2016) Securing video integrity using decentralized trusted timestamping on the bitcoin blockchain. *Proceedings of Mediterr. Conf. Inf. Syst.*

Groth, J., & Maller, M. (2017) Snarky signatures: Minimal signatures of knowledge from simulation-extractable SNARKs. Lecture Notes in Computer Science, 581–612.

Hakak, S., Khan, W., Gilkar, G., Imran, M., & Guizani, N. (2020). Securing smart cities through blockchain technology: Architecture, requirements, and challenges. *IEEE Network*, *34*(1), 8–14. doi:10.1109/MNET.001.1900178

Hao, Z., Mao, D., Zhang, B., Zuo, M., & Zhao, Z. (2020). A novel visual analysis method of food safety risk traceability based on blockchain. *International Journal of Environmental Research and Public Health*, *17*(17), 2300. doi:10.3390/ijerph17072300 PMID:32235378

Harris, C. (2019) Consensus-based secret sharing in blockchain smart contracts. In Int Work Big Data Inf Secur (pp. 79–84). doi:10.1109/IWBIS.2019.8935853

Hu, R. (2019) *Visual blockchain using Merkle tree* [Master's Thesis]. Auckland University of Technology.

Hu, R., & Yan, W. (2020) Design and implementation of visual blockchain with Merkle tree. In Handbook of Research on Multimedia Cyber Security (pp. 282 – 295). doi:10.4018/978-1-7998-2701-6.ch014

Kalbo, N., Mirsky, Y., Shabtai, A., & Elovici, Y. (2020). The security of IP-based video surveillance systems. *Sensors (Basel)*, *20*(17), 4806. doi:10.339020174806 PMID:32858840

Khan, M., & Salah, K. (2018). IoT security: Review, blockchain solutions, and open challenges. *Future Generation Computer Systems*, *82*, 395–411. doi:10.1016/j.future.2017.11.022

Khan, P., Byun, Y., & Park, N. (2020). A data verification system for CCTV surveillance cameras using blockchain technology in smart cities. *Electron*, *9*(3).

Khrais, L. (2020) The combination of IoT-sensors in appliances and block-chain technology in smart cities energy solutions. In *Proceedings of Int Conf Adv Comput Commun Syst* (pp. 1373–1378). 10.1109/ICACCS48705.2020.9074362

Li, X., Jiang, P., Chen, T., Luo, X., & Wen, Q. (2020). A survey on the security of blockchain systems. *Future Generation Computer Systems*, *107*, 841–853. doi:10.1016/j.future.2017.08.020

Liao, S., Wu, J., Li, J., Bashir, A., & Yang, W. (2021). Securing collaborative environment monitoring in smart cities using Blockchain enabled software-defined Internet of drones. *IEEE Internet Things (Mag)*, *4*(1), 12–18. doi:10.1109/IOTM.0011.2000045

Majdoubi, D., Bakkali, H., & Sadki, S. (2020). Towards smart blockchain-based system for privacy and security in a smart city environment. *Proceedings of Int. Conf. Cloud Comput. Artif. Intell. Technol. Appl (CloudTech).* 10.1109/CloudTech49835.2020.9365905

Maxwell, G., Poelstra, A., Seurin, Y., & Wuille, P. (2019). Simple Schnorr multi-signatures with applications to Bitcoin. *Designs, Codes and Cryptography*, *87*(9), 2139–2164. doi:10.100710623-019-00608-x

Neven, G., Smart, N., & Warinschi, B. (2009). Hash function requirements for Schnorr signatures. *J. Math Cryptol*, *3*(1), 69–87. doi:10.1515/JMC.2009.004

Novotny, P., Zhang, Q., Hull, R., Baset, S., Laredo, J., Vaculin, R., Ford, D. L., & Dillenberger, D. N. (2018). Permissioned blockchain technologies for academic publishing. *Information Services & Use*, *38*(3), 159–171. doi:10.3233/ISU-180020

Panarello, A., Tapas, N., Merlino, G., Longo, F., & Puliafito, A. (2018). Blockchain and IoT integration: A systematic survey. *Sensors (Basel), 18*(8), 2575. doi:10.339018082575 PMID:30082633

Pedersen, T. (1992). Non-interactive and information-theoretic secure verifiable secret sharing. Lect Notes Comput Sci, 576, 129–140. doi:10.1007/3-540-46766-1_9

Pramod, N., & Sankaran, S. (2019) Blockchain-based framework for driver profiling in smart cities. In *Proceedings of Int. Symp. Adv. Networks Telecommun. Syst.* (pp. 1–6). 10.1109/ANTS47819.2019.9117923

Raman, R., & Varshney, L. (2018). *Distributed storage meets secret sharing on the blockchain*. Inf Theory Appl Work ITA. doi:10.1109/ITA.2018.8503089

Shyu, S., & Chen, Y. (2008). Threshold secret image sharing by Chinese remainder theorem. In *Proceedings of IEEE Asia-Pacific Serv Comput Conf.* (pp. 1332–1337). 10.1109/APSCC.2008.223

Singhal, B., Dhameja, G., & Panda, P. (2018). *Beginning blockchain – A beginner's guide to building blockchain solutions*. Jenson Books Inc.

Taylor, P., Dargahi, T., Dehghantanha, A., Parizi, R., & Choo, K. (2020). A systematic literature review of blockchain cyber security. *Digit. Commun. Networks, 6*(2), 147–156. doi:10.1016/j.dcan.2019.01.005

Tian, S., Liu, Y., Zhang, Y., & Zhao, Y. (2021) A byzantine fault-tolerant raft algorithm combined with Schnorr signature. In *Proceedings of Int. Conf. Ubiquitous Inf. Manag. Commun* (pp. 1–5). 10.1109/IMCOM51814.2021.9377376

Treiblmaier, H., Rejeb, A., & Strebinger, A. (2020). Blockchain as a driver for smart city development: Application fields and a comprehensive research agenda. *Smart Cities, 3*(3), 853–872. doi:10.3390martcities3030044

Xu, R., Nikouei, S., Nagothu, D., Fitwi, A., & Chen, Y. (2020). BlendSPS: A Blockchain-enabled decentralized smart public safety system. *Smart Cities, 3*(3), 928–951. doi:10.3390martcities3030047

Yu, G. (2020). Simple Schnorr signature with Pedersen commitment as key. IACR Cryptol ePrint Arch, 61.

Chapter 13
The Integration of Blockchain With IoT in Smart Appliances:
A Systematic Review

A. Anitha
Vellore Institute of Technology, India

Haritha T.
Vellore Institute of Technology, India

ABSTRACT

An amazing utilization of internet of things (IoT)-based smart technologies has been observed in recent years. Internet of things (IoT) became smarter and autonomous, leading the way by interconnecting and interacting with the environment. Existing technologies such as big data, AI, and cloud computing have been supported by IoT but with some limitations. Blockchain is the better solution to access and share data with the decentralized system. IoT spreads its applications in healthcare, industrial internet of things, supply chain management, traffic management, etc., out of which smart cities take special attention in recent days by integrating with IoT to create a sustainable ecosystem by using information and communication technology (ICT). Smart cities can offer various smart applications for boosting the life prominence of society. These characteristics of blockchain are beneficial to the creation of smart cities. In this regard, the authors have conducted a systematic review of articles. This study highlights the integration of blockchain with IoT.

DOI: 10.4018/978-1-7998-9274-8.ch013

INTRODUCTION

With the advancement of numerous technologies and the emergence of novel concepts such as Big Data, Cloud Computing, Cyber Physical Systems (CPS), Artificial Intelligence (AI), and so on, the future seems bright, the modern industry has been promoted to a new level (Wan et al., 2019). The expanding global population, along with the rapid growth of inhabitants in cities, has resulted in a slew of social, technical, organizational, and economic issues that threaten cities' environmental and economic viability. (Bhushan et al., 2020). Most of the sectors are introducing "smart" concepts in the environment to improvise the visible and invisible assets. Smart cities offer different applications like Smart e-commerce, smart e-voting, smart healthcare systems, smart transportation, smart governance, and so on are all examples of smart technologies. Smart cities, in this context, strive to improve inhabitants' quality of life and establish a sustainable urbanization by incorporating sophisticated information and communication technologies (ICT) into local administration and services. (Wong et al.,2020). To integrate the various technologies to enhance the infrastructure of cities. In this context, IoT combined with different new concepts like Cloud computing, big data, AI. Internet of Things (IoT) smart devices become smart and autonomous. IoT devices will be interconnected with the environment in order to collect information and mechanize certain data. Internet of Things plays a critical role in transforming contemporary cities into smart cities, electrical grids into smart grids, and homes into smart homes. According to numerous study projections, by 2020, the number of interconnected massive gadgets is expected to increase from 20 to 50 billion. (Ana et al., 2018). Traditionally, the data collection over internet of things (IoT) devices are stored on central server. It may lead to privacy challenges, lack of data loss and security. Unlike traditional methods, blockchain technology facilitates digital assets without any mediators or third-party entities. Further, threats attempting to acquire access to personal information or control over the entire system are protected by blockchain-based technologies. (Bhushan et al., 2020).

Many researchers are highlighted different directions to addressed blockchain IoT applications in smart cities. The research directions are addressed various attacks, authentication issues in various applications such as healthcare sector (Thomas et al, 2019), e-voting system. It is related to patient data leaks at some manner during cyber security attacks (Yaqoob et al., 2021) To overcome that challenges block chain can assist streamline operations in healthcare system. The authors Provides fundamental security goals based on blockchain technology in smart homes (Dorri et al.,2021). Helo & Hao (2019) discussed about supply chain management for transaction details. Taş & Tanrover (2020) discussed the challenges involved in accomplishing blockchain-based e-voting. Due to cyber-security concerns, remote voting is considered impracticable. Several examples of diverse applications are shown in Table 1.

Table 1. Addressed various issues on blockchain technology

Author & year	Smart environment	Technology employed	Contributions of the paper
(Xiaoqi et al, 2017)		Blockchain security	Addressed about various security attacks in blockchain and analyzed the vulnerabilities.
(Zheng et al,2017)		Blockchain challenges	Overviewed the blockchain features and challenges such as privacy leakage, anonymity, decentralization. Explore various consensus mechanisms to enhance security.
(Khan. et al., 2018)		Blockchain security solution	Discussed about IoT security issues and solution for various attacks using authorization, authenticate, data secure communications and enhance integrity in blockchain technology.
(Wubing et al., 2018)	Blockchain applications		Surveyed on various applications in different areas and presents future directions for researchers in blockchain applications
(Muhammad et al., 2018)	IoT based Blockchain		Surveyed various features, techniques and applications addressed in IoT based blockchain. highlighted challenges in IoT and different solutions are comparatively provided both industry and research areas.
(Nallapaneni et al., 2018)		Security challenges in blockchain	Surveyed various security challenges with distributed ledger based blockchain. Identify various interoperability, storing, privacy emerging issues.
(Panarello et al., 2018)		Integration of blockchain with IoT	Discussed about various application areas with development solution grant as data privacy and integrity, designed a system of tamperproof records.
(Hu et al.,2019)	Blockchain based smart parking	Blockchain open sources(bcos) and smart contract technology	To provide privacy of its users without reliable third-party entity. Smart contracts are allowing to search vacant. BCOS provides a one-time pad mechanism and group signature.
(Jeong et.al ., 2019)	Network security in smart cities	Block chain-based framework for security	Surveyed various technologies and different network architecture, open issues and challenges in lack of security with smart cities. To overcome that blockchain is introduced.
(Xie *et al.*, 2019)	Smart cities issues		Surveyed urban problems in the smart cities, aims to enhance the quality of life with blockchain features such as trust free, transparency, security.
(Rathore et al., 2019)	Security attacks	Software defined networking	To overcome single point of failure attacks and early detection of attacks in IoT using software defined networking, fog computing and edge computing with blockchain architecture.
(Carmen et al., 2019)	Blockchain with smart cities	Self-sovereign identity model and smart contracts	Discussion on administration infrastructure in smart cities using digital technology, identify security threats in various aspects and corresponding challenges.
(Wang et al., 2019)		Smart contracts	Surveyed the study about smart contracts, open challenges, different attacks, smart contract-based framework for future research directions.
(Ferrag et al., 2019)		Blockchain IoT attacks.	A reviewed about blockchain protocols in IoT networks with different applications, identify major various attacks and analyzed in detailed.

Continued on following page

Table 1. Continued

Author & year	Smart environment	Technology employed	Contributions of the paper
(Hassan et al., 2019)	Privacy issues	Blockchain privacy implementation	Dedicated to the integration of privacy-preserving technologies into Internet of Things (IoT) applications that run on the blockchain.
(Wang et al., 2020)		Various protocols	To reduce the possibility of criminal attacks and illegal issues in the blockchain platform under the condition of preserving privacy protection.
(Fei et al., 2020)			Surveyed access control, data storing, trusted third party, automatic payment system views to understand the benefits with blockchain IoT system.
(Gade et al., 2020)	Blockchain with smart cities	Blockchain with security	Discussed about literature of blockchain technology in smart cities and proposed safe, reliable and secure transaction using frameworks.
(Ahmed et al., 2020)		Smart city challenges	In the smart city, it addresses several obstacles and issues including as immutability, decentralization, interoperability, trust, and privacy. To investigate the problems with blockchain features.
(Saqib et al., 2020)	Blockchain security with smart cities	Three layers of architecture projects	Discussed blockchain characteristics, benefits, three layers of architecture, challenges are identified in smart cities.
(Wang et al., 2021)		Smart contracts security	Discussed about security and vulnerabilities attacks in blockchain based smart contract system. Focusing on various challenges of security, challenges.

Introduction to Internet of Things (IoT)

The term internet of things IoT was introduced by Kevin Ashton (Sadawi et al., 2021). who coined the express "the Internet of Things" to illustrate the network connecting objects in the physical world to the Internet. The IoT plays a central role in turning current cities into smart cities, electrical grids into smart grids and houses into smart homes, and this is only the beginning. According to various research reports, the number of connected devices is predicted to reach anywhere from 20 to 50 (Ana et al., 2018) billion by 2020, mainly the interconnectedness of vast devices. The IoT was firstly most stimulating to business and manufacturing, where its application is sometimes known as machine-to-machine (M2M), but the importance is filling our environment with smart devices and transforming it into related to nearly everyone. some tech analyst predicts that total there will be 41.6 billion connected IoT devices by 2025. IoT are growing faster sector and building automation system in enterprises, businesses, Industrial internet of things and smart city etc. the benefits of the IoT in business is quickness and efficiency, they can access more data about their own products and ability to make changes as a result. Industrial internet of things or fourth industrial revolution aim is to use a grouping of sensors, wireless networks, big data, AI and analytics to measure and optimize industrial processes. The benefits of IoT in consumers are to make our homes or environment as smarter, more measurable. Smart speakers like amazon echo and goggle home etc. smart lighting when we are in out automatic light is activated in home. Home security system, self-driving cars etc. Security and privacy are one the major issues with the IoT, while sensors are collecting extremely sensitive data. For example, take our smart home who visits you and passes by your house- thanks to your smart doorbell, listen music -thanks to smart speakers etc. In smart city, IoT devices are produces large amount of real time data and require processing. In IoT advancing technology will produce large amount of data in the context of smart cities have major challenges such as bandwidth, latency, scalability, security and privacy issues (Singh, 2021). IoT has its applications in smart home security system, cyber security and lot more. Application of IoT in smart home security system as discussed by (Anitha, 2017), and cyber security system as discussed by (Anitha, 2016). Cyber defense using Artificial system using IoT was disucced by (Anitha et.al,2016). (Anitha,2017) also discussed about the garbage monitoring system using IoT.

Basics of Block Chain

Blockchain, at times referred to as Distributed Ledger Technology (DLT), makes the record of any digital asset unalterable and transparent through the use of decentralization and cryptographic hashing. It is a structure that stores every transactional record known as blocks; each block has cryptographic hash value using former value links are formed as chain. Each block is cryptography linked to previous one, forms blockchain. Blockchain is an especially shows potential and innovative technology because it helps reduce risk, stamps out fraud and brings transparency in a scalable way for numerous uses. For simple example a comparison for understanding blockchain technology is a Google Doc. When we create a document and share it with a collection of people, the document is distributed instead of copied or transferred. This creates a decentralized distribution chain that gives everyone access to the document at the same time. No one is locked out awaiting changes from another party, while all modifications to the doc are being recorded in real-time, making changes completely transparent. Of course, blockchain is more complex than Google doc but it creates some fundamental ideas of the technology.

Blockchain was described in 1991 by the research scientist Stuart Haber and W. Scott Stornetta. He works on computational practical solution on digital documents as a result, not supported for back dated and tampered. In 2008, blockchain technology was combined with several other technologies implemented with modern crypto currencies. In 2008, Nakamoto jumpstarted with expansion of digital currencies using electronic cash system. Resolve the problem of double spending using Proof of Work (Nakamoto, 2008). The blockchain idea was combined with several other technologies and computing concepts to create modern crypto currencies: electronic cash protected through cryptographic mechanisms instead of a central authority (Yaga et al., 2019). Block chains are tampered evident and tamper resistant digital ledgers implemented in a distributed fashion (i.e., without a central repository) and usually without a central authority (i.e., a bank, company or government). In 2008, blockchain came into existence with the name of bitcoin for digital concurrencies. Bitcoin is based on the blockchain.

The following example explains how the blockchain technology works. Let's pretend that X and Y are two entities in a blockchain-based parking system, and that X is paying the parking authority, Y. This transaction is represented online as a block that contains information like as the block number, proof of work, previous block, and transaction records, and this block is broadcast to all network entities. The block is verified by the other entities, and if more than half of them approve it, the transaction is validated and put to the chain. The fee is then sent from entity X's account to authority Y's account. (Biswas & Muthukkumarasamy, 2016).

Block Structure

A blockchain is a collection of interconnected data chunks. A ledger page can be used to compare each block. Each block is made up of a variety of different elements. It's possible to tell the difference between the block's head (block header) and his body (block body)
Block header: The head of the block is divided into six components.

Table 2. Components of block structure

Headers attributes	Description
The version number of the software	Validation rule for protocol signals
The previous block's hash.	Hash of previous block, to create new block
The Merkle tree's root hash.	All transactions can be aggregated in a hash
Since 1970–01–01 T00: 00 UTC, the time has been measured in seconds.	A time stamp
The current difficulty's goal	Compact representation of the current hashing target.
The Nonce	A 32-bit field starts at zero and works its way up to every hash function.

The block body can be thought of as a truck's loading area. It contains all transactions that the block has confirmed. The miners successfully verify and validate a transaction before it is included in a block. All the transactions can be validated by nodes in the network, after this the new block can be created with the help cryptographically hash pointer and added into the existing blockchain in the network (Bhushan et al., 2020).

Figure 1. The blockchain structure display chained blocks with header and body fields

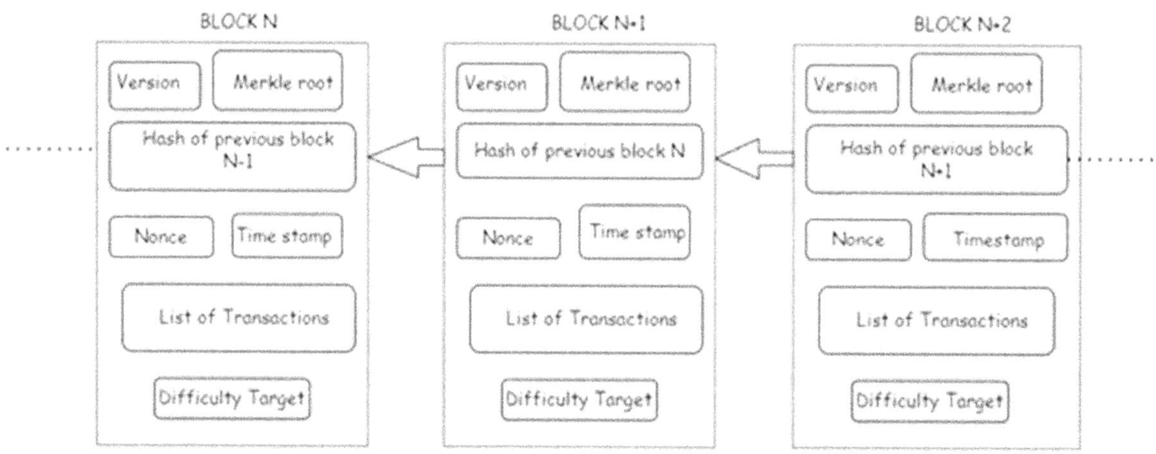

Types of Blockchain

On the basis of control mechanisms and authentication, blockchain can be divided into four categories: public, private, consortium, and hybrid.

Figure 2. Various categories of Blockchain

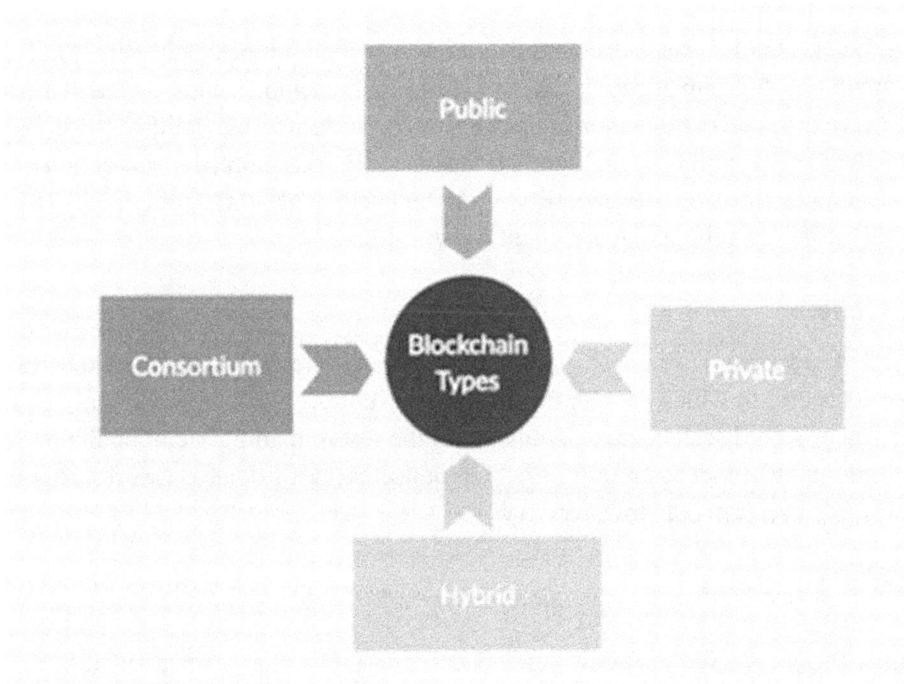

Public Blockchain

A public blockchain is a permissionless, non-restrictive distributed ledger technology. It's an open-source platform any one can perform operations in blockchain. A node or user which is a part of the public blockchain is authorized to access current and past records, verify transactions or do proof-of-work for an incoming block, and do mining. The most basic use of public blockchain is for mining and exchanging cryptocurrencies. In terms of solving such challenges, Proof-of-Work (PoW) consensus procedures are the most efficient. (Bhushan et al., 2020). Advantages are trustable, open and transparent for in public blockchain. However, proof of work is time consuming process for huge amount of data and lower transaction rate can occur. An example of public blockchain is Bitcoin, Ethereum and Litecoin.

Private Blockchain

A private blockchain is a permissioned or restricted blockchain that can only be used in a closed network. Private blockchains are typically utilized within a company or organization where just a few people are allowed to participate in a blockchain network. The controlling organization determines the level of security, authorizations, permissions, and accessibility. The mining process on a private blockchain is controlled by a single person or a dedicated team, limiting access to unknown or new users until they are invited by some governing authority. (Khan & Murad et al.,2016) Practical Byzantine Fault Tolerance (PBFT) (Helo & Hao 2019). However Private blockchain networks are controlled by central authority. Voting, supply chain management, digital identity, asset ownership, and other applications use private blockchain networks. Hyperledger projects are one example.

Consortium Blockchain

A consortium blockchain is a semi-decentralized kind in which a blockchain network is managed by multiple organizations. A group of people takes on the task of reaching a consensus and blocking validation choices. More than one organization can act as node and do mining, validate the transactions. The major drawbacks are vulnerability against tampering attacks. Furthermore, the transactions are controlled by blockchain immutability and irreversibility of the blockchain network from threats (Bhushan et al., 2020). Normally used in banks and government organization.

Hybrid Blockchain

The term "hybrid blockchain" refers to a combination of private and public blockchains. Users may regulate who has access to which data on the blockchain in this hybrid network. Only a portion of the blockchain's data or records can be made public, with the rest remaining secret on the private network. Through multiple public blockchain, all users are connected in a private network. This improves the blockchain network's security and transparency.

Consensus Algorithms

It's a mechanism that allows all blockchain network users to reach a shared understanding (consensus) on the current state of the ledger's data and trust unknown peers in a distributed environment. The challenge, however, is in developing a system that allows the trusty generals to achieve an agreement in a trustless environment by identifying diverse answers through consensus. In the applications of blockchain can solve the double spending and byzantine problems, the method verifying transactions by many distributed nodes in the network. Consensus algorithms are major scenario in blockchain technology and analyzed how to make a performance better (Mingxiao et.al, 2017). Therefore, various algorithms are ensuring among ledgers as shown in Table 2.

Table 3. Consensus algorithms

Consensus algorithm	Author & year	Description	Limitations
Proof of work	(Wang, W. et al., 2019; Memon et al., 2019)	Solves the computational puzzles to figure out the result	High energy consumption and long processing time
Proof of stake	(Saleh,2019; (Andreina, 2017)	Getting chances for block validation based on stake system, virtual currency is held, implemented protocols to enhance secure.	Currency paid strategy
Proof of capacity	(Bhushan et al., 2020)	consume memory space for nodes	More space requires more stakes
Proof of activity	(Bentov, 2014)	Fast transactions based on stake	Scalability issues
Proof of burn	(Bhushan et al., 2020)	Crypto coin spending solution	Resource wastage
Delegated PoS	(Larimer, 2014.)	'Voting' is used by every node with a stake in the network.	Limitations on the amount of people who can own a token
Proof of Elapsed Time	(Chen, et al., 2017)	To estimate its waiting time, each node creates a random number.	Every time, the same nodes are elected as the Leader.
Practical Byzantine Fault Tolerance	(Wang, X. et al., 2018; Gramoli, 2020).	Voting based decisions are consider to prove the data integrity, identify attacks in bitcoin and Ethereum.	High consumption overhead
RAFT	(Chaudhary et al., 2018)	To elects leader based on voting consensus and perform logs for consistency in BFT environment	low security

CHALLENGES OF IOT AND BLOCK CHAIN

Apart from various technologies, implementation and deployment in smart cities also face quite a few technological challenges are

- In smart cities have numerous data collection and analysis from various entities. To improvise the security, reliability and data integrity are utmost importance in city management.
- Complexity issues: now a day's number of devices is increasing exponentially in smart cities. In traditional system collection of all devices data is becomes more complex and creates network issues. It takes long processing time for transferring the data stored.
- Data Security: One more challenge is security and transparency while transferring the collection of data such as public personal data, organization data and IoT data. To improvise the smart city administration and provides security. which aims to prevent unauthorized access.
- Trusted environment: it is essential to sharing data, analyzed and processed data in IoT systems. While transferring the data through network may cause disruption to its network (Minhaj Ahmad Khan & Khaled Salah 2018).
- Data privacy is identified as key challenge in the smart cities, this is because the IoT devices in smart cities capture huge amount data are storing and further analyzed the sensitive information

leads to privacy concerns. Data can collect without the user permissions can cause major issue. Without taking the permissions of data owners may rise transparency issue. Anonymity is one more challenge in IoT devices for path communication and track back the user locations (Sookhak, Tang & Yu, 2018). Again, a systemic study can explore the various threats in smart city privacy concerns and various challenges such as zero of knowledge proof, data minimization, differential privacy, private information retrieval (Eckhoff &Wagner 2018).

- Multi agent collaboration which means operated single entity in IoT system when peer communications were collaborated mutual trust cost is high (Fei Chen et al., 2020).

Due to these flaws, blockchain technology can enable transparent city management, assure data integrity, support cooperative decision-making among individuals and organizations (e.g., universities, hospitals, corporations, national and local government), and stimulate the adoption of a democratized smart city.

NEED FOR INTEGRATING IOT WITH BLOCK CHAIN

Blockchain is an emerging technology with many advantages in an increasingly digital world. The record of transactions is kept in a peer-to-peer network. There are no central authority needs to confirm the transactions. The transactions are verified by the participants in peer-to-peer networks. There is no need for any central authority for any kind of trade settlement, voting, or money transactions. The challenges of IoT blockchain technology are

- Security: Security is the main concern for all kinds of online activities. Blockchain provides a high level of security which makes it impossible to breach for anyone because of the decentralized nature of Blockchain. As finding single point of failure is deadly task, in network to overall security should be enhanced (Bhushan et al., 2020). One of the contributions are cryptographic block structure. Blocks are connected through previous hash value embedded in block header of the block using hash algorithm. If any hackers want to access the data in any block, they can be recalculated with the previous successive hash value in blocks it is not possible task. If any recalculated hash values, cannot altered or added any data in to the blockchain because every content or transaction details make a copy in the blockchain network (Swan, 2015).
- Decentralized system: Transactions are inherently authorized or trusted in a traditional centralized system via central trusted intermediaries. In blockchain technology, transactions are done without third party entities. There is no centralized system. All types of blockchain system are fully decentralized and all the participants in the network can have copy of transaction details.
- Transparency: The blockchain technology is very transparent as everything is visible to all the participants in network. One can see each and every transaction in decentralized network which makes it very open technology. Everyone has copy of all the transactions in the network (Majeed et al., 2021).
- Immutability: blockchain has made immutable and secure using cryptography. All the transactions are signed and verified cryptographically. Data blocks are securely linked via hash value. One which requires multiple parties to make their own independent decisions.
- P2P exchanges: In IoT all the data can communicate through nodes or gateways and route to store in cloud server (Gui et al., 2016). There are different paradigms are used in nodes at same levels,

as fog computing. To enhance the peer-to-peer communication in blockchain without any gateways (Suárez-Albela et al.,2017).

- Privacy: Another feature of cryptographic structure using public and private keys in blockchain. To improvise the authentication process, only authorized persons can allow in blockchain (Fernández-Caramés, Tiago & Fraga-Lamas, Paula. 2018).

RECENT RESEARCH DIRECTIONS ON INTEGRATING BLOCK CHAIN WITH IOT

The wide spread and expansion of IoT applications in various areas carried significant issues. The IoT deployment contains heterogenous devices which has uniquely identifiable, low power and limited processing (Minhaj Ahmad Khan and Khaled Salah 2018). IoT can integrated with different technologies such as information and communication technology, automated sensors, cloud computing, artificial intelligence and data analytics. The integration of such technologies increasing the complexity of IoT expansion (Sadawi et al.,2021). As a result, all the technologies are adopted traditional architecture (central server), while authenticating a huge amount of data from devices causes trust factors. The structure might have faces data privacy. Integrity, tamper records issues with IoT. Also, the central server architecture has features in data to tamper or modified the records according to their interest in the entities while transferring the data in network. IoT can produce massive amount of data that needs to be stored, communicating and processed data that requires high energy consumption and storage capacity. Privacy and security issues are extremely important factor in transit and processed the data. Lastly IoT has increasingly using cloud computing for storage and upgraded the data. IoT centralized severs have huge amount of device data that performs tasks such as authentication, analyzing and storing. To build the trusted environment in constrained IoT devices. Based on above addressed security, privacy, storage, trust issues, blockchain technology was proposed in IoT environment to the solution of all the challenges in IoT paradigm. Blockchain providing all the capable solutions for facing challenges in IoT such as decentralized platform, building trustful environment, security using hashing techniques, privacy implemented cryptographic keys, traceability and transparency (Sadawi et al.,2021). Some of security issues are addressed in IoT such as energy efficiency, availability of services, single point failures, IoT protocols and standards in gateway communications. Security attacks are categorized into various levels. To provide blockchain solutions, secured communications while eliminating the standard protocols and key management distribution instead of using asymmetric key pair are connected to the network. Light weight protocols are used for memory space. In IoT one more challenge is trusted environment to develop that authorized identity relationship, transparency in every transaction, traceability using blockchain. Data authentication and integrity are bringing the secure transmission of data, every data should be recorded in distributed ledger. Authentication, authorization and privacy are implemented in decentralized authentication process, authorization and privacy can be ensured by using smart contracts (Minhaj Ahmad Khan & Khaled Salah 2018). The integration of promising technologies like IoT with different technologies such as cloud computing has verified invaluable. Whereas features of blockchain are key benefits to enhance the IoT deployment like security, privacy, reliability, autonomy etc. Additionally, IoT interactions are enabled with blockchain such as IoT-IoT which has low latency communication between IoT through blockchain. IoT-blockchain which has all the transactions are recorded in immutable ledger and develop autonomy of IoT devices.

Lastly hybrid design to implement which data is stored in blockchain rest of the data is stored in fog computing it is used for potential changes. Authors provides the feasibility of blockchain nodes with IoT applications (Ana Reyna et al., 2018). Data management issues in IoT based on service-oriented approaches (Missier et al., 2017), to handling large amount of data using IoT mobility (Li, T et al., 2012) additionally suggest that partially decentralized using clusters, sub servers to enable scalability (Lu, T. et al.,2015). Blockchain features for data management issues are leverage the auditability of sharing the information without any third parties. The proposed solution is built on three ways a) data stored through cloud using decentralized hash tables b) data stored through blockchain c) IoT end devices data (Shafagh et al., 2017). Another feature to implement IoT data management issues are using immutability and auditability of storing data. all the data are distributed databases it brings tamper proof records in blockchain (Liang et al., 2017). Maintaining and storing the data in medical records using frameworks for hosting all the data through blockchain. To reduce the storage requirements for storing all the transactional details in off chain (Missier et al., 2017). Using Ethereum blockchain frameworks for securing and transferring the data in blockchain (Azaria et al., 2016). Another goal for IoT based blockchain is security, as expansion of IoT networks to provide protection against threats. IoT edge nodes increases points of failures attacks for example DDos (distributed denial of service) (Suo et al., 2012). Within the IoT edge device, the no of corrupted nodes is together spoiling the IoT services for example botnet attacks (Kolias et al., 2017). At center point of failures are Availability threats, confidentiality and authorization (Gubbi et al., 2013):(Sicari et al., 2018). Some confidentiality rises issues like analyzed routing protocols, traffic information and identity spoofing techniques (Farooq et al., 2015). To enable the blockchain in IoT devices using public blockchains without tokenization to protect from DDos attacks (Ali et al., 2019). Authors proposed blockchain based smart city solutions are decentralized platform for immutable records, using Ethereum blockchain with smart contracts as top of decentralized records (Biswas and Muthukkumarasamy, 2016).

To improvise the data integrity in blockchain overlay, decentralized hash tables are implemented to store all the data (Yuan, Yong. 2016). One more feature is confidentiality, each transaction is stored in key pairs. IoT devices can be managed by blockchain based public key interfaces and commands are managed through smart contracts in IoT devices using blockchain unique addresses (Axon, & Goldsmith, 2017). The service offered by blockchain is trustless environment, In IoT ecosystem, centralized server or third-party entities are managed by data communication between peer-to-peer networks. The assumption of third-party entities lead to cracks the privacy and security issues. To overcome the IoT trust, solutions are allowed multiple users to store and copy of all the transactions in peer-to-peer to network using permissionless blockchain (Zyskind et al., 2015). Author proposes the trustless IoT architecture where IoT connected devices are register themselves directly in IoT blockchain, for securing, storing and processing without need for any intermediataries.to enhance the end on communications with trust less environment (Biswas and Muthukkumarasamy, 2016). In IoT produces large amount of data and requires real time processing, it may rise bandwidth, latency, and scalability issues to solution of the issues integrated with blockchain, to reducing bandwidth, achieving low latency and improving the security and privacy implemented hybrid architecture with the help of proof of work mechanism (Singh, 2021).

To overcome the data sharing, data integrity, authentication and access control privacy issues with the integration of blockchain and IoT using frameworks such as data sharing framework, data integrity framework, authentication and access control to get better efficiency, to avoid privacy in IoT system, private blockchain and consortium mechanisms are involved. Avoid third party entities through decentralized system (Y. Yu et al., 2018).

APPLICATIONS OF BLOCK CHAIN IN SMART CITIES

There are various aspects of smart cities such as smart healthcare, smart transportation, smart grid, Supply chain management, financial systems and data center networks. In this section, we surveyed existing blockchain efforts in each of these aforementioned aspects. This will give readers an understanding of how blockchain technology is being used in the realm of smart cities.

Figure 3. Emerging smart city applications.

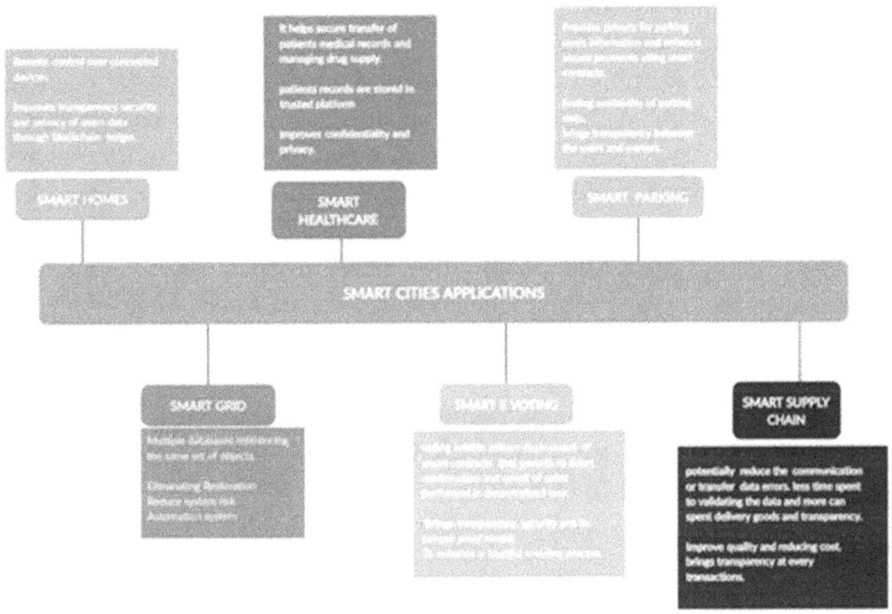

Smart Healthcare

The existing healthcare system is integrated with IoT based system. IoT based medical systems have various areas with end-to-end connectivity. In case of emergencies IoT based health care systems can monitor the real time situations. But gathering person's medical data and storing all the health information are drawbacks. A typical healthcare network is made up of a collection of hospitals that are owned, managed, and supported by a single entity (Chaudhary et al., 2018). These centralized healthcare networks, however, are vulnerable to a single point of failure. Furthermore, because to the rapid urbanization of the world's population, traditional healthcare systems are finding it difficult to meet public demand. The conflict between finite resources and ever-increasing demand necessitates the development of an efficient, intelligent, and long-term healthcare system. Blockchain technology is the best-known approach for achieving the appropriate level of decentralization in healthcare networks while also improving security. Existing systems, such as healthcare data management systems, can be significantly improved using blockchain technology. It consists of key features, opportunities and benefits of blockchain to adopt various healthcare industries (Yaqoob, Ibrar et al., 2021). To enabling sharing, transparency and immutable ledger in healthcare blockchain technology with better outcomes at lowest cost as well as increases privacy in patients' data (Treiblmaier et al., 2020). A blockchain technology enabled trusted exchange of health information moving towards to provide insight of patient treatments, proper drugs, secure food diet, medical devices, quality and better safety precautions. Using smart contracts for exchanging the patient information taking decision making process (Transaction & MPI, 2016).

Smart Grid

Smart grid improves efficiency, economics and resiliency of electric grids, facilitate the use of the distribution grid to enable platform for modern energy services. Discussed the employment of blockchain in demand-side management of the smart grid. The study aims huge industrial energy users and presents a novel power trade structure based upon the P2P blockchain network for the contract, day-ahead, adjustment, and balancing markets. They formulate an optimal load management problem for a particular industrial user to minimize the operational cost (Dang et al., 2019). Blockchain applications in the smart grid such as real time operational monitoring of the power grid and cyber-physical security of the smart grid and effective consumption analytics (Musleh et al., 2019). Authors presented cyber security layer frameworks for smart grid. Smart cities enable to secure access of energy consumption at lower cost, strengthen privacy, every transaction is stored in peer-to-peer network (Treiblmaier et al., 2020). Promoting energies is the part of smart grid, the production of local energy and avoid transport energy loses blockchain aims to decentralized, distributed ledger and which is owned and updated by all users in the transactions. Peer to peer communications are exchanges through smart contracts. In smart city has several micro grids, every micro grid is own blockchain to exchange the energy and reduce the energy loses (Nehai & Guérard, 2017).

Smart Supply Chain

In smart city, a supply chain is a set of connections between a company and its traders to produce and deal out a specific product to the final buyer. Smart Supply Chain aims at bringing awareness for efficient decision-making, by tapping the data gathered by IoT devices and providing detailed visibility of products / items, all the way from manufacturer to retailer. Authors implemented blockchain based supply chain management system to enhance the tracking and parcel details, it brings transparency and all transaction record history are immutable within the ecosystem. Introduces Ethereum platform for frameworks to develop blockchain system in logistics (Helo, & Hao, 2019). Again, blockchain in supply chain to determining the problems that arising from lack of trust, quality issues etc. blockchain is implemented in supply chain as decentralized system, data sharing, all the product details are recorded, transparency, quality control system. Authors introduces different layers are one for location and gathering the information, second one for keeping various kinds of data in secure manner, thirdly sharing and privacy of data, aims to maintain quality and efficiency, finally implemented frameworks using smart contracts to enhance the quality products (Chen et al., 2017). To reduce the pollution within the metropolitan boundaries and to raise customer satisfactions with tamperproof records (Treiblmaier et al., 2020).

Smart E-Voting

In smart city, Electronic Voting Machine (also known as EVM) is voting using electronic means to either aid or takes care of the responsibilities of casting and counting votes. With the EVM, instead of issuing a ballot paper, the polling officer will press the Ballot Button which enables the voter to cast their vote. Remote voting system has some drawbacks like time consuming and slow. It was the foremost type of electoral scam found in India. Once the voting schedule is over, the result of the election cannot be declared immediately. All the ballot boxes are collected at a specific location and then the counting of the votes is carried out which is a lot of time taking process. To overcome problems electronic voting system has easy, efficient, no supply costs, quick and easy polling, accurate and no waiting for results. (Kashif Mehboob et al., 2020) Discussed about voting system to fill the gap use blockchain hashes and block size generations. In blockchain voter's addresses are stored in unique assets to provide secure, convenient process for voting and accurate results in e voting system. Authors presented different analysis for the solution of e-voting system in blockchain. (Kshetri, & Voas, 2018) Presented voting system in blockchain with different challenges and benefits are to reduce voter fraud and increases voters to cast their votes in online it and using all the voters' data are cryptographically secure manner, created to voter's personal ids using blockchain. Authors presented different case studies to provide security, accurate and public trust. (Shahzad & Crowcroft, 2019) Discussed about electronic voting process in blockchain, paper-based voting system has some flaws for security and privacy. Some existing literatures enhance different voting process solution in blockchain electronic voting. Authors introduced creation of blocks and sealing of blocks using hashing techniques. Means voting polling process, to secure the data after completion of time using sealing the data into blocks and results are declared by adjustable methods. Traditional ballot-based voting move towards to electronic voting system to develop the transparency, immutable records, trust less environment with the help of blockchain technology. It aims to anonymity, uniqueness of voter identity, authentication and gives instant accuracy results. To implement the blockchain based framework for secure e-voting system (Al-Abri & Raya 2019).

Smart Homes

A smart home is integrated with IoT with human comforts for their residences or living environment that provides quality life, security etc. IoT is major platform of smart homes that interconnectedness of various smart devices like smart phones, smart tv etc. (Younghun Lee et al., 2020) smart homes refer to inhabited or living the environment furnished with technologies that can automatically control devices. Smart home gateways are configured with central authority. To enhance the smart home gateways with blockchain, decentralized, peer to peer communication is configured. authors implemented those solutions are confidentiality, data integrity, authentication with different data from gateways. Proposed architecture is implemented with Ethereum frameworks along with encryption algorithm. A smart home-based survey was conducted by (Moniruzzaman et al., 2020) to understand that adaption of blockchain in smart home ecosystem. One of the key challenges in smart home applications to enhance the security and privacy have possible solution is blockchain technology. majorly focuses on Data marketplace, access control management, payment services, smart city services, homecare systems. Blockchain has flexibility feature to adopt and to integrate different smart home devices, explores frameworks with four phases are data source layer that generates data from devices, blockchain network layer that cryptographically validated, application layer that integrated blockchain with different smart home applications, client layer such as retailer shop etc. (WEI SHE et al., 2019) described that privacy protection in smart home system device level information from various threatens. Existing traditional smart home system is using homomorphic consortium blockchain, to increase the level of privacy protection and encrypt sensitive data using paillier encryption algorithm. To enhance the security blockchain based models are analyzed.

Smart Governance

As cities are industrialized on the source of smart city model, governments across all regions are using e-governance to support democracy, citizen involvement and public welfare. The aim of smart or e-governance is to make the system more transparent and citizens more informed. The three characteristics of e governance are reductionist doctrine: This reductionist approach assumes that complex problems I.e., energy, transportation, water. Secondly separation of power with public actors and political ideas. Lastly democracy system I.e., voting system. To addressed all the issues moving forward from urban governance to e governance which is easier and transparency to citizens (Razaghi and Finger, 2018). Authors discussed about various use cases and technical details to reintroduce the government with blockchain technology addressed different technical and adoption challenges are fully preserving privacy protection and increases scalability. In blockchain all the data are distributed in the network it brings transparency and secure data storing availability (Clavin et al., 2020).

Smart Transportation

The rapid development of modern sensing, communication, analyzing and computing techniques ha potentially increasing the intelligent transportation system. The fast-growing technology such as Internet of things and cloud computing make it possible for maximum amount of the data, analyses and decisions are managed by centralized authorities. Two crucial issues are security and trust in intelligent transportation system due to centralized authorities. To enhance the security, trust, transparency through decentralized system with blockchain based intelligent transportation system. To develop a trusted architecture using parallel intelligent transportation system (Yuan & Yong. (2016)).

Smart Parking

In smart parking system, IoT collected, analyzed and processed data from different sources. Cloud acts as intermediate between the things and applications. The integration of internet of things with cloud computing technologies can rise the opportunities such as storage capacity, computational power, scalability, availability, interoperability and communication resources. Smart parking system addresses real time availability of parking spaces to booking a slot using a mobile application (Khanna, & Anand, 2016). Smart parking system addressed the trust and data integrity issues among several existing highlights. To resolve the data integrity and third-party trust issues with the concept of decentralized system. And also parking service provider offers parking as a service and updates, parking spaces and offer in integrated system. blockchain network contains a public ledger with the valid transactions only, a consensus mechanism is used for verify the transactions. The parking user is the participant who request for a car park. In blockchain based integrated smart parking system, every parking area has a local copy of ledger. Data generated parking availability spaces by the parking sensor. Data is related to Parking service provider set a price of parking based on time (Ahmed et al., 2019). Some existing smart parking systems are threatening the drivers (lack of privacy) and revealing information about their visited locations. some failures and attacks in availability of parking services and doesn't bring transparency in booking slots using centralized server. Consortium blockchain is created to overcome the parking system securely and transparently, where all the parking offers are recorded in immutable ledger. Using private information retrieval techniques to improve privacy issues and randomizable signature, time locked anonymous payment techniques using smart contract to produce the secure and privacy pay methods, trust less manner for parking services (Amiri et al., 2019). Also added reputation management scheme to rate the parking lots based on blockchain technology, it is useful to identify quality parking lot services based on reputation score. In this phase, driver users are using different public key pair to submit their rating every time (Zhang et al. 2020).

CONCLUSION

The explosive development in the world's population together with the rapid urbanisation process tends to endanger the environmental and economical sustainability of cities. To this end, the concept of "Smart City" is proposed that use modern Information and Communication Technology (ICT) in a smart manner so as to build a sustainable urban environment and improve the citizens life. However, there are increasing security challenges in smart cities. These challenges can be successfully addressed by the use of blockchain technology, due to its good properties such as transparency, immutability and decentralization. In this paper, the possibilities and benefits of applying blockchain technology to smart cities along with its trade-offs are presented through a comprehensive survey. The paper begins with some related recently published surveys and background knowledge of blockchain with IoT in smart cities. Further, the paper aims to integrate the two areas by exploring of blockchain in various smart communities such as healthcare, transportation, smart grid, supply chain management, financial systems and data center networks. This survey is expected to provide as a knowledge base and systematic guideline for future research in applying blockchain technology to smart cities.

REFERENCES

Ahmed, S., Rahman, M. S., & Rahaman, M. S. (2019, March). A blockchain-based architecture for integrated smart parking systems. In *2019 IEEE International Conference on Pervasive Computing and Communications Workshops (PerCom Workshops)*. IEEE.

Ahmed, S., Shah, M. A., & Wakil, K. (2020). Blockchain as a trust builder in the smart city domain: A systematic literature review. *IEEE Access: Practical Innovations, Open Solutions*, *8*, 92977–92985.

Aithal, P. S. (2020). Blockchain Technology: A Driving Force in Smart Cities Development. *International Journal of Applied Engineering and Management Letters*, *4*(2), 237–252.

Al-Abri, Khalique, & Goldsmith. (2019). *Designing of a Framework for Electronic Voting Systems using Blockchain Technology.* International Business Information Management Conference (34th IBIMA), Madrid, Spain.

Al Amiri, W., Baza, M., Banawan, K., Mahmoud, M., Alasmary, W., & Akkaya, K. (2019, December). Privacy-preserving smart parking system using blockchain and private information retrieval. In *2019 International Conference on Smart Applications, Communications and Networking (SmartNets)*. IEEE.

Al Amiri, W., Baza, M., Banawan, K., Mahmoud, M., Alasmary, W., & Akkaya, K. (2020, January). *Towards secure smart parking system using blockchain technology. In 2020 IEEE 17th Annual Consumer Communications & Networking Conference (CCNC)*. IEEE.

Al Sadawi, A., Hassan, M. S., & Ndiaye, M. (2021, March). A Review on the Integration of Blockchain and IoT. *2020 International Conference on Communications, Signal Processing, and their Applications (ICCSPA)*, 1-6.

Ali, M. S., Dolui, K., & Antonelli, F. (2017, October). IoT data privacy via blockchains and IPFS. *Proceedings of the seventh international conference on the internet of things*, 1-7.

Ali, M. S., Vecchio, M., Pincheira, M., Dolui, K., Antonelli, F., & Rehmani, M. H. (2018). Applications of blockchains in the Internet of Things: A comprehensive survey. *IEEE Communications Surveys and Tutorials*, *21*(2), 1676–1717.

Anitha, A. (2017a). Home security system using internet of things. In *IOP conference series: materials science and engineering*. IOP Publishing.

Anitha, A. (2017b). Garbage monitoring system using IoT. In *IOP Conference Series: Materials Science and Engineering*. IOP Publishing.

Anitha, A., & Kalra, S. (2016). A Cyber defence using artificial home automation system using IoT. *International Journal of Pharmacy and Technology*, *8*, 25358–25364.

Anitha, A., Paul, G., & Kumari, S. (2016). A cyber defence using artificial intelligence. *International Journal of Pharmacy and Technology*, *8*(4), 25325–25357.

Axon, L., & Goldsmith, M. (2017, July). PB-PKI: A Privacy-aware Blockchain-based PKI. SECRYPT, 311-318.

Azaria, A., Ekblaw, A., Vieira, T., & Lippman, A. (2016, August). Medrec: Using blockchain for medical data access and permission management. In *2016 2nd international conference on open and big data (OBD)*. IEEE.

Bentov, I., Lee, C., Mizrahi, A., & Rosenfeld, M. (2014). Proof of activity: Extending bitcoin's proof of work via proof of stake. ACM SIGMETRICS Perform. *Evaluation Review*, *42*(3), 34–37.

Bhushan, B., Khamparia, A., Sagayam, K. M., Sharma, S. K., Ahad, M. A., & Debnath, N. C. (2020). Blockchain for smart cities: A review of architectures, integration trends and future research directions. *Sustainable Cities and Society*, *61*, 102360.

Biswas, K., & Muthukkumarasamy, V. (2016, December). Securing smart cities using blockchain technology. *2016 IEEE 18th international conference on high performance computing and communications; IEEE 14th international conference on smart city; IEEE 2nd international conference on data science and systems (HPCC/SmartCity/DSS)*, 1392-1393.

Chaudhary, R., Jindal, A., Aujla, G. S., Kumar, N., Das, A. K., & Saxena, N. (2018). Lscsh: Lattice-based secure cryptosystem for smart healthcare in smart cities environment. *IEEE Communications Magazine*, *56*(4), 24–32.

Chen, F., Xiao, Z., Cui, L., Lin, Q., Li, J., & Yu, S. (2020). Blockchain for Internet of things applications: A review and open issues. *Journal of Network and Computer Applications*, 102839.

Chen, L., Xu, L., Shah, N., Gao, Z., Lu, Y., & Shi, W. (2017, November). On security analysis of proof-of-elapsed-time (poet). In *International Symposium on Stabilization, Safety, and Security of Distributed Systems* (282-297). Springer.

Chen, S., Shi, R., Ren, Z., Yan, J., Shi, Y., & Zhang, J. (2017, November). A blockchain-based supply chain quality management framework. In *2017 IEEE 14th International Conference on e-Business Engineering (ICEBE)*. IEEE.

Chen, W., Xu, Z., Shi, S., Zhao, Y., & Zhao, J. (2018, December). A survey of blockchain applications in different domains. *Proceedings of the 2018 International Conference on Blockchain Technology and Application*, 17-21.

Clavin, J., Duan, S., Zhang, H., Janeja, V. P., Joshi, K. P., Yesha, Y., ... Li, J. D. (2020). Blockchains for Government: Use Cases and Challenges. *Digital Government: Research and Practice*, *1*(3), 1–21.

Dang, C., Zhang, J., Kwong, C. P., & Li, L. (2019). Demand side load management for big industrial energy users under blockchain-based peer-to-peer electricity market. *IEEE Transactions on Smart Grid*, *10*(6), 6426–6435.

Dorri, A. (2017). *Blockchain for IoT Security and Privacy: The Case Study of a Smart Home*. Academic Press.

Eckhoff, D., & Wagner, I. (2017). Privacy in the smart city—Applications, technologies, challenges, and solutions. *IEEE Communications Surveys and Tutorials*, *20*(1), 489–516.

Emmanuel, M., & Rayudu, R. (2016). Communication technologies for smart grid applications: A survey. *Journal of Network and Computer Applications*, *74*, 133–148.

Fan, T., & Chen, Y. (2010, September). A scheme of data management in the Internet of Things. In *2010 2nd IEEE International Conference on Network Infrastructure and Digital Content*. IEEE.

Farooq, M. U., Waseem, M., Mazhar, S., Khairi, A., & Kamal, T. (2015). A review on internet of things (IoT). *International Journal of Computers and Applications*, *113*(1), 1–7.

Fernández-Caramés, T. M., & Fraga-Lamas, P. (2018). A Review on the Use of Blockchain for the Internet of Things. *IEEE Access: Practical Innovations, Open Solutions*, *6*, 32979–33001.

Ferrag, M. A., Derdour, M., Mukherjee, M., Derhab, A., Maglaras, L., & Janicke, H. (2018). Blockchain technologies for the internet of things: Research issues and challenges. *IEEE Internet of Things Journal*, *6*(2), 2188–2204.

Gramoli, V. (2020). From blockchain consensus back to Byzantine consensus. *Future Generation Computer Systems*, *107*, 760–769.

Gubbi, J., Buyya, R., Marusic, S., & Palaniswami, M. (2013). Internet of Things (IoT): A vision, architectural elements, and future directions. *Future Generation Computer Systems*, *29*(7), 1645–1660.

Gui, T., Ma, C., Wang, F., & Wilkins, D. E. (2016, March). *Survey on swarm intelligence-based routing protocols for wireless sensor networks: An extensive study. In 2016 IEEE international conference on industrial technology (ICIT)*. IEEE.

Hakak, S., Khan, W. Z., Gilkar, G. A., Imran, M., & Guizani, N. (2020). Securing smart cities through blockchain technology: Architecture, requirements, and challenges. *IEEE Network*, *34*(1), 8–14.

Hassan, M. U., Rehmani, M. H., & Chen, J. (2019). Privacy preservation in blockchain based IoT systems: Integration issues, prospects, challenges, and future research directions. *Future Generation Computer Systems*, *97*, 512–529.

Helo, P., & Hao, Y. (2019). Blockchains in operations and supply chains: A model and reference implementation. *Computers & Industrial Engineering*, *136*, 242–251.

Hu, J., He, D., Zhao, Q., & Choo, K. K. R. (2019). Parking management: A blockchain-based privacy-preserving system. *IEEE Consumer Electronics Magazine*, *8*(4), 45–49.

Ibba, S., Pinna, A., Seu, M., & Pani, F. E. (2017, May). CitySense: blockchain-oriented smart cities. *Proceedings of the XP2017 Scientific Workshops*, 1-5.

Jo, J. H., Sharma, P. K., Sicato, J. C. S., & Park, J. H. (2019). Emerging technologies for sustainable smart city network security: Issues, challenges, and countermeasures. *Journal of Information Processing Systems*, *15*(4), 765–784.

Khan, K. M., Arshad, J., & Khan, M. M. (2020). Investigating performance constraints for blockchain based secure e-voting system. *Future Generation Computer Systems*, *105*, 13–26.

Khan, M., Silva, B. N., & Han, K. (2016). Internet of things based energy aware smart home control system. *IEEE Access: Practical Innovations, Open Solutions*, *4*, 7556–7566.

Khan, M. A., & Salah, K. (2018). IoT security: Review, blockchain solutions, and open challenges. *Future Generation Computer Systems*, *82*, 395–411.

Khan, M. A., & Salah, K. (2018). IoT security: Review, blockchain solutions, and open challenges. *Future Generation Computer Systems*, *82*, 395–411.

Khanna, A., & Anand, R. (2016, January). IoT based smart parking system. In *2016 International Conference on Internet of Things and Applications (IOTA)* (pp. 266-270). IEEE.

Kolias, C., Kambourakis, G., Stavrou, A., & Voas, J. (2017). DDoS in the IoT: Mirai and other botnets. *Computer*, *50*(7), 80–84.

Kshetri, N., & Voas, J. (2018). Blockchain-enabled e-voting. *IEEE Software*, *35*(4), 95–99.

Kumar, N. M., & Mallick, P. K. (2018). Blockchain technology for security issues and challenges in IoT. *Procedia Computer Science*, *132*, 1815–1823.

Larimer, D. (2014). Delegated proof-of-stake (dpos). *Bitshare Whitepaper, 81*, 85.

Lee, Y., Rathore, S., Park, J. H., & Park, J. H. (2020). A blockchain-based smart home gateway architecture for preventing data forgery. *Human-centric Computing and Information Sciences*, *10*(1), 1–14.

Li, T., Liu, Y., Tian, Y., Shen, S., & Mao, W. (2012, November). A storage solution for massive iot data based on nosql. In *2012 IEEE International Conference on Green Computing and Communications*. IEEE.

Li, W., Andreina, S., Bohli, J. M., & Karame, G. (2017). Securing proof-of-stake blockchain protocols. In *Data Privacy Management, Cryptocurrencies and Blockchain Technology* (pp. 297–315). Springer.

Li, X., Jiang, P., Chen, T., Luo, X., & Wen, Q. (2020). A survey on the security of blockchain systems. *Future Generation Computer Systems*, *107*, 841–853.

Liang, X., Zhao, J., Shetty, S., & Li, D. (2017, October). Towards data assurance and resilience in IoT using blockchain. In *MILCOM 2017-2017 IEEE Military Communications Conference (MILCOM)* (pp. 261–266). IEEE.

Lu, T., Fang, J., & Liu, C. (2015, September). A unified storage and query optimization framework for sensor data. In *2015 12th Web Information System and Application Conference (WISA)*. IEEE.

Majeed, U., Khan, L. U., Yaqoob, I., Kazmi, S. A., Salah, K., & Hong, C. S. (2021). Blockchain for IoT-based smart cities: Recent advances, requirements, and future challenges. *Journal of Network and Computer Applications*, *181*(9), 1–33. doi:10.1016/j.jnca.2021.103007

McGhin, T., Choo, K. K. R., Liu, C. Z., & He, D. (2019). Blockchain in healthcare applications: Research challenges and opportunities. *Journal of Network and Computer Applications*, *135*, 62–75.

Memon, R., Li, J., & Ahmed, J. (2019). Simulation Model for Blockchain Systems Using Queuing Theory. *Electronics (Basel)*, *8*(2), 234. doi:10.3390/electronics8020234

Mingxiao, D., Xiaofeng, M., Zhe, Z., Xiangwei, W., & Qijun, C. (2017, October). *A review on consensus algorithm of blockchain. In 2017 IEEE international conference on systems, man, and cybernetics (SMC)*. IEEE.

Missier, P., Bajoudah, S., Capossele, A., Gaglione, A., & Nati, M. (2017, October). Mind my value: a decentralized infrastructure for fair and trusted iot data trading. *Proceedings of the Seventh International Conference on the Internet of Things*, 1-8.

Moniruzzaman, M., Khezr, S., Yassine, A., & Benlamri, R. (2020). Blockchain for smart homes: Review of current trends and research challenges. *Computers & Electrical Engineering*, *83*, 106585.

Musleh, A. S., Yao, G., & Muyeen, S. M. (2019). Blockchain applications in smart grid–review and frameworks. *IEEE Access: Practical Innovations, Open Solutions*, *7*, 86746–86757.

Nakamoto, S. (2008). Bitcoin: A peer-to-peer electronic cash system. *Decentralized Business Review*, 21260.

Nehai, Z., & Guérard, G. (2017, May). Integration of the blockchain in a smart grid model. *The 14th International Conference of Young Scientists on Energy Issues (CYSENI) 2017*, 127-134.

Panarello, A., Tapas, N., Merlino, G., Longo, F., & Puliafito, A. (2018). Blockchain and iot integration: A systematic survey. *Sensors (Basel)*, *18*(8), 2575.

Rathore, S., Kwon, B. W., & Park, J. H. (2019). BlockSecIoTNet: Blockchain-based decentralized security architecture for IoT network. *Journal of Network and Computer Applications*, *143*, 167–177.

Razaghi, M., & Finger, M. (2018). Smart governance for smart cities. *Proceedings of the IEEE*, *106*(4), 680–689.

Reyna, A., Martín, C., Chen, J., Soler, E., & Díaz, M. (2018). On blockchain and its integration with IoT. Challenges and opportunities. *Future Generation Computer Systems*, *88*, 173–190.

Rotună, C., Gheorghiță, A., Zamfiroiu, A., & Smada Anagrama, D. (2019). Smart City Ecosystem Using Blockchain Technology. *Informações Econômicas*, *23*(4), 41–50.

Saleh, F. (2019). *Blockchain Without Waste: Proof-of-Stake. SSRN Scholarly Paper ID 3183935*. Social Science Research Network.

Shafagh, H., Burkhalter, L., Hithnawi, A., & Duquennoy, S. (2017, November). Towards blockchain-based auditable storage and sharing of iot data. *Proceedings of the 2017 on cloud computing security workshop*, 45-50.

Shahzad, B., & Crowcroft, J. (2019). Trustworthy electronic voting using adjusted blockchain technology. *IEEE Access: Practical Innovations, Open Solutions*, *7*, 24477–24488.

Sharma, P. K., & Park, J. H. (2018). Blockchain based hybrid network architecture for the smart city. *Future Generation Computer Systems*, *86*, 650–655.

She, W. E. I., Gu, Z. H., Lyu, X. K., Liu, Q. I., Tian, Z., & Liu, W. (2019). Homomorphic consortium blockchain for smart home system sensitive data privacy preserving. *IEEE Access: Practical Innovations, Open Solutions*, *7*, 62058–62070.

Sicari, S., Rizzardi, A., Cappiello, C., Miorandi, D., & Coen-Porisini, A. (2018). Toward data governance in the internet of things. In *New advances in the internet of things* (pp. 59–74). Springer.

Singh, C. (2021). Blockchain and IOT integrated Smart City Architecture. *Turkish Journal of Computer and Mathematics Education*, *12*(9), 62–69.

Sookhak, M., Tang, H., & Yu, F. R. (2018, June). Security and Privacy of Smart Cities: Issues and Challenge. *2018 IEEE 20th International Conference on High Performance Computing and Communications; IEEE 16th International Conference on Smart City; IEEE 4th International Conference on Data Science and Systems (HPCC/SmartCity/DSS)*, 1350-1357.

Suárez-Albela, M., Fernández-Caramés, T. M., Fraga-Lamas, P., & Castedo, L. (2017). A practical evaluation of a high-security energy-efficient gateway for IoT fog computing applications. *Sensors (Basel)*, *17*(9), 1978.

Suo, H., Wan, J., Zou, C., & Liu, J. (2012, March). *Security in the internet of things: a review. In 2012 international conference on computer science and electronics engineering*. IEEE.

Swan, M. (2015). *Blockchain: Blueprint for a new economy*. O'Reilly Media, Inc.

Taş, R., & Tanrıöver, Ö. Ö. (2020). A Systematic Review of Challenges and Opportunities of Blockchain for E-Voting. *Symmetry*, *12*(8), 1328.

Treiblmaier, H., Rejeb, A., & Strebinger, A. (2020). Blockchain as a driver for smart city development: Application fields and a comprehensive research agenda. *Smart Cities*, *3*(3), 853–872.

Valera, A. J. J., Zamora, M. A., & Skarmeta, A. F. (2010, January). An architecture based on internet of things to support mobility and security in medical environments. In *2010 7th IEEE consumer communications and networking conference*. IEEE.

Wan, J., Li, J., Imran, M., & Li, D. (2019). A blockchain-based solution for enhancing security and privacy in smart factory. *IEEE Transactions on Industrial Informatics*, *15*(6), 3652–3660.

Wang, D., Zhao, J., & Wang, Y. (2020). A Survey on Privacy Protection of Blockchain: The Technology and Application. *IEEE Access: Practical Innovations, Open Solutions*, *8*, 108766–108781.

Wang, S., Ouyang, L., Yuan, Y., Ni, X., Han, X., & Wang, F. Y. (2019). Blockchain-enabled smart contracts: Architecture, applications, and future trends. *IEEE Transactions on Systems, Man, and Cybernetics. Systems*, *49*(11), 2266–2277.

Wang, W., Hoang, D. T., Hu, P., Xiong, Z., Niyato, D., Wang, P., ... Kim, D. I. (2019). A survey on consensus mechanisms and mining strategy management in blockchain networks. *IEEE Access: Practical Innovations, Open Solutions*, *7*, 22328–22370.

Wang, X., WeiLi, J., & Chai, J. (2018, December). The research on the incentive method of consortium blockchain based on practical byzantine fault tolerant. In *2018 11th international symposium on computational intelligence and design (ISCID)* (2, 154-156). IEEE.

Wang, Z., Jin, H., Dai, W., Choo, K. K. R., & Zou, D. (2021). Ethereum smart contract security research: Survey and future research opportunities. *Frontiers of Computer Science*, *15*(2), 1–18.

Wong, P. F., Chia, F. C., San Kiu, M., & Lou, E. C. (2020). *Potential integration of blockchain technology into smart sustainable city (SSC) developments: a systematic review*. Smart and Sustainable Built Environment.

Xie, J., Tang, H., Huang, T., Yu, F. R., Xie, R., Liu, J., & Liu, Y. (2019). A survey of blockchain technology applied to smart cities: Research issues and challenges. *IEEE Communications Surveys and Tutorials*, *21*(3), 2794–2830.

Yaga, D., Mell, P., Roby, N., & Scarfone, K. (2019). *Blockchain technology overview*. arXiv preprint arXiv:1906.11078.

Yaqoob, I., Salah, K., Jayaraman, R., & Al-Hammadi, Y. (2021). Blockchain for healthcare data management: Opportunities, challenges, and future recommendations. *Neural Computing & Applications*, 1–16.

Yong, Y., & Fei-Yue, W. (2016). *Towards blockchain-based intelligent transportation systems*. Academic Press.

Yu, B., Wright, J., Nepal, S., Zhu, L., Liu, J., & Ranjan, R. (2018). Iotchain: Establishing trust in the internet of things ecosystem using blockchain. *IEEE Cloud Computing*, *5*(4), 12–23.

Yu, Y., Li, Y., Tian, J., & Liu, J. (2018). Blockchain-based solutions to security and privacy issues in the internet of things. *IEEE Wireless Communications*, *25*(6), 12–18.

Yue, X., Wang, H., Jin, D., Li, M., & Jiang, W. (2016). Healthcare data gateways: Found healthcare intelligence on blockchain with novel privacy risk control. *Journal of Medical Systems*, *40*(10), 1–8.

Zhang, C., Zhu, L., Xu, C., Zhang, C., Sharif, K., Wu, H., & Westermann, H. (2020). BSFP: Blockchain-enabled smart parking with fairness, reliability and privacy protection. *IEEE Transactions on Vehicular Technology*, *69*(6), 6578–6591.

Zheng, Z., Xie, S., Dai, H., Chen, X., & Wang, H. (2017, June). *An overview of blockchain technology: Architecture, consensus, and future trends. In 2017 IEEE international congress on big data (BigData congress)*. IEEE.

Zyskind, G., Nathan, O., & Pentland, A. (2015). *Enigma: Decentralized computation platform with guaranteed privacy.* arXiv preprint arXiv:1506.03471.

Compilation of References

Joon, V., Chandra, A., & Bhattacharya, M. (2009). Household energy consumption pattern and socio-cultural dimensions associated with it: A case study of rural Haryana, India. *Biomass and Bioenergy*, *33*(11), 1509–1512. doi:10.1016/j.biombioe.2009.07.016

Wang, S., Ouyang, L., Yuan, Y., Ni, X., Han, X., & Wang, F. (2019, November). Blockchain-Enabled Smart Contracts: Architecture, Applications, and Future Trends. *IEEE Transactions on Systems, Man, and Cybernetics. Systems*, *49*(11), 2266–2277. doi:10.1109/TSMC.2019.2895123

Alexaki, S., Alexandris, G., Katos, V., & Petroulakis, N. E. (2018). Blockchain-based Electronic Patient Records for Regulated Circular Healthcare Jurisdictions. *IEEE 23rd International Workshop on Computer Aided Modeling and Design of Communication Links and Networks (CAMAD),* 1-6. 10.1109/CAMAD.2018.8514954

Celebi, E., & Fuller, J. D. (2007). A Model for Efficient Consumer Pricing Schemes in Electricity Markets. *IEEE Transactions on Power Systems*, *22*(1), 60–67. doi:10.1109/TPWRS.2006.888956

McNeil, M. A., & Letschert, V. E. (2010). Modeling diffusion of electrical appliances in the residential sector. *Energy and Building*, *42*(6), 783–790. doi:10.1016/j.enbuild.2009.11.015

Rajput, S., Singh, A., Khurana, S., Bansal, T., & Shreshtha, S. (2019). Blockchain Technology and Cryptocurrencies. *2019 Amity International Conference on Artificial Intelligence (AICAI)*, 909-912. 10.1109/AICAI.2019.8701371

Rouhani, S., & Deters, R. (2017). Performance analysis of ethereum transactions in private blockchain. *8th IEEE International Conference on Software Engineering and Service Science (ICSESS),* 70-74. 10.1109/ICSESS.2017.8342866

Swan, L. G., & Ugursal, V. I. (2009). Modeling of end-use energy consumption in the residential sector: A review of modeling techniques. *Renewable & Sustainable Energy Reviews*, *13*(8), 1819–1835. doi:10.1016/j.rser.2008.09.033

Chen, P., Jiang, B., & Wang, C. (2017). Blockchain-based payment collection supervision system using pervasive Bitcoin digital wallet. *IEEE 13th International Conference on Wireless and Mobile Computing, Networking and Communications (WiMob),* 139-146. 10.1109/WiMOB.2017.8115844

Nishio, K-I., & Asano, H. (2006). *A residential end-use demand model for analyzing the energy conservation potential of new energy efficient technologies*. Academic Press.

Fernandez-Vazquez, S., Rosillo, R., De La Fuente, D., & Priore, P. (2019). Blockchain in FinTech: A Mapping Study. *Sustainability*, *11*(22), 6366. doi:10.3390u11226366

Golmohamadi, H., & Keypour, R. (2017). Retail Energy Management in Electricity Markets: Structure, Challenges and Economic Aspects- a Review. *Technology and Economics of Smart Grids and Sustainable Energy, 2*(1).

Hasselgren, A., Kralevska, K., Gligoroski, D., Pedersen, S. A., & Faxvaag, A. (2020). Blockchain in healthcare and health sciences—A scoping review. *International Journal of Medical Informatics, 134*, 104040. doi:10.1016/j.ijmedinf.2019.104040 PMID:31865055

Newsham, G. R., & Bowker, B. G. (2010). The effect of utility time-varying pricing and load control strategies on residential summer peak electricity use: A review. *Energy Policy, 38*(7), 3289–3296. doi:10.1016/j.enpol.2010.01.027

Khan, F. A., Asif, M., Ahmad, A., Alharbi, M., & Aljuaid, H. (2020). Blockchain technology, improvement suggestions, security challenges on smart grid and its application in healthcare for sustainable development. *Sustainable Cities and Society, 55*, 102018. doi:10.1016/j.scs.2020.102018

Young, D. (2008). When do energy-efficient appliances generate energy savings? Some evidence from Canada. *Energy Policy, 36*(1), 34–46. doi:10.1016/j.enpol.2007.09.011

Vineetha, C. P., & Babu, C. A. (2014). Smart grid challenges, issues and solutions. *2014 International Conference on Intelligent Green Building and Smart Grid (IGBSG)*. 10.1109/IGBSG.2014.6835208

Fischer, C. (2008). Feedback on household electricity consumption: A tool for saving energy? *Energy Efficiency, 1*(1), 79–104. doi:10.100712053-008-9009-7

Hölbl, M., Kompara, M., Kamišalić, A., & Nemec Zlatolas, L. (2018). A Systematic Review of the use of Blockchain in Healthcare. *Symmetry, 10*(10), 470. doi:10.3390ym10100470

Sianaki, O., Hussain, O., & Tabesh, A. (2010). A Knapsack problem approach for achieving efficient energy consumption in smart grid for endusers' life style. *2010 IEEE Conference on Innovative Technologies for an Efficient and Reliable Electricity Supply, CITRES 2010*, 159 - 164. 10.1109/CITRES.2010.5619873

Sempértegui, R., Bautista, J., Cubero, R., & Pereira, J. (2002). Models And Procedures for Electric Energy Distribution Planning. A Review. *IFAC Proceedings Volumes, 35*(1), 395-400. 10.3182/20020721-6-ES-1901.01217

Mozina, C. (2007). Undervoltage Load Shedding. *IEEE Power Systems Conference: Advanced Metering, Protection, Control, Communication, and Distributed Resources*, 39-54.

Lakra & Kirar. (2015). Load Shedding techniques for System with Cogeneration: A Review. *Electrical and Electronics Engineering: An International Journal, 4*(3).

Verayiah, R. (2014). *Review of Under-voltage Load Shedding Schemes in Power System Operation*. Przegląd Elektrotechniczny.

Pferschy, U. (1999). Dynamic Programming Revisited: Improving Knapsack Algorithms. *Computing, 63*(4), 419–430. doi:10.1007006070050042

Feng, Q., He, D., Zeadally, S., Khan, M. K., & Kumar, N. (2018). A survey on privacy protection in blockchain systems. *Journal of Network and Computer Applications*.

Wood, G., & Newborough, M. (2007). Energy-use information transfer for intelligent homes: Enabling energy conservation with central and local displays. *Energy and Building, 39*(4), 495–503. doi:10.1016/j.enbuild.2006.06.009

Barriers to the adoption of efficient electricity technologies: A case study of India. (1998). *Fuel and Energy Abstracts, 39*(3), 196.

Tribis, Y., El Bouchti, A., & Bouayad, H. (2018). Supply Chain Management based on Blockchain: A Systematic Mapping Study. *MATEC Web of Conferences, 200*, 20. 10.1051/matecconf/201820000020

Helo, P., & Shamsuzzoha, A. H. M. (2020). Real-time supply chain— A blockchain architecture for project deliveries. *Robotics and Computer-integrated Manufacturing*, *63*, 101909. doi:10.1016/j.rcim.2019.101909

Klüser, R. (2009). Current challenges for efficient electricity grids. *Poiesis & Praxis: International Journal of Ethics of Science and Technology Assessment*, *6*(3–4), 265–271. doi:10.100710202-009-0070-x

GNetworkArchitecture – A high-level perspective. (2016). Huawei Technologies.

Protecting Patient Data the New Scope Cybersecurity for Healthcare. (2019, Summer). *Elsevier Journal of Allied Health.*

Yamamoto, Y., Suzuki, A., Fuwa, Y., & Sato, T. (2008). Decision-making in electrical appliance use in the home. *Energy Policy*, *36*(5), 1679–1686. doi:10.1016/j.enpol.2008.01.022

Aydinalp, M., Ismet Ugursal, V., & Fung, A. S. (2002). Modeling of the appliance, lighting, and space-cooling energy consumptions in the residential sector using neural networks. *Applied Energy*, *71*(2), 87–110. doi:10.1016/S0306-2619(01)00049-6

Narayanaswami, Nooyi, Govindaswamy, & Viswanathan. (2019). Blockchain anchored supply chain automation. *IBM Journal of Research and Development, 63*(2-3), 7:1-7:11.

FaruquiA.SergiciS.SharifA. (2009). *The Impact of Informational Feedback on Energy Consumption — A Survey of the Experimental Evidence*. doi:10.2139/ssrn.1407701

Twesige. (2015). *Bitcoin. A simple explanation of Bitcoin and Block Chain technology*. GRIN Verlag.

Aslanoglou, C., Konstantopoulos, M., Chondros, N., & Roussopoulos, M. (2020). Take Back your Friends with DCS: A Decentralized Connectivity Service for private social communication apps. *2020 IEEE International Conference on Decentralized Applications and Infrastructures (DAPPS)*, 133-138. 10.1109/DAPPS49028.2020.00017

Wood, G., & Newborough, M. (2003). Dynamic energy-consumption indicators for domestic appliances: Environment, behaviour and design. *Energy and Building*, *35*(8), 821–841. doi:10.1016/S0378-7788(02)00241-4

Shahnaz, A., Qamar, U., & Khalid, A. (2019). Using Blockchain for Electronic Health Records. *IEEE Access: Practical Innovations, Open Solutions*, *7*, 147782–147795. doi:10.1109/ACCESS.2019.2946373

Ziel, F., Steinert, R., & Husmann, S. (2015). Efficient modeling and forecasting of electricity spot prices. *Energy Economics*, *47*, 98–111. doi:10.1016/j.eneco.2014.10.012

Adekunle. (2015). Islamic Education in Nigeria. From Rote Learning to E-Learning. *European Conference on E-Learning, Academic Conferences International Limited, 422.*

Aggarwal, A., Gupta, I., Garg, N., & Goel, A. (2019). Deep Learning Approach to Determine the Impact of Socio Economic Factors on Bitcoin Price Prediction. *2019 Twelfth International Conference On Contemporary Computing (IC3)*. 10.1109/IC3.2019.8844928

Ahmed, S., Rahman, M. S., & Rahaman, M. S. (2019, March). A blockchain-based architecture for integrated smart parking systems. In *2019 IEEE International Conference on Pervasive Computing and Communications Workshops (PerCom Workshops)*. IEEE.

Ahmed, S., Shah, M. A., & Wakil, K. (2020). Blockchain as a trust builder in the smart city domain: A systematic literature review. *IEEE Access: Practical Innovations, Open Solutions*, *8*, 92977–92985.

Aithal, P. S. (2020). Blockchain Technology: A Driving Force in Smart Cities Development. *International Journal of Applied Engineering and Management Letters*, *4*(2), 237–252.

Al Amiri, W., Baza, M., Banawan, K., Mahmoud, M., Alasmary, W., & Akkaya, K. (2019, December). Privacy-preserving smart parking system using blockchain and private information retrieval. In *2019 International Conference on Smart Applications, Communications and Networking (SmartNets)*. IEEE.

Al Amiri, W., Baza, M., Banawan, K., Mahmoud, M., Alasmary, W., & Akkaya, K. (2020, January). *Towards secure smart parking system using blockchain technology. In 2020 IEEE 17th Annual Consumer Communications & Networking Conference (CCNC)*. IEEE.

Al Sadawi, A., Hassan, M. S., & Ndiaye, M. (2021, March). A Review on the Integration of Blockchain and IoT. *2020 International Conference on Communications, Signal Processing, and their Applications (ICCSPA)*, 1-6.

Al-Abri, Khalique, & Goldsmith. (2019). *Designing of a Framework for Electronic Voting Systems using Blockchain Technology.* International Business Information Management Conference (34th IBIMA), Madrid, Spain.

Alammary, A., Alhazmi, S., Almasri, M., & Gillani, S. (2019). Blockchain-Based Applications in Education: A Systematic Review. *Applied Sciences (Basel, Switzerland)*, *9*(12), 2400. doi:10.3390/app9122400

Albayati, H., Kim, S. K., & Rho, J. J. (2021). Accepting financial transactions using blockchain technology and cryptocurrency: A customer perspective approach. *Technology in Society*, *62*, 101320. doi:10.1016/j.techsoc.2020.101320

Al-Husban, A. A. (2014). Accounting Information Systems. Al-Yazuri Publishing House.

Alick, K., Davies, A., Leopoldseder, M., & Niemeyer, A. (2017). *Blockchain technology for supply chains--A must or a maybe?* McKinsey & Company. https://www.mckinsey.com/business-functions/operations/our-insights/blockchain-technology-for-supply-chainsa-must-or-a-maybe

Ali, M. S., Dolui, K., & Antonelli, F. (2017, October). IoT data privacy via blockchains and IPFS. *Proceedings of the seventh international conference on the internet of things*, 1-7.

Ali, M. S., Vecchio, M., Pincheira, M., Dolui, K., Antonelli, F., & Rehmani, M. H. (2018). Applications of blockchains in the Internet of Things: A comprehensive survey. *IEEE Communications Surveys and Tutorials*, *21*(2), 1676–1717.

Alketbi, A., Nasir, Q., & Abu Talib, M. (2018). Blockchain for government services — Use cases, security benefits and challenges. 15th Learning and Technology Conference (L&T). doi:10.1109/LT.2018.836849410.1109/LT.2018.8368494

Allessie, D., Sobolewski, M., & Vaccari, L. (2019). Blockchain for digital government. JRC Science for Policy Report: Vol. EUR 29677. 10.2760/93808

Allison, I. (2016, October 12). Shipping giant Maersk tests blockchain-powered bill of lading. *International Business Times UK*. https://www.ibtimes.co.uk/shipping-giant-maersk-tests-blockchain-powered-bills-lading-1585929

Alotaibi, B. (2019). Utilizing blockchain to overcome cyber security concerns in the internet of things: A review. *IEEE Sensors Journal*, *19*(23), 10953–10971. doi:10.1109/JSEN.2019.2935035

Alshattnawi. (2017). Smart Water Distribution Management System Architecture Based on Internet of Things and Cloud Computing. *International Conference on New Trends.* ieeexplore.ieee.org

Anagnostopoulos, I. (2018). Fintech and Regtech: Impact on regulators and banks. *Journal of Economics and Business*, *100*, 7–25. doi:10.1016/j.jeconbus.2018.07.003

Anandarajan. (2018). *Protecting the Internet of medical things: A situational crime-prevention approach. In Cogent Medicine*. Taylor & Francis.

Anitha, A. (2017a). Home security system using internet of things. In *IOP conference series: materials science and engineering*. IOP Publishing.

Anitha, A. (2017b). Garbage monitoring system using IoT. In *IOP Conference Series: Materials Science and Engineering*. IOP Publishing.

Anitha, A., & Kalra, S. (2016). A Cyber defence using artificial home automation system using IoT. *International Journal of Pharmacy and Technology*, *8*, 25358–25364.

Anitha, A., Paul, G., & Kumari, S. (2016). A cyber defence using artificial intelligence. *International Journal of Pharmacy and Technology*, *8*(4), 25325–25357.

Ant Group Co. Ltd. (2020). *Prospectus of Ant Technology Group Co. Ltd.* Available online: https://www.antgroup.com

Ante, L. (2020). Smart Contracts on the Blockchain – A Bibliometric Analysis and Review. SSRN Electronic Journal, 10, 1–48. doi:10.2139/ssrn.3576393 doi:10.2139srn.3576393

Antonopoulos, A. M. (2014). Mastering Bitcoin: Unlocking Digital Cryptocurrencies. In *Google Books*. O'Reilly Media, Inc. https://books.google.com/books?hl=en&lr=&id=IXmrBQAAQBAJ&oi=fnd&pg=PR4&dq=Antonopoulos+AM.+Mastering+Bitcoin:+unlocking+digital+cryptocurrencies.+%E2%80%9C+O%E2%80%99Reilly+Media

Anwer, M., Saad, A., & Ashfaque, A. (2020) Security of IoT using block chain: A review. In *Proceedings of International Conference on Information Science and Communication Technology* (pp. 1–5). 10.1109/ICISCT49550.2020.9079943

Araz, O. M., Choi, T., Olson, D. L., & Sibel, S. F. (2020). Role of Analytics for Operational Risk Management in the Era of Big Data. *Decision Sciences, 51*(6), 1320–1346.

ATLANT. (n.d.). Atlant CTO Attends Cityscape's Real Estate Summit in Dubai. https://blog.atlant.io/denis-donin-atlant-cto-attends-cityscapes-real-estate-summit-in-dubai-e65a47e6a79f

Axon, L., & Goldsmith, M. (2017, July). PB-PKI: A Privacy-aware Blockchain-based PKI. SECRYPT, 311-318.

Azaria, A., Ekblaw, A., Vieira, T., & Lippman, A. (2016, August). Medrec: Using blockchain for medical data access and permission management. In *2016 2nd international conference on open and big data (OBD)*. IEEE.

Bauerle, N. (n.d.). How does blockchain technology work? https://www.coindesk.com/learn/blockchain-101/how-does-blockchain-technology-work

Beatley, T. (2006). Green Urbanism in European Cities. *The Human Metropolis: People and Nature in the 21st-Century City, 297.*

Beck, R., & Müller-Bloch, C. (2017). Blockchain as radical innovation: a framework for engaging with distributed ledgers. *Proceedings of the 50th Hawaii International Conference on System Sciences*, 5390–5399.

Beijing Blockchain Technology Application Association. (2020). *Regulatory Technology Blue Book: China Regulatory Technology Development Report (2020).* Author.

Benčić, F. M., P. Skočir, & Žarko, I. P. (2019). DLTags: DLT and Smart Tags for Decentralized, PrivacyPreserving, and Verifiable Supply Chain Management. *IEEE Access, 7*, 46198–46209. doi:10.1109/ACCESS.2019.2909170

Benítez-Martínez, F. L., Hurtado-Torres, M. V., & Romero-Frías, E. (2021). A neural blockchain for a tokenizable e-Participation model. *Neurocomputing*, *423*, 703–712. doi:10.1016/j.neucom.2020.03.116

Bentov, I., Lee, C., Mizrahi, A., & Rosenfeld, M. (2014). Proof of activity: Extending bitcoin's proof of work via proof of stake. ACM SIGMETRICS Perform. *Evaluation Review*, *42*(3), 34–37.

Bhushan, B., Khamparia, A., Sagayam, K. M., Sharma, S. K., Ahad, M. A., & Debnath, N. C. (2020). Blockchain for smart cities: A review of architectures, integration trends and future research directions. *Sustainable Cities and Society, 61*, 102360.

Bing, T. (2019). *The application prospect of blockchain technology in the financial industry*. People Venture Blockchain.

BIS. PWC. (2020). *Central Bank Digital Currencies (CBDC) interbank or wholesale projects that are furthest in their development as of 2020, by country*. PwC CBDC Global Index. Retrieved from: https://www.statista.com/statistics/1229776/countries-with-most-mature-interbank-cbdg-worldwide/

Biswas, K., & Muthukkumarasamy, V. (2016, December). Securing smart cities using blockchain technology. *2016 IEEE 18th international conference on high performance computing and communications; IEEE 14th international conference on smart city; IEEE 2nd international conference on data science and systems (HPCC/SmartCity/DSS)*, 1392-1393.

Biswas, S., Pawar, M., Badole, S., Galande, N., & Rathod, S. (2021). Cryptocurrency Price Prediction Using Neural Networks and Deep Learning. *2021 7Th International Conference On Advanced Computing And Communication Systems (ICACCS)*. 10.1109/ICACCS51430.2021.9441872

Bitansky, N., Canetti, R., Chiesa, A., & Tromer, E. (2012) From extractable collision resistance to succinct non-interactive arguments of knowledge, and back again. In *Proceedings of Innovations in Theoretical Computer Science Conference* (pp 326–349). 10.1145/2090236.2090263

Boettke, P. (2016). *Special Collections at the GMU Library – October is Nobel Celebration Month | Main | 2016 Southern Economic Association Meetings*. https://www.coordinationproblem.org/2016/10/and-the-nobel-goes-to-oliver-hart-and-bengt-holmstrom.html

Bollaert, H., de Silanes, F. L., & Schwienbacher, A. (2021). Fintech and access to finance. *Journal of Corporate Finance, 68*, 101941. Advance online publication. doi:10.1016/j.jcorpfin.2021.101941

Bruyn, A. S. (2017). Blockchain an introduction Research paper. https://beta.vu.nl/nl/Images/werkstuk-bruyn_tcm235-862258.pdf

Buchanan, W. J., Li, S., & Asif, R. (2017). Lightweight cryptography methods. *Journal of Cyber Security Technology, 1*(3-4), 187–201. doi:10.1080/23742917.2017.1384917

Burstall, R., & Clarke, B. (2019, January 14). *Blockchain, IP, and the Pharma Industry – How Distributed Ledger Technologies can help Secure the Pharma Supply Chain – Blockchain*. blockchain.bakermckenzie.com https://blockchain.bakermckenzie.com/2019/01/14/blockchain-ip-and-the-pharma-industry-how-distributed-ledger-technologies-can-help-secure-the-pharma-supply-chain/

CAICT. (2021a). *Market size of the digital economy in China in selected years from 2005 to 2020*. White Paper on China's Digital Economy Development. Retrieved from: https://www.statista.com/statistics/1250080/china-digital-economy-size/

CAICT. (2021b). *Share of the digital economy in the GDP of China from 2015 to 2020*. White Paper on China's Digital Economy Development. Retrieved from: https://www.statista.com/statistics/1250092/china-digital-economy-gdp-share/

CAICT. (2021c). *Value added share of digital economy in China from 2016 to 2020, by industry*. White Paper on China's Digital Economy Development. Retrieved from: https://www.statista.com/statistics/1250100/china-digital-economy-penetration-rate-by-industry/

Caragliu, A., Del Bo, C., & Nijkamp, P. (2011). Smart cities in europe. *Journal of Urban Technology, 18*(2), 65–82.

Casino, F., Dasaklis, T. K., & Patsakis, C. (2019). A systematic literature review of blockchainbased applications: Current status, classification and open issues. *Telematics and Informatics, 36*, 55–81. doi:10.1016/j.tele.2018.11.006

Catania. (2014). An approach for monitoring and Smart planning of urban solid waste management using the smart-M3 platform. *Proceedings of 15th Conference of Open.* ieeexplore.ieee.org

CBRC General Office. (2021). *Notice on further promoting the high-quality development of financial services for small and micro enterprises in 2021.* Retrieved from: http://www.gov.cn/zhengce/zhengceku/2021-04/26/content_5602140.htm

CCID Industry and Information Research Institute. (2021). *Blockchain development trends in China in 2021.* Retrieved from: https://baijiahao.baidu.com/s?id=1693991735268111467&wfr=spider&for=pc

Chang, S. E., Chen, Y., & Lu, M. (2019). Supply chain reengineering using blockchain technology: A case of smart contract based tracking process. *Technological Forecasting and Social Change*, *144*, 1–11. doi:10.1016/j.techfore.2019.03.015

Chang, Y., Iakovou, E., & Shi, W. (2020). Blockchain in global supply chains and cross border trade: A critical synthesis of the stateoftheart, challenges and opportunities. *Null*, *58*(7), 2082–2099. doi:10.1080/00207543.2019.1651946

Chaudhary, R., Jindal, A., Aujla, G. S., Kumar, N., Das, A. K., & Saxena, N. (2018). Lscsh: Lattice-based secure cryptosystem for smart healthcare in smart cities environment. *IEEE Communications Magazine*, *56*(4), 24–32.

Chen, S., Shi, R., Ren, Z., Yan, J., Shi, Y., & Zhang, J. (2017, November). A blockchain-based supply chain quality management framework. In *2017 IEEE 14th International Conference on e-Business Engineering (ICEBE).* IEEE.

Chen, W., Xu, Z., Shi, S., Zhao, Y., & Zhao, J. (2018). A Survey of Blockchain Applications in Different Domains. *Proceedings of the 2018 International Conference on Blockchain Technology and Application - ICBTA 2018.* 10.1145/3301403.3301407

Chen, F., Xiao, Z., Cui, L., Lin, Q., Li, J., & Yu, S. (2020). Blockchain for Internet of things applications: A review and open issues. *Journal of Network and Computer Applications*, 102839.

Cheng, B. (2015). Building a big data platform for smart cities: Experience and lessons from Santander. In *2015 IEEE International Congress on Big Data (BigData Congress'15).* IEEE.

Cheng, S., Wang, L., Ao, N., & Han, Q. (2020). A selective video encryption scheme based on coding characteristics. *Symmetry*, *12*(3), 332. doi:10.3390ym12030332

Chen, L., Xu, L., Shah, N., Gao, Z., Lu, Y., & Shi, W. (2017, November). On security analysis of proof-of-elapsed-time (poet). In *International Symposium on Stabilization, Safety, and Security of Distributed Systems* (282-297). Springer.

Chen, W., Xu, Z., Shi, S., Zhao, Y., & Zhao, J. (2018, December). A survey of blockchain applications in different domains. *Proceedings of the 2018 International Conference on Blockchain Technology and Application*, 17-21.

China Insurance Regulatory Commission. (2017). *Circular of CIRC on Printing and Distributing the 'Construction Plan of the Second Generation Project of Compensation'.* Retrieved from: http://www.gov.cn/xinwen/2017-09/20/content_5226368.htm

China Patent Protection Association. (2020). *2020 Global Patent Authorization report in blockchain field.* Retrieved from: http://www.199it.com/archives/1077801.html

China Strategic Emerging Industry. (2018). The world's first cross-border blockchain remittance was born in Hong Kong. *China Strategic Emerging Industry*, *14*, 12–13.

Chinese IoT Blockchain Branch. (2021). *2021 China industry blockchain ecological map.* Retrieved from: https://mp.weixin.qq.com/s/zkwrOTSFPaM74pW5bUO_hg

Choi, Wen, Sun, & Chung. (2019). The mean-variance approach for global supply chain risk analysis with air logistics in the blockchain technology era. *Transportation Research Part E: Logistics and Transportation Review, 127*, 178–191.

Choi, T., Guo, S., Liu, N., & Shi, X. (2020). Optimal pricing in ondemandserviceplatformoperations with hired agents and risksensitive customers in the blockchain era. *European Journal of Operational Research, 284*(3), 1031–1042. doi:10.1016/j.ejor.2020.01.049

Choi, T.-M. (2020). Creating all-win by blockchain technology in supply chains: Impacts of agents' risk attitudes towards cryptocurrency. *The Journal of the Operational Research Society, 0*, 1–16. doi:10.1080/01605682.2020.1800419

Chu & Yong. (2019, November 28). *Fake degrees: A worrying trend in Malaysia.* Retrieved from https://www.thestar.com.my/news/nation/2019/05/06/a-worrying-trend-in-msia/

Cioroianu, I., Corbet, S., & Larkin, C. (2021). The differential impact of corporate blockchain-development as conditioned by sentiment and financial desperation. *Journal of Corporate Finance, 66*, 101814. doi:10.1016/j.jcorpfin.2020.101814

Clavin, J., Duan, S., Zhang, H., Janeja, V. P., Joshi, K. P., Yesha, Y., ... Li, J. D. (2020). Blockchains for Government: Use Cases and Challenges. *Digital Government: Research and Practice, 1*(3), 1–21.

CoinMarketCap. (2021, October 11). *Cryptocurrency Prices, Charts And Market Capitalizations.* https://coinmarketcap.com

Cong, L. W., & He, Z. (2019). Blockchain disruption and smart contracts. *Review of Financial Studies, 32*(5), 1754–1797. doi:10.1093/rfs/hhz007

Corluka, D., & Lindh, U. (2017). Blockchain - A New Technology That Will Transform the Real Estate Market. Thesis, 74.

Corten, P. A. (2018). Implementation of Blockchain Powered Smart Contracts in Governmental Services (Issue March). https://repository.tudelft.nl/islandora/object/uuid:87709465-b9a1-48da-9ba5-eba98bc263d7

CrunchBase, SCMP. (2020). *Total funding of the leading fintech and blockchain startups in China as of June 2020 (in million U.S. dollars).* China internet report 2020.

CryptoLUX. (n.d.). *Lightweight cryptography.* https://www.cryptolux.org/index.php/LightweightCryptography

Cyberspace Administration of China. (2019). *Blockchain Information Service Management Regulations.* Retrieved from: http://www.cac.gov.cn/2019-01/10/c_1123971164.htm

Cyberspace Administration of China. (2019). *On the notice of the first batch of domestic blockchain information service record number of the announcement.* Retrieved from: http://www.cac.gov.cn/2019-03/30/c_1124305122.htm?from=singlemessage&isappinstalled=0

Czymmek, V., Moller, C., Harders, L., & Hussmann, S. (2021). Deep Learning Approach for high Energy efficient Real-Time Detection of Weeds in Organic Farming. *2021 IEEE International Instrumentation And Measurement Technology Conference (I2MTC).* 10.1109/I2MTC50364.2021.9459943

D'Agostino, R., & Schmidt, T. (2020). EMOTIONET: A Multi-Convolutional neural Network hierarchical approach to facial and Emotional classification using TPUs. *2020 IEEE International Conference on Engineering, Technology and Innovation (ICE/ITMC).* 10.1109/ICE/ITMC49519.2020.9198366

da Silva, R. G., Dal Molin Ribeiro, M. H., Fraccanabbia, N., Mariani, V. C., & dos Santos Coelho, L. (2020). Multi-step ahead bitcoin price FORECASTING based on VMD and ensemble learning methods. *2020 International Joint Conference on Neural Networks (IJCNN).* 10.1109/IJCNN48605.2020.9207152

Dai, F., Shi, Y., Meng, N., Wei, L., & Ye, Z. (2017). From Bitcoin to cybersecurity: A comparative study of blockchain application and security issues. 2017 4th International Conference on Systems and Informatics, ICSAI 2017, (61471129), 975–979. 10.1109/ICSAI.2017.8248427

Dang, C., Zhang, J., Kwong, C. P., & Li, L. (2019). Demand side load management for big industrial energy users under blockchain-based peer-to-peer electricity market. *IEEE Transactions on Smart Grid, 10*(6), 6426–6435.

Daniel, F., & Guida, L. (2019). A service-oriented perspective on blockchain smart contracts. IEEE Internet Computing, 23(1), 46–53. doi:10.1109/MIC.2018.2890624 doi:10.1109/MIC.2018.2890624

Deepak, K., Badiger, A., Akshay, J., Awomi, A., Deepak, G., & Kumar, N. (2020) Blockchain-based management of video surveillance systems: A survey. In *Proceedings of Int. Conf. Adv. Comput. Commun. Syst* (pp. 1256–1258) 10.1109/ICACCS48705.2020.9074197

Deloitte. (2016). Blockchain Opportunity Contents. Deloitte.

Deloitte. (2017, September 20). *Using Blockchain to Drive Supply Chain Transparency and Innovation*. Deloitte United States. https://www2.deloitte.com/us/en/pages/operations/articles/blockchain-supply-chain-innovation.html

Demestichas. (2015). Intelligent 5G Networks: Managing 5G Wireless\/Mobile Broadband. *IEEE Vehicular Technology Magazine, 10*(3), 41–50. doi:10.1109/mvt.2015.2446419

Dempere, J. (2019). *Factors Affecting the Return and Volatility of Major Cryptocurrencies. In 2019 Sixth HCT Information Technology Trends*. ITT. doi:10.1109/ITT48889.2019.9075117

Dey, S., & Sarkar, S. (2021). A theoretical investigation on the distinguishers of Salsa and ChaCha. *Discrete Applied Mathematics, 302*, 147–162. doi:10.1016/j.dam.2021.06.017

Di Vaio, Assunta, & Varriale. (2020). Blockchain technology in supply chain management for sustainable performance: Evidence from the airport industry. *International Journal of Information Management, 52*, 102014.

Dimitriou, T. (2020). Efficient, coercion-free and universally verifiable blockchain-based voting. *Computer Networks, 174*, 107234. doi:10.1016/j.comnet.2020.107234

Ding, L. (2019). Improved related-cipher attack on Salsa20 stream cipher. *IEEE Access: Practical Innovations, Open Solutions, 7*, 30197–30202. doi:10.1109/ACCESS.2019.2892647

Ding, L., Liu, C., Zhang, Y., & Ding, Q. (2019). A new lightweight stream cipher based on chaos. *Symmetry, 11*(7), 853. doi:10.3390ym11070853

Docusign. (n.d.). eSignature Plans and Pricing. https://www.docusign.com/products-and-pricing

Dorri, A. (2017). *Blockchain for IoT Security and Privacy: The Case Study of a Smart Home*. Academic Press.

Duong, T., Chepurnoy, A., Fan, L., & Zhou, H. (2018) TwinsCoin: A cryptocurrency via proof-of-work and proof-of-stake. In *Proceedings of the ACM Workshop on Blockchains, Cryptocurrencies, and Contracts, Co-located with ASIA CCS* (pp 1–13). 10.1145/3205230.3205233

Dziembowski, S., Faust, S., Kolmogorov, V., & Pietrzak, K. (2015). Proofs of space. Lecture Notes in Computer Science, 9216(616160), 585–605. doi:10.1007/978-3-662-48000-7_29

Eastmoney Choice database. (2021). Retrieved from: http://choice.eastmoney.com/

Eckhoff, D., & Wagner, I. (2017). Privacy in the smart city—Applications, technologies, challenges, and solutions. *IEEE Communications Surveys and Tutorials, 20*(1), 489–516.

E-commerce Definition - What is Ecommerce. (n.d.). https://www.shopify.com/encyclopedia/what-is-ecommerce

ECRYPT. (n.d.). *eSTREAM: the ECRYPT stream cipher project*. https://www.ecrypt.eu.org/stream/

Efanov, D., & Roschin, P. (2018). The AllPervasiveness of the Blockchain Technology. *8th Annual International Conference on Biologically Inspired Cognitive Architectures, BICA 2017 (Eighth Annual Meeting of the BICA Society), 123*, 116–121.

El Maouchi, M., Ersoy, O., & Erkin, Z. (2018b). TRADE: A Transparent, Decentralized Traceability System for the Supply Chain. *ERCIM - The European Research Consortium for Informatics and Mathematics*. doi:10.18420/blockchain2018_01

Elbahrawy, J., Lovejoy, J., Ouyang, A., & Perez, J. (2020). Analysis of Bitcoin improvement proposal 340 — Schnorr signatures. MIT Report, 1–15.

Emmanuel, M., & Rayudu, R. (2016). Communication technologies for smart grid applications: A survey. *Journal of Network and Computer Applications, 74*, 133–148.

Ernst & Young. (2016). UK FinTech census: The voice of FinTech, HM treasury. *EY and Innovate Finance*. Retrieved from: http://www.ey.com/Publication/vwLUAssets/EY-UK-FinTechCensus-2017/$FILE/EY-UK-FinTech-Census-2017.pdf

Ersoy, P., & Börühan, G. (2015). Intelligent transportation systems and their applications in road transportation industry in Turkey. *The International Conference on Logistics & Sustainable Transport 2015*.

Fake degrees in Malaysia offered online from S$4,000. (n.d.). Retrieved from https://www.todayonline.com/world/fake-degrees-malaysia-offered-online-s4000

Fan, T., & Chen, Y. (2010, September). A scheme of data management in the Internet of Things. In *2010 2nd IEEE International Conference on Network Infrastructure and Digital Content*. IEEE.

Farooq, M. U., Waseem, M., Mazhar, S., Khairi, A., & Kamal, T. (2015). A review on internet of things (IoT). *International Journal of Computers and Applications, 113*(1), 1–7.

Fattahi, S., Makanju, A., & Fard, A. (2020) SIMBA: An efficient simulator for blockchain applications. In *Proceedings of IEEE/IFIP Int. Conf. Dependable Syst. Networks* (pp. 51–52). 10.1109/DSN-S50200.2020.00028

Fernández-Caramés, T. M., Blanco-Novoa, O., Froiz-Míguez, I., & Fraga-Lamas, P.Fernández-Caramés. (2019). Towards an autonomous industry 4.0 warehouse: A UAV and blockchain-based system for inventory and traceability applications in big data-driven supply chain management. *Sensors (Basel), 19*(10), 10. Advance online publication. doi:10.339019102394 PMID:31130644

Fernández-Caramés, T. M., & Fraga-Lamas, P. (2018). A Review on the Use of Blockchain for the Internet of Things. *IEEE Access: Practical Innovations, Open Solutions, 6*, 32979–33001.

Ferrag, M. A., Derdour, M., Mukherjee, M., Derhab, A., Maglaras, L., & Janicke, H. (2018). Blockchain technologies for the internet of things: Research issues and challenges. *IEEE Internet of Things Journal, 6*(2), 2188–2204.

Fill, H., & Haerer, F. (2018) Knowledge Blockchains: Applying Blockchain technologies to enterprise modelling. In *Proceedings of the Hawaii International Conference on System Sciences* (pp 4045–4054). 10.24251/HICSS.2018.509

Finance, F. (2020). Fintech helps securities business transform and restructure. *China Securities Journal*. Retrieved from: https://baijiahao.baidu.com/s?id=1654762300223125200&wfr=spider&for=pc

Fish Farming - The IoT Marketplace. (n.d.). https://www.the-iot-marketplace.com/solutions/fish-farming

Fosso Wamba, S., Kala Kamdjoug, J. R., Epie Bawack, R., & Keogh, J. G. (2019). Bitcoin, Blockchain and Fintech: A systematic review and case studies in the supply chain. *Production Planning and Control, 31*(2-3), 115–142. doi:10.1080/09537287.2019.1631460

Frentrup, M., & Theuvsen, L. (2009). Information technologies and transparency in agri-food supply chains: empirical results from the German pig and dairy production. *Georg-August University Goettingen*. Department of Agricultural Economics and Rural Development.

Frisby, D. (2014) Who is Satoshi Nakamoto? In Bitcoin: The Future of Money (pp. 85–149). Academic Press.

Fu, Y., & Zhu, J. (2019). Big production enterprise supply chain endogenous risk management based on blockchain. *IEEE Access: Practical Innovations, Open Solutions*, *7*, 15310–15319. doi:10.1109/ACCESS.2019.2895327

Gabay, D., Akkaya, K., & Cebe, M. (2020). Privacy-preserving authentication scheme for connected electric vehicles using blockchain and zero knowledge proofs. *IEEE Transactions on Vehicular Technology*, *69*(6), 5760–5772. doi:10.1109/TVT.2020.2977361

Gallo, P., Pongnumkul, S., & Nguyen, U. (2018) BlockSee: Blockchain for IoT video surveillance in smart cities. In *Proceedings of IEEE Int. Conf. Environ. Electr. Eng., IEEE Ind. Commer. Power Syst. Eur.* 10.1109/EEEIC.2018.8493895

Gao, W., Hatcher, W. G., & Yu, W. (2018). A Survey of Blockchain: Techniques, Applications, and Challenges. *2018 27th International Conference on Computer Communication and Networks (ICCCN)*.

Gao, Z., Xu, L., Chen, L., Zhao, X., Lu, Y., & Shi, W. (2018). CoC: A Unified Distributed Ledger Based Supply Chain Management System. *Journal of Computer Science and Technology*, *33*(2), 237–248. doi:10.100711390-018-1816-5

Gatteschi, V., Lamberti, F., Demartini, C., Pranteda, C., & Santamaria, V. (2018). To blockchain or not to blockchain: That is the question. IT Professional, 20(2), 62–74. doi:10.1109/MITP.2018.021921652 doi:10.1109/MITP.2018.021921652

General Office of Ministry of Industry and Information Technology. (2018). *"Circular of the General Office of the Ministry of Industry and Information Technology on the Application of the Pilot and Demonstration Projects of Information Consumption.* Retrieved from: http://www.gov.cn/zhengce/zhengceku/2021-03/04/content_5590162.htm

General Office of the State Council. (2017). *Guiding Opinions of the General Office of the State Council on Actively Promoting Innovation and Application of Supply Chain*. Retrieved from: http://www.gov.cn/zhengce/content/2017-10/13/content_5231524.htm

General Office of the State Council. (2017). *Opinions of the General Office of the State Council on Innovating Management, Improving Services, Cultivating and Strengthening New Driving Forces of Economic Development, and Accelerating the Transition Between Old and New Driving Forces of Economic Development.* Retrieved from: http://www.gov.cn/zhengce/content/2017-01/20/content_5161614.htm

General Postal Office. (2017). *Guiding Opinions of State Post Bureau on Promoting the Construction of "One Belt And One Road" in Postal Service*. Retrieved from: http://www.spb.gov.cn/zc/flfgjzc_1/201801/t20180112_1467509.html

Georgoula, I., Pournarakis, D., Bilanakos, C., Sotiropoulos, D. N., & Giaglis, G. M. (2015). Using time-series and sentiment analysis to detect the determinants of bitcoin prices. SSRN *Electronic Journal*. doi:10.2139/ssrn.2607167

Gipp, B., Kosti, J., & Breitinger, C. (2016) Securing video integrity using decentralized trusted timestamping on the bitcoin blockchain. *Proceedings of Mediterr. Conf. Inf. Syst.*

Golosova, J., & Romanovs, A. (2018). The advantages and disadvantages of the blockchain technology. 6th Workshop on Advances in Information, Electronic and Electrical Engineering (AIEEE). doi:10.1109/AIEEE.2018.859225310.1109/AIEEE.2018.8592253

Gomathi, S., Soni, M., Dhiman, G., Govindaraj, R., & Kumar, P. (2021). A survey on applications and security issues of blockchain technology in business sectors. Materials Today: Proceedings. Advance online publication. doi:10.1016/j.matpr.2021.02.088 doi:10.1016/j.matpr.2021.02.088

Gomber, P., Koch, J. A., & Siering, M. (2017). Digital Finance and FinTech: Current research and future research directions. *Journal of Business Economics*, *87*(5), 537–580. doi:10.100711573-017-0852-x

Graglia. (2018). Blockchain and property in 2018: At the end of the beginning. Academic Press.

Gramoli, V. (2020). From blockchain consensus back to Byzantine consensus. *Future Generation Computer Systems*, *107*, 760–769.

Grid Control | PRIME Alliance. (n.d.). https://www.prime-alliance.org/?page_id=312

Groth, J., & Maller, M. (2017) Snarky signatures: Minimal signatures of knowledge from simulation-extractable SNARKs. Lecture Notes in Computer Science, 581–612.

Gubbi, J., Buyya, R., Marusic, S., & Palaniswami, M. (2013). Internet of Things (IoT): A vision, architectural elements, and future directions. *Future Generation Computer Systems*, *29*(7), 1645–1660.

Guegan, D. (2018). The Digital World : II – Alternatives to the Bitcoin Blockchain ? Halshs.

Gui, T., Ma, C., Wang, F., & Wilkins, D. E. (2016, March). *Survey on swarm intelligence-based routing protocols for wireless sensor networks: An extensive study. In 2016 IEEE international conference on industrial technology (ICIT).* IEEE.

Guler, M. G. (2015). *2.1.7. Technology Infrastructure Analysis H1.* Yildiz Technical University.

Guo, Y. M., Huang, Z. L., Guo, J., Guo, X.-R., Li, H., Liu, M.-Y., Ezzeddine, S., & Nkeli, M. J. (2021). A bibliometric analysis and visualization of blockchain. *Future Generation Computer Systems*, *116*, 316–332. doi:10.1016/j.future.2020.10.023

Gurtu, A., & Johny, J. (2019). Potential of blockchain technology in supply chain management: A literature review. *International Journal of Physical Distribution & Logistics Management*, *49*(9), 881–900. doi:10.1108/IJPDLM-11-2018-0371

Hakak, S., Khan, W., Gilkar, G., Imran, M., & Guizani, N. (2020). Securing smart cities through blockchain technology: Architecture, requirements, and challenges. *IEEE Network*, *34*(1), 8–14. doi:10.1109/MNET.001.1900178

Hall. (n.d.). *Online cheats: The rise of 'fake' degrees in Malaysia.* Retrieved from https://thepienews.com/analysis/documentary-highlights-malaysian-students-with-fake-degrees/

Hao, W., Ma, S.L., Dai, H.N., Imran, M., & Wang, T.S. (2020). Blockchain-based data privacy management with Nudge theory in open banking. *Future Generation Computer Systems*, *110*.

Hao, Z., Mao, D., Zhang, B., Zuo, M., & Zhao, Z. (2020). A novel visual analysis method of food safety risk traceability based on blockchain. *International Journal of Environmental Research and Public Health*, *17*(17), 2300. doi:10.3390/ijerph17072300 PMID:32235378

Harris, C. (2019) Consensus-based secret sharing in blockchain smart contracts. In Int Work Big Data Inf Secur (pp. 79–84). doi:10.1109/IWBIS.2019.8935853

Hassan, M. U., Rehmani, M. H., & Chen, J. (2019). Privacy preservation in blockchain based IoT systems: Integration issues, prospects, challenges, and future research directions. *Future Generation Computer Systems*, *97*, 512–529.

HBSGB Transport. (2013-2015). *National Smart Transport Systems Strategy Document 2013-123 and Action Plan.* Draft. Author.

Helo, P., & Hao, Y. (2019). Blockchains in operations and supply chains: A model and reference implementation. *Computers & Industrial Engineering*, *136*, 242–251.

Hernández-Muñoz, J. M., Vercher, J. B., Muñoz, L., Galache, J. A., Presser, M., Luis A Hernández, G., & Pettersson, J. (2011). Smart cities at the forefront of the future internet. In *The Future Internet Assembly* (Vol. 6656, pp. 447–462). Springer.

Hollands, R. G. (2008). Will the real smart city please stand up? Intelligent, progressive, or entrepreneurial? *City*, *12*(3), 303–320.

Hou, H. (2017). The application of blockchain technology in E-government in China. 26th International Conference on Computer Communication and Networks (ICCCN). doi:10.1109/ICCCN.2017.803851910.1109/ICCCN.2017.8038519

Hu, R. (2019) *Visual blockchain using Merkle tree* [Master's Thesis]. Auckland University of Technology.

Hu, R., & Yan, W. (2020) Design and implementation of visual blockchain with Merkle tree. In Handbook of Research on Multimedia Cyber Security (pp. 282 – 295). doi:10.4018/978-1-7998-2701-6.ch014

Hughes, A., Park, A., Kietzmann, J., & ArcherBrown, C. (2019). Beyond Bitcoin: What blockchain and distributed ledger technologies mean for firms. *Business Horizons, 62*(3), 273–281.

Huhmo, M. (2018). Blockchain technology, Bitcoin as a case. In Oulu University of Applied Sciences. Oulu University of Applied Sciences.

Hu, J., He, D., Zhao, Q., & Choo, K. K. R. (2019). Parking management: A blockchain based privacy-preserving system. *IEEE Consumer Electronics Magazine*, *8*(4), 45–49.

Iansiti, M., & Lakhani, K. (2018, March 6). The Truth About Blockchain. *Harvard Business Review*. https://hbr.org/2017/01/the-truth-about-blockchain

Ibba, S., Pinna, A., Seu, M., & Pani, F. E. (2017, May). CitySense: blockchain-oriented smart cities. *Proceedings of the XP2017 Scientific Workshops*, 1-5.

IBM Institute for Business Value. (2017, January). *Building trust in government*. IBM. https://www.ibm.com/thought-leadership/institute-business-value/report/blockchain-for-government

IBM. (2017, March 5). *Maersk and IBM Unveil Supply Chain Solution on Blockchain*. https://www-03.ibm.com/press/us/en/pressrelease/51712.wss#feeds

Infopulse. (2019, October 7). *Blockchain in Supply Chain Management: Key Use Cases and Benefits*. Medium. https://medium.com/@infopulseglobal_9037/blockchain-in-supply-chain-management-key-use-cases-and-benefits-6c6b7fd43094

Iraqi laws and legislation, Pub. L. No. 184 (1984). http://wiki.dorar-aliraq.net/iraqilaws/law/18200.html

Irreversible Transactions - Bitcoin Wiki. (n.d.). *En.bitcoin.it*. https://en.bitcoin.it/wiki/Double-spending

Işıkdağ, Ü. (2011). Barriers to E-Procurement in Turkish AEC Industry. *CIB W078 2011 Conference*. https://www.academia.edu/5469207/Barriers_to_E_Procurement_in_Turkish_AEC_Industry

Iskakova, A., Kumar Nunna, H. S. V. S., & Siano, P. (2020). Ethereum blockchain-based peer-to-peer energy trading platform. PECon 2020 - 2020 IEEE International Conference on Power and Energy, 327–331. 10.1109/PECon48942.2020.9314591

ITU-T. (2014). *Smart Sustainable Cities: An Analysis of Definitions*. Tech. Report. FG-SSC-10/2014.

Ivanov, D., Dolgui, A., & Sokolov, B. (2019). The impact of digital technology and Industry 4.0 on the ripple effect and supply chain risk analytics. *Null*, *57*(3), 829–846. doi:10.1080/00207543.2018.1488086

J. AlJaroodi, & Mohamed, N. (2019). Blockchain in Industries: A Survey. *IEEE Access, 7*, 36500–36515. doi:10.1109/ACCESS.2019.2903554

Jamil, F., Hang, L., Kim, K., & Kim, D. (2019). A Novel Medical Blockchain Model for Drug Supply Chain Integrity Management in a Smart Hospital. *Electronics (Basel)*, *8*(5), 505. doi:10.3390/electronics8050505

Jan, B., Farman, H., Khan, M., Talha, M., & Din, I. U. (2019). Designing a smart transportation system: An internet of things and big data approach. *IEEE Wireless Communications*, *26*(4), 73–79.

Jayachandran, P. (2017). The difference between public and private blockchain. IBM. https://www.ibm.com/blogs/blockchain/2017/05/the-difference-between-public-and-private-blockchain/

Jiang, S. R., Li, Y. Z., Wang, S. Y., & Lin, Z. (2021). Blockchain competition: The tradeoff between platform stability and efficiency. *European Journal of Operational Research*. Advance online publication. doi:10.1016/j.ejor.2021.05.031

Jiao, L., Li, Y., & Hao, Y. (2020). A guess-and-determine attack on SNOW-V stream cipher. *The Computer Journal*, *63*(12), 1789–1812. doi:10.1093/comjnl/bxaa003

Johansson, J., & Nilsson, C. (2018). How the blockchain technology can enhance sustainability for contractors within the construction industry. Chalmers University of Technology. https://publications.lib.chalmers.se/records/fulltext/255155/255155.pdf

Jo, J. H., Sharma, P. K., Sicato, J. C. S., & Park, J. H. (2019). Emerging technologies for sustainable smart city network security: Issues, challenges, and countermeasures. *Journal of Information Processing Systems*, *15*(4), 765–784.

Juma, H., Shaalan, K., & Kamel, I. (2019). A Survey on Using Blockchain in Trade Supply Chain Solutions. *IEEE Access*, 1. doi:10.1109/ACCESS.2019.2960542

Junejo, A. Z., Memon, M. M., Junejo, M. A., Talpur, S., & Memon, R. M. (2020). Blockchains Technology Analysis: Applications, Current Trends and Future Directions—An Overview. *Intelligent Computing and Innovation on Data Science Lecture Notes in Networks and Systems*, 411-419.

Junis, F. (2019). A Revisit on Blockchain-based Smart Contract Technology. Academic Press.

Kalbo, N., Mirsky, Y., Shabtai, A., & Elovici, Y. (2020). The security of IP-based video surveillance systems. *Sensors (Basel)*, *20*(17), 4806. doi:10.339020174806 PMID:32858840

Kalis, R. (2018). Using blockchain to validate audit trail data in private business applications. Universiteit van Amsterdam.

Kamble, S. S., Gunasekaran, A., & Sharma, R. (2020). Modeling the blockchain enabled traceability in agriculture supply chain. *International Journal of Information Management, 52*, 101967.

Kanter & Litow. (2009). *Informed and interconnected: A manifesto for smarter cities.* https://papers.ssrn.com/sol3/papers.cfm?abstract_id=1420236

Karamitsos, I., Papadaki, M., & Al Barghuthi, N. B. (2018). Design of the Blockchain Smart Contract: A Use Case for Real Estate. Journal of Information Security, 09(03), 177–190. doi:10.4236/jis.2018.93013 doi:10.4236/jis.2018.93013

Karami, Z., & Kashef, R. (2020). Smart transportation planning: Data, models, and algorithms. *Transportation Engineering*, *2*, 100013. doi:10.1016/j.treng.2020.100013

Khan, P., Byun, Y., & Park, N. (2020). A data verification system for CCTV surveillance cameras using blockchain technology in smart cities. *Electron*, *9*(3).

Khan, K. M., Arshad, J., & Khan, M. M. (2018). Secure digital voting system based on blockchain technology. *International Journal of Electronic Government Research*, *14*(1), 53–62. doi:10.4018/IJEGR.2018010103

Khan, K. M., Arshad, J., & Khan, M. M. (2020). Investigating performance constraints for blockchain based secure e-voting system. *Future Generation Computer Systems*, *105*, 13–26. doi:10.1016/j.future.2019.11.005

Khan, M., & Salah, K. (2018). IoT security: Review, blockchain solutions, and open challenges. *Future Generation Computer Systems*, *82*, 395–411. doi:10.1016/j.future.2017.11.022

Khan, M., Silva, B. N., & Han, K. (2016). Internet of things based energy aware smart home control system. *IEEE Access: Practical Innovations, Open Solutions*, *4*, 7556–7566.

Khanna, A., & Anand, R. (2016, January). IoT based smart parking system. In *2016 International Conference on Internet of Things and Applications (IOTA)* (pp. 266-270). IEEE.

Khazaei, S., & Rezaei-Aliabadi, M. (2018). A rigorous security analysis of a decentralized electronic voting protocol in the universal composability framework. *Journal of Information Security and Applications, 43*, 99-109.

Khosla, D., Sharma, M., Sharma, A., Budhiraja, A., & Singh, S. (2019). Blockchain based supply chain management: An overview. *International Journal of Control and Automation*, *12*, 424–430.

Khrais, L. (2020) The combination of IoT-sensors in appliances and block-chain technology in smart cities energy solutions. In *Proceedings of Int Conf Adv Comput Commun Syst* (pp. 1373–1378). 10.1109/ICACCS48705.2020.9074362

Kikitamara, S. (2017). Digital Identity Management on Blockchain for Open Model Energy System. Radboud University.

Kim, H., Bock, G., & Lee, G. (2021). Predicting Ethereum prices with machine learning based on Blockchain information. *Expert Systems with Applications*, *184*, 115480. doi:10.1016/j.eswa.2021.115480

Kimmich, M. (2019, November 5). *Devices & Technologies - Blockchain*. https://www.juniperresearch.com/researchstore/devices-technology/blockchain-research-report

Kirit, N., & Sarkar, P. (2017). EscrowChain: Leveraging ethereum blockchain as escrow in real estate. International Journal of Innovative Research in Computer and Communication Engineering, 5(10).

Knutson, P. (2020). *Cryptanalytic attack on Salsa20*. https://github.com/catsymptote/Salsacryptanalysis

Kolias, C., Kambourakis, G., Stavrou, A., & Voas, J. (2017). DDoS in the IoT: Mirai and other botnets. *Computer*, *50*(7), 80–84.

Konya intelligent transportation system applications. (2019). Retrieved from https://akillisehir.konya.bel.tr/uygulama/atus-akilli-toplu-ulasim-sistemi https://atus.konya.bel.tr

Kopylash, V. (2018). An Ethereum-based Real Estate Application with Tampering-resilient Document Storage. University of Tartu.

Ko, T., Lee, J., & Ryu, D. (2018). Blockchain Technology and Manufacturing Industry: Real-Time Transparency and Cost Savings. *Sustainability*, *10*(11), 4274. doi:10.3390u10114274

Krishnamurthy, R., Rathee, G., & Jaglan, N. (2020). An enhanced security mechanism through blockchain for E-polling/counting process using IoT devices. *Wireless Networks*, *26*(4), 2391–2402. doi:10.100711276-019-02112-5

Kristoufek, L. (2021). Tethered, or Untethered? On the interplay between stablecoins and major cryptoassets. *Finance Research Letters*, *101991*. Advance online publication. doi:10.1016/j.frl.2021.101991

Kshetri, N. (2018). Blockchain's roles in meeting key supply chain management objectives. *International Journal of Information Management*, *39*, 80–89. doi:10.1016/j.ijinfomgt.2017.12.005

Kshetri, N. (2021). Blockchain and sustainable supply chain management in developing countries. *International Journal of Information Management, 60*, 102376. doi:10.1016/j.ijinfomgt.2021.102376

Kshetri, N., & Voas, J. (2018). Blockchain-enabled e-voting. *IEEE Software, 35*(4), 95–99.

Kumar, D., & Rath, S. K. (2020). Predicting the Trends of Price for Ethereum Using Deep Learning Techniques. In S. Dash, C. Lakshmi, S. Das, & B. Panigrahi (Eds.), *Artificial Intelligence and Evolutionary Computations in Engineering Systems. Advances in Intelligent Systems and Computing* (Vol. 1056). Springer. doi:10.1007/978-981-15-0199-9_9

Kumar, N. M., & Mallick, P. K. (2018). Blockchain technology for security issues and challenges in IoT. *Procedia Computer Science, 132*, 1815–1823. doi:10.1016/j.procs.2018.05.140

Kuyumcu, Y. (2017). *Bicycle as a Means of Transportation and Bicycle Path Proposal for Urban Transportation in Çorum* [Master Thesis]. Bahçeşehir University.

Lai, F., & Carsten, O. (2012). What benefit does Intelligent Speed Adaptation deliver: A close examination of its effect on vehicle speeds. *Accident; Analysis and Prevention, 48*, 4–9. doi:10.1016/j.aap.2010.01.002 PMID:22664662

Larimer, D. (2014). Delegated proof-of-stake (dpos). *Bitshare Whitepaper, 81*, 85.

Larriba, A. M., Sempere, J. M., & López, D. (2020). A two authorities electronic vote scheme. *Computers & Security, 97*, 101940. doi:10.1016/j.cose.2020.101940

Lee, Y., Rathore, S., Park, J. H., & Park, J. H. (2020). A blockchain-based smart home gateway architecture for preventing data forgery. *Human-centric Computing and Information Sciences, 10*(1), 1–14.

Li, G., Li, L., Choi, T.-M., & Sethi, S. P. (2020). Green supply chain management in Chinese firms: Innovative measures and the moderating role of quick response technology. *Journal of Operations Management, 66*, 958–988.

Liang, X., Zhao, J., Shetty, S., & Li, D. (2017, October). Towards data assurance and resilience in IoT using blockchain. In *MILCOM 2017-2017 IEEE Military Communications Conference (MILCOM)* (pp. 261–266). IEEE.

Liao, S., Wu, J., Li, J., Bashir, A., & Yang, W. (2021). Securing collaborative environment monitoring in smart cities using Blockchain enabled software-defined Internet of drones. *IEEE Internet Things (Mag), 4*(1), 12–18. doi:10.1109/IOTM.0011.2000045

Li, J., Boonaert, J., Doniec, A., & Lozenguez, G. (2021). Multi-models machine learning methods for traffic flow estimation from Floating Car Data. *Transportation Research Part C, Emerging Technologies, 132*, 103389. doi:10.1016/j.trc.2021.103389

Li, J., Jin, M., Wang, L., Qin, B., & Wang, K. (2016). MDS clinical diagnostic criteria for Parkinson's disease in China. *Journal of Neurology, 264*(3), 476–481. doi:10.100700415-016-8370-2 PMID:28025665

Li, J., Wang, X., Huang, Z., Wang, L., & Xiang, Y. (2019). Multi-level multi-secret sharing scheme for decentralized e-voting in cloud computing. *Journal of Parallel and Distributed Computing, 130*, 91–97. doi:10.1016/j.jpdc.2019.04.003

Li, L., Freeman, G., & Wohn, D. Y. (2021). The Interplay of Financial Exchanges and Offline Interpersonal Relationships through Digital Peer-to-Peer Payments. *Telematics and Informatics, 63*, 63. doi:10.1016/j.tele.2021.101671

Lin, Z., Yu, S., Feng, X., & Lü, J. (2018). Cryptanalysis of a chaotic stream cipher and its improved scheme. *International Journal of Bifurcation and Chaos in Applied Sciences and Engineering, 28*(07), 1850086. doi:10.1142/S0218127418500864

Li, T., Liu, Y., Tian, Y., Shen, S., & Mao, W. (2012, November). A storage solution for massive iot data based on nosql. In *2012 IEEE International Conference on Green Computing and Communications*. IEEE.

Litke, A., Anagnostopoulos, D., & Varvarigou, T. (2019). Blockchains for Supply Chain Management: Architectural Elements and Challenges Towards a Global Scale Deployment. *Logistics*, *3*(1), 5. Advance online publication. doi:10.3390/logistics3010005

Liu, S. P. (2019). *Digital Economy New Engine (Blockchain Technology) Innovation Application Summit.* Retrieved from: https://www.cebnet.com.cn/20190419/102566686.html

Liu, Y., & Wang, Q. (2017). An E-voting Protocol Based on Blockchain. *IACR Cryptol. ePrint Arch.*, *2017*, 1043.

Liu, T., Wang, X., & Woo, W. T. (2019). The road to currency internationalization: Global perspectives and Chinese experience. *Emerging Markets Review*, *38*, 38. doi:10.1016/j.ememar.2018.11.003

Li, W., Andreina, S., Bohli, J. M., & Karame, G. (2017). Securing proof-of-stake blockchain protocols. In *Data Privacy Management, Cryptocurrencies and Blockchain Technology* (pp. 297–315). Springer.

Li, X., Jiang, P., Chen, T., Luo, X., & Wen, Q. (2020). A survey on the security of blockchain systems. *Future Generation Computer Systems*, *107*, 841–853.

Li, X., Jiang, P., Chen, T., Luo, X., & Wen, Q. (2020). A survey on the security of blockchain systems. *Future Generation Computer Systems*, *107*, 841–853. doi:10.1016/j.future.2017.08.020

Li, Y., Li, T., & Liu, H. (2017). Recent advances in feature selection and its applications. *Knowledge and Information Systems*, *53*(3), 551–577. doi:10.100710115-017-1059-8

Li, Z., Liang, C., Hong, Y., & Zhang, Z. (2021). How do on-demand ridesharing services affect traffic congestion? The moderating role of urban compactness. *Production and Operations Management*, poms.13530. doi:10.1111/poms.13530

Lnes, S., Ubacht, J., & Janssen, M. (2017). Blockchain in government: Benefits and implications of distributed ledger technology for information sharing. Government Information Quarterly, 34(3), 355–364.

Long, Y. A., Zhang, J., & Ai, R. (2019). Research on supply chain finance System Optimization based on Blockchain Technology. *Southwest Finance*, *2019*(1), 74–81.

Loup's. (2017). *The design of ChaCha20.* https://loup-vaillant.fr/tutorials/chacha20-design

Lu, T., Fang, J., & Liu, C. (2015, September). A unified storage and query optimization framework for sensor data. In *2015 12th Web Information System and Application Conference (WISA)*. IEEE.

Lu, T., Chen, Y., Hechtman, B., Wang, T., & Anderson, J. (2021). Large-Scale Discrete Fourier Transform on TPUs. *IEEE Access: Practical Innovations, Open Solutions*, *9*, 93422–93432. doi:10.1109/ACCESS.2021.3092312

Lu, T., Marin, T., Zhuo, Y., Chen, Y., & Ma, C. (2020). Accelerating MRI Reconstruction on TPUs. *2020 IEEE High Performance Extreme Computing Conference (HPEC)*. 10.1109/HPEC43674.2020.9286192

Lu, Y. (2021). Consistency, breakthrough and structure of the blockchain regulation in China. *Journal of Nanjing University of Posts and Telecommunications*, *2*, 24–36.

Lyócsa, Š., Molnár, P., Plíhal, T., & Širaňová, M. (2020). Impact of macroeconomic news, regulation and hacking exchange markets on the volatility of bitcoin. *Journal of Economic Dynamics & Control*, *119*, 103980. doi:10.1016/j.jedc.2020.103980

M., P., Sharma, A., V., V., Bhardwaj, V., Sharma, A., Iqbal, R., & Kumar, R. (2020). Prediction of the price of Ethereum blockchain cryptocurrency in an industrial finance system. *Computers & Electrical Engineering, 81*, 106527. doi:10.1016/j.compeleceng.2019.106527

Ma, F. (2021). Design of running training assistance system based on blockchain technology in wireless network. Eurasip Journal on Wireless Communications and Networking, 2021(1). 10.1186/s13638-021-01897-4

Madarro-Capó, E. J., Legón-Pérez, C. M., Rojas, O., & Sosa-Gómez, G. (2021). Information Theory Based Evaluation of the RC4 Stream Cipher Outputs. *Entropy (Basel, Switzerland)*, *23*(7), 896. doi:10.3390/e23070896 PMID:34356437

Majdoubi, D., Bakkali, H., & Sadki, S. (2020). Towards smart blockchain-based system for privacy and security in a smart city environment. *Proceedings of Int. Conf. Cloud Comput. Artif. Intell. Technol. Appl (CloudTech)*. 10.1109/CloudTech49835.2020.9365905

Majeed, U. K., Khan, L. U., Yaqoob, I., Kazmi, S. M. A., Salah, K., & Hong, C. S. (2021). Blockchain for IoT-based smart cities: Recent advances, requirements, and future challenges. *Journal of Network and Computer Applications*, *181*, 103007. doi:10.1016/j.jnca.2021.103007

Malik, G., Tayal, D. K., & Vij, S. (2018). An Analysis of the Role of Artificial Intelligence in Education and Teaching. *Advances in Intelligent Systems and Computing Recent Findings in Intelligent Computing Techniques*, 407-417.

Mao, D., Wang, F., Hao, Z., & Li, H. (2018). Credit Evaluation System Based on Blockchain for Multiple Stakeholders in the Food Supply Chain. *International Journal of Environmental Research and Public Health*, *15*(8), 1627. Advance online publication. doi:10.3390/ijerph15081627 PMID:30071695

Martijn, D. (2017). Blockchain: Towards Disruption in the Real Estate Sector: An exploration on the impact of blockchain technology in the real estate management process. https://repository.tudelft.nl/islandora/object/uuid%3Ab6ec7ece-e879-4ae3-8232-d8144ac2642d?collection=education

Mavroeidis, V., Vishi, K., D, M., & Jøsang, A. (2018). The impact of quantum computing on present cryptography. *International Journal of Advanced Computer Science and Applications*, *9*(3). Advance online publication. doi:10.14569/IJACSA.2018.090354

Maxwell, G., Poelstra, A., Seurin, Y., & Wuille, P. (2019). Simple Schnorr multi-signatures with applications to Bitcoin. *Designs, Codes and Cryptography*, *87*(9), 2139–2164. doi:10.100710623-019-00608-x

McClintock. (2016). *Big Data Brings Big Changes to Product Design*. https://www.ptc.com/en/cad-software-blog/big-data-brings-big-changes-to-product-design

McGhin, T., Choo, K. K. R., Liu, C. Z., & He, D. (2019). Blockchain in healthcare applications: Research challenges and opportunities. *Journal of Network and Computer Applications*, *135*, 62–75.

McKinsey & Company. (2017). *Artificial intelligence takes shape*. McKinsey & Company. https://www.mckinsey.com/quarterly/the-magazine/2017-issue-4-mckinsey-quarterly#

McNally, S., Roche, J., & Caton, S. (2018). Predicting the price of bitcoin using machine learning. *2018 26th Euromicro International Conference on Parallel, Distributed and Network-Based Processing (PDP)*. 10.1109/PDP2018.2018.00060

Memon, R., Li, J., & Ahmed, J. (2019). Simulation Model for Blockchain Systems Using Queuing Theory. *Electronics (Basel)*, *8*(2), 234. doi:10.3390/electronics8020234

Meric, E. B. (2018). Intelligent transportation systems (AUS) and development agencies. *Journal of Intelligent Transportation Systems and Applications*, *1*(2), 33–55.

Midasium Contracts. (n.d.). The Blockchain of Real Estate. http://midasium.herokuapp.com/smart-contracts

Mingxiao, D., Xiaofeng, M., Zhe, Z., Xiangwei, W., & Qijun, C. (2017, October). *A review on consensus algorithm of blockchain. In 2017 IEEE international conference on systems, man, and cybernetics (SMC)*. IEEE.

Min, H. (2019). Blockchain technology for enhancing supply chain resilience. *Business Horizons*, *62*(1), 35–45. doi:10.1016/j.bushor.2018.08.012

Ministry of Commerce, Ministry of Finance. (2017). *Circular of the General Office of the Ministry of Commerce and the General Office of the Ministry of Finance on the Construction of Supply Chain System*. Retrieved from: http://www.mofcom.gov.cn/article/h/redht/201708/20170802627302.shtml

Ministry of Commerce, the National Development and Reform Commission, the Ministry of Land and Resources, the Ministry of Transport & the State Post Bureau. (2017). *Notice on the issuance of the 13th Five-Year Plan for the Development and Development of Trade and Goods*. Retrieved from: http://www.mofcom.gov.cn/article/b/d/201702/20170202511705.shtml

Ministry of Education. (2018). *Notice of the Ministry of Education on the issuance of the Action Plan for Educational Informatization 2.0*. Retrieved from: http://www.moe.gov.cn/srcsite/A16/s3342/201804/t20180425_334188.html

Ministry of Industry and Information Technology of the People's Republic of China. (2018). *White Paper on China's Blockchain Industry 2018*. Retrieved from: https://www.miit.gov.cn/n1146290/n1146402/n1146445/c6180238/part/6180297.pdf

Ministry of Industry and Information Technology, Office of the Central Cyberspace Affairs Commission. (2021). Guidance on accelerating the application of blockchain technology and industrial development. Joint Letter Of Ministry of Industry and Information Technology.

Ministry of Industry and Information Technology. (2016a). *China Blockchain Technology and Application Development White Paper (2016)*. Retrieved from: http://www.shujuju.cn/lecture/detail/2134

Ministry of Industry and Information Technology. (2016b). *Circular of the Ministry of Industry and Information Technology on the issuance of the Development Plan for Software and Information Technology Service Industry (2016-2020)*. Retrieved from: http://www.cac.gov.cn/2017-01/17/c_1120330752.htm

Ministry of Industry and Information Technology. (2017). *Circular of the Ministry of Industry and Information Technology on the issuance of the Three-Year Action Plan for the Development of Cloud Computing (2017-2019)*. Retrieved from: https://www.miit.gov.cn/jgsj/xxjsfzs/zlgh/art/2020/art_fb1e14b54f234fc7b4f52c062b9d3d08.html

Ministry of Industry and Information Technology. (2018a). *Notice of Ministry of Industry and Information Technology of China*. Retrieved from: https://www.miit.gov.cn/jgsj/xxjsfzs/gzdt/art/2020/art_ebe80855c8234ba7bd1b7cf7ffee5813.html

Ministry of Industry and Information Technology. (2018b). *China Blockchain Technology and Application Development White Paper (2018)*. Retrieved from: https://www.miit.gov.cn/n1146290/n1146402/n1146445/c6180238/part/6180297.pdf

Mire, S. (2018). 37 Startups using blockchain to transform real estate [market map]. https://www.disruptordaily.com/blockchain-market-map-real-estate/

Missier, P., Bajoudah, S., Capossele, A., Gaglione, A., & Nati, M. (2017, October). Mind my value: a decentralized infrastructure for fair and trusted iot data trading. *Proceedings of the Seventh International Conference on the Internet of Things*, 1-8.

Mohanta, B. K., Panda, S. S., & Jena, D. (2018). An overview of smart contract and use cases in blockchain technology. 9th International Conference on Computing, Communication and Networking Technologies (ICCCNT). doi:10.1109/ICCCNT.2018.849404510.1109/ICCCNT.2018.8494045

Moniruzzaman, M., Khezr, S., Yassine, A., & Benlamri, R. (2020). Blockchain for smart homes: Review of current trends and research challenges. *Computers & Electrical Engineering*, *83*, 106585.

Mu, Geng, Lv, Qian, & Qing. (2020). Blockchain research Group of Digital Currency Research Institute, People's Bank of China: Existing blockchains are not suitable for high concurrency scenarios such as traditional retail payments. *Chinese Finance, 4.*

Mulár, O. (2018). Blockchain technology in the enterprise environment. Masaryk University. http://search.ebscohost.com/login.aspx?direct=true&db=izh&AN=131823111&lang=pt-br&site=ehost-live

Musleh, A. S., Yao, G., & Muyeen, S. M. (2019). Blockchain applications in smart grid–review and frameworks. *IEEE Access: Practical Innovations, Open Solutions*, *7*, 86746–86757.

Nakamoto, S. (2008). Bitcoin: A Peer-to-Peer Electronic Cash System. 10.1162/ARTL_a_00247

Nakamoto, S. (2008). Bitcoin: A peer-to-peer electronic cash system. *Decentralized Business Review*, 21260.

Nakamoto, S. (2008). *Bitcoin: A Peer-to-Peer Electronic Cash System.* https://bitcoin.org/bitcoin.pdf

Nakamoto. (2020). *Bitcoin: A Peer-to-Peer Electronic Cash System.* Retrieved from https://bitcoin.org/bitcoin.pdf

Narayanaswami, Nooyi, Govindaswamy, & Viswanathan. (n.d.). Blockchain anchored supply chain automation. *IBM Journal of Research and Development, 63*(2/3), 7:1–7:11. doi:10.1147/JRD.2019.2900655

National Internet Finance Association of China, China Banking Association, Payment & Clearing Association of China. (2021). *Notice of the prevention of virtual currency trading speculation risk.* Retrieved from: https://baijiahao.baidu.com/s?id=1700098234762596743&wfr=spider&for=pc

National Internet Finance Association of China. (2020a). *Risk tips for participating in speculation on overseas virtual currency trading platforms.* Retrieved from: https://baijiahao.baidu.com/s?id=1662828395598126938&wfr=spider&for=pc

National Internet Finance Association of China. (2020b). *Research Report on Blockchain Financial Application and Development in China.* Retrieved from: https://wenku.baidu.com/view/0193c7d6b6360b4c2e3f5727a5e9856a5712267e.html

Nehai, Z., & Guérard, G. (2017, May). Integration of the blockchain in a smart grid model. *The 14th International Conference of Young Scientists on Energy Issues (CYSENI) 2017*, 127-134.

Neven, G., Smart, N., & Warinschi, B. (2009). Hash function requirements for Schnorr signatures. *J. Math Cryptol*, *3*(1), 69–87. doi:10.1515/JMC.2009.004

Nguyen, G.-T., & Kim, K. (2018). A survey about consensus algorithms used in blockchain. Journal of Information Processing Systems, 14(1), 101–128. 10.3745/JIPS

Nguyen, L. T. Q., Hoang, T. G., Do, L. H., Ngo, X. T., Nguyen, P. H. T., Nguyen, G. D. L., & Nguyend, G. N. T. (2021). The role of blockchain technology-based social crowdfunding in advancing social value creation. *Technological Forecasting and Social Change*, *170*, 120898. doi:10.1016/j.techfore.2021.120898

Ning, Y. J. (2021). *Speech on "Industrial Blockchain 2.0:2021 New Opportunities for Digital Economy"*. Retrieved from: https://new.qq.com/omn/20210128/20210128A08O3A00.html

Norta, A., Fernandez, C., & Hickmott, S. (2018). Commercial Property Tokenizing with Smart Contracts. Proceedings of the International Joint Conference on Neural Networks. doi:10.1109/IJCNN.2018.848953410.1109/IJCNN.2018.8489534

North, D. (1990). *Institutions, Institutional Change and Economic Performance.* Cambridge University. doi:10.1017/CBO9780511808678

Novotny, P., Zhang, Q., Hull, R., Baset, S., Laredo, J., Vaculin, R., Ford, D. L., & Dillenberger, D. N. (2018). Permissioned blockchain technologies for academic publishing. *Information Services & Use*, *38*(3), 159–171. doi:10.3233/ISU-180020

O'Leary, D. E. (2017). Configuring blockchain architectures for transaction information in blockchain consortiums: The case of accounting and supply chain systems. *Intelligent Systems in Accounting, Finance & Management, 24*(4), 138–147. doi:10.1002/isaf.1417

Olaniyi, E. O., Oyedotun, O. K., & Adnan, K. (2015). Heart Diseases Diagnosis Using Neural Networks Arbitration. *International Journal of Intelligent Systems and Applications, 7*(12), 75–82. doi:10.5815/ijisa.2015.12.08

Oyamada, M., & Nakadai, S. (2017). Relational mixture of experts: Explainable demographics prediction with behavioral data. *2017 IEEE International Conference on Data Mining (ICDM).* 10.1109/ICDM.2017.45

P, N., Tom, R., Gupta, P., Shanthini, A., John, V., & Sharma, V. (2021). Prediction of Bitcoin Price Using Bi-LSTM Network. *2021 International Conference on Computer Communication and Informatics (ICCCI).* doi:10.1109/ICCCI50826.2021.9402427

P. P. T. & S. K. L. (2019). Smart City Services - Challenges and Approach. *2019 International Conference on Machine Learning, Big Data, Cloud and Parallel Computing (COMITCon)*, 553-558. doi: 10.1109/COMITCon.2019.8862243

Pal, A., Tiwaria, C. K., & Haldar, N. (2021). Blockchain for business management: Applications, challenges and potentials. *The Journal of High Technology Management Research, 32*(2), 100414. doi:10.1016/j.hitech.2021.100414

Palaniappan, S., & Awang, R. (2008). Intelligent heart disease prediction system using data mining techniques. *2008 IEEE/ACS International Conference on Computer Systems and Applications.* 10.1109/AICCSA.2008.4493524

Panarello, A., Tapas, N., Merlino, G., Longo, F., & Puliafito, A. (2018). Blockchain and iot integration: A systematic survey. *Sensors (Basel), 18*(8), 2575.

Panarello, A., Tapas, N., Merlino, G., Longo, F., & Puliafito, A. (2018). Blockchain and IoT integration: A systematic survey. *Sensors (Basel), 18*(8), 2575. doi:10.339018082575 PMID:30082633

Panetta, K. (2017, October 3). *Smarter With Gartner.* https://www.gartner.com/smarterwithgartner/gartner-top-10-strategic-technology-trends-for-2018/

Panetta, K. (2018, October 15). *Smarter With Gartner.* https://www.gartner.com/smarterwithgartner/gartner-top-10-strategic-technology-trends-for-2019/

Panetta, K. (2019, October 21). *Smarter With Gartner.* https://www.gartner.com/smarterwithgartner/gartner-top-10-strategic-technology-trends-for-2020/

Panja, S., & Roy, B. (2021). A secure end-to-end verifiable e-voting system using blockchain and cloud server. *Journal of Information Security and Applications, 59*, 102815. doi:10.1016/j.jisa.2021.102815

Pant, R. R., Prakash, G., & Farooquie, J. A. (2015). A Framework for Traceability and Transparency in the Dairy Supply Chain Networks. *Operations Management in Digital Economy, 189*, 385–394.

Papa, R., Gargiulo, C., & Galderisi, A. (2013). Towards an urban planners' perspective on smart city. *TeMA J. Land Use. Mobil. Environ., 6*(1), 5–17.

Park, J., Kim, H., Kim, G., & Ryou, J. (2021). Smart contract data feed framework for privacy-preserving oracle system on blockchain. Computers, 10(1), 1–12. 10.3390/computers10010007

Peck, M. E. (2017). Blockchain world - Do you need a blockchain? This chart will tell you if the technology can solve your problem. *IEEE Spectrum, 54*(10), 38–60. doi:10.1109/MSPEC.2017.8048838

Pedersen, T. (1992). Non-interactive and information-theoretic secure verifiable secret sharing. Lect Notes Comput Sci, 576, 129–140. doi:10.1007/3-540-46766-1_9

People's Bank of China, Ministry of Industry and Information Technology & China Banking Regulatory Commission (CBRC). (2013). *Notice on Bitcoin Risk Prevention*. Retrieved from: http://www.gov.cn/gzdt/2013-12/05/content_2542751.htm

People's Bank of China, Office of the Central Cyberspace Affairs Commission, Ministry of Industry and Information Technology, State Administration for Market Regulation, China Insurance Regulatory Commission, China's Securities Regulatory Commission & China's Bank Regulatory Commission. (2017). *Notice on Preventing the Risk of Token Issuance Financing*. Retrieved from: http://www.pbc.gov.cn/goutongjiaoliu/113456/113469/3374222/index.html

People's Bank of China. (2014). *Set up a special research group on legal digital currency*. Annual Report of the People's Bank of China 2014. Retrieved from: http://www.pbc.gov.cn/eportal/fileDir/image_public/UserFiles/chubanwu/upload/File/中国人民银行年报2014-最终定稿.pdf

People's Bank of China. (2015). *Some key issues in the field of digital currency were investigated*. Annual Report of the People's Bank of China 2015. Retrieved from: http://www.pbc.gov.cn/chubanwu/114566/115296/3087023/3086996/2016062115305560130.pdf

People's Bank of China. (2020a). *Blockchain technology financial application evaluation rules*. Retrieved from: https://www.mpaypass.com.cn/Download/202007/30153205.html

People's Bank of China. (2020b). *Notice to promote the application of blockchain technical specifications*. Retrieved from: https://www.jinse.com/news/blockchain/772215.html

People's Bank of China. (2020c). *Technical security specification for financial distributed ledger*. Retrieved from: https://www.sohu.com/a/375656914_120057347

People's Bank of China. (2021). *White paper of "The development of Chinese digital RMB"*. Retrieved from: http://www.gov.cn/xinwen/2021-07/16/content_5625569.htm

Peyton, A. (2018). Emirates NBD goes blockchain for its "Cheque Chain." https://www.fintechfutures.com/2018/04/emirates-nbd-goes-blockchain-for-its-cheque-chain/

Pham. (2014). Smart cities in Japan. An assessment on the potential for EU-Japan Cooperation and Business Development, Tokyo.

Philipp, R., Prause, G., & Gerlitz, L. (2019). Blockchain and Smart Contracts for Entrepreneurial Collaboration in Maritime Supply Chains. *Transport and Telecommunication Journal, 20*(4), 365–378.

Poirier, C. C. (1999). Advanced Supply Chain Management: How to Build a Sustained Competitive Advantage. In *Google Books*. Berrett-Koehler Publishers. https://books.google.co.in/books?hl=en&lr=&id=66XhLR3Ih_MC&oi=fnd&pg=PR7&dq=C.+C.+Poirier

Potestad-Ordóñez, F. E., Valencia-Barrero, M., Baena-Oliva, C., Parra-Fernández, P., & Jiménez-Fernández, C. J. (2020). Breaking Trivium Stream Cipher Implemented in ASIC Using Experimental Attacks and DFA. *Sensors (Basel), 20*(23), 6909. doi:10.339020236909 PMID:33287234

Pramod, N., & Sankaran, S. (2019) Blockchain-based framework for driver profiling in smart cities. In *Proceedings of Int. Symp. Adv. Networks Telecommun. Syst.* (pp. 1–6). 10.1109/ANTS47819.2019.9117923

Proof of Stake - Bitcoin Wiki. (n.d.). *En.bitcoin.it*. https://en.bitcoin.it/wiki/Proof_of_Stake

Proof of work - Bitcoin Wiki. (2019). *Bitcoin.it*. https://en.bitcoin.it/wiki/Proof_of_work

PuHuiMingYan financial research institute. (2018). *Application scenario of block chain in supply chain finance and the first block chain ABS case.* Retrieved from: https://mp.weixin.qq.com/s/2TbAxw7G_MD4OmEkfZ2npQ

Purbarani, S. C., & Jatmiko, W. (2018). Performance comparison of bitcoin prediction in big data environment. *2018 International Workshop on Big Data and Information Security (IWBIS).* 10.1109/IWBIS.2018.8471691

Qiang, W., & Min, S., & Rong, R.L. (2020). Is China the world's blockchain leader? Evidence, evolution and outlook of China's blockchain research. *Journal of Cleaner Production*, 264.

Qi-Long, C., Rong-Hua, Y., & Fei-Long, L. (2019). A Blockchain-based Housing Rental System. CTISC, 184–190. 10.5220/0008097201840190

Queiroz, M. M., Telles, R., & Bonilla, S. H. (2020). Blockchain and supply chain management integration: A systematic review of the literature. *Supply Chain Management*, *25*(2), 241–254. doi:10.1108/SCM-03-2018-0143

Rahman, M. A., Hossain, M. S., Islam, M. S., Alrajeh, N. A., & Muhammad, G. (2020). Secure and provenance enhanced Internet of health things framework: A blockchain managed federated learning approach. *IEEE Access: Practical Innovations, Open Solutions*, *8*, 205071–205087. doi:10.1109/ACCESS.2020.3037474 PMID:34192116

Raman, R., & Varshney, L. (2018). *Distributed storage meets secret sharing on the blockchain.* Inf Theory Appl Work ITA. doi:10.1109/ITA.2018.8503089

Raschka, S. (2018, December 3). *Model Evaluation, Model Selection, and Algorithm Selection in Machine Learning.* Retrieved from https://arxiv.org/abs/1811.12808v2

Rathee, G., Iqbal, R., Waqar, O., & Bashir, A. K. (2021). On the Design and Implementation of a Blockchain Enabled E-Voting Application Within IoT-Oriented Smart Cities. *IEEE Access: Practical Innovations, Open Solutions*, *9*, 34165–34176. doi:10.1109/ACCESS.2021.3061411

Rathore, S., Kwon, B. W., & Park, J. H. (2019). BlockSecIoTNet: Blockchain-based decentralized security architecture for IoT network. *Journal of Network and Computer Applications*, *143*, 167–177.

Razaghi, M., & Finger, M. (2018). Smart governance for smart cities. *Proceedings of the IEEE*, *106*(4), 680–689.

Remote-Sensing. (n.d.). https://www.nrcan.gc.ca/maps-tools-publications/satellite-imagery-air-photos/remote-sensing-tutorials/fundamentals-remote-sensing-introduction/9363

Rentberry. (2018). A Decentralized Global Home Rental Platform Whitepaper. https://rentberry.cryptonomos.com/#whitepaper

Reyna, A., Martín, C., Chen, J., Soler, E., & Díaz, M. (2018). On blockchain and its integration with IoT. Challenges and opportunities. *Future Generation Computer Systems*, *88*, 173–190.

Risius, M., & Spohrer, K. (2017). A Blockchain Research Framework. *Business & Information Systems Engineering*, *59*(6), 385–409. doi:10.100712599-017-0506-0

Roh, C. G., & Jeon, H. (2021). Decision-Making Process for Demand Response Public Transportation Service Design—A Case Study in Incheon, Korea. *Sustainability*, *13*(10), 5574. doi:10.3390u13105574

Rong, Z.X. & Yu, L.M. (2020). The application of blockchain technology in the banking industry. *China Finance, 10.*

Rotună, C., Gheorghiță, A., Zamfiroiu, A., & Smada Anagrama, D. (2019). Smart City Ecosystem Using Blockchain Technology. *Informações Econômicas*, *23*(4), 41–50.

Saberi, S., Kouhizadeh, M., Sarkis, J., & Shen, L. (2018). Blockchain technology and its relationships to sustainable supply chain management. *International Journal of Production Research, 57*(7), 2117–2135. doi:10.1080/00207543.2018.1533261

Saha. (2017). A Remote Sensing Approach to Smart City Development in India: Case of Bhopal City, Madhya Pradesh. *Conference: the Special Collection.* Doi:10.1145/3055219.3055232

Salam, I., Ooi, T. H., Xue, L., Yau, W. C., Pieprzyk, J., & Phan, R. C. W. (2021). Random Differential Fault Attacks on the Lightweight Authenticated Encryption Stream Cipher Grain-128AEAD. *IEEE Access: Practical Innovations, Open Solutions, 9*, 72568–72586. doi:10.1109/ACCESS.2021.3078845

Saleh, F. (2019). *Blockchain Without Waste: Proof-of-Stake. SSRN Scholarly Paper ID 3183935.* Social Science Research Network.

Sanka, A. I., Irfan, M., Huang, I., & Cheung, R. C. C. (2021). A survey of breakthrough in blockchain technology: Adoptions, applications, challenges and future research. *Computer Communications, 169*, 179–201. doi:10.1016/j.comcom.2020.12.028

Santana. (2017). Software Platforms for Smart Cities: Concepts, Requirements, Challenges, and a Unified Reference Architecture. ACM Comput. Surv., 50(6).

Sattarov, O., Jeon, H., Oh, R., & Lee, J. (2020). Forecasting Bitcoin Price Fluctuation by Twitter Sentiment Analysis. *2020 International Conference On Information Science And Communications Technologies (ICISCT).* 10.1109/ICISCT50599.2020.9351527

Scherer, M. (2017). Performance and Scalability of Blockchain Networks and Smart Contracts. UMEA. http://www.diva-portal.org/smash/record.jsf?pid=diva2%3A1111497&dswid=8567

Scott, W. R. (2008). *Institutions and organizations: Ideas and interests* (3rd ed.). SAGE Publications.

Securities, M. (2017). *Driven by the dual track of blockchain policy and practice, the Fintech industry is heating up.* Minsheng Securities Institute.

Seebacher, S., & Schüritz, R. (2017). Blockchain technology as an enabler of service systems: A structured literature review. Lecture Notes in Business Information Processing, 279(April), 12–23. doi:10.1007/978-3-319-56925-3_2 doi:10.1007/978-3-319-56925-3_2

Sengupta, J., Kubendran, R., Neftci, E., & Andreou, A. (2020). High-speed, real-time, spike-based object tracking and path prediction on google edge tpu. *2020 2nd IEEE International Conference on Artificial Intelligence Circuits and Systems (AICAS).* 10.1109/AICAS48895.2020.9073867

Serafini, G., Yi, P., Zhang, Q., Brambilla, M., Wang, J., Hu, Y., & Li, B. (2020). Sentiment-Driven Price Prediction of the Bitcoin based on Statistical and Deep Learning Approaches. *2020 International Joint Conference On Neural Networks (IJCNN).* 10.1109/IJCNN48605.2020.9206704

Shafagh, H., Burkhalter, L., Hithnawi, A., & Duquennoy, S. (2017, November). Towards blockchain-based auditable storage and sharing of iot data. *Proceedings of the 2017 on cloud computing security workshop*, 45-50.

Shahzad, B., & Crowcroft, J. (2019). Trustworthy electronic voting using adjusted blockchain technology. *IEEE Access: Practical Innovations, Open Solutions, 7*, 24477–24488.

Shanker, M. (2019). Use Case: Smart Contract for Lease Agreements using Blockchain Technology. International Journal of Scientific Research in Computer Science and Engineering, 7(6), 1–09.

Sharma, P. K., N. Kumar, & Park, J. H. (n.d.). BlockchainBased Distributed Framework for Automotive Industry in a Smart City. *IEEE Transactions on Industrial Informatics, 15*(7), 4197–4205. doi:10.1109/TII.2018.2887101

Sharma, P. K., & Park, J. H. (2018). Blockchain based hybrid network architecture for the smart city. *Future Generation Computer Systems*, *86*, 650–655.

Sheel, A., & Nath, V. (2019). Effect of blockchain technology adoption on supply chain adaptability, agility, alignment and performance. *Management Research Review*, *42*(12), 1353–1374. doi:10.1108/MRR-12-2018-0490

Shen, C., & Pena-Mora, F. (2018). Blockchain for Cities—A Systematic Literature Review. *IEEE Access: Practical Innovations, Open Solutions*, *6*, 76787–76819. doi:10.1109/ACCESS.2018.2880744

Sheng, A., & He, J. (2020). Distributed Evolution Strategies Using TPUs for Meta-Learning. *2020 IEEE Symposium Series On Computational Intelligence (SSCI)*. 10.1109/SSCI47803.2020.9308334

She, W. E. I., Gu, Z. H., Lyu, X. K., Liu, Q. I., Tian, Z., & Liu, W. (2019). Homomorphic consortium blockchain for smart home system sensitive data privacy preserving. *IEEE Access: Practical Innovations, Open Solutions*, *7*, 62058–62070.

Shin, M., Mohaisen, D., & Kim, J. (2021). Bitcoin Price Forecasting via Ensemble-based LSTM Deep Learning Networks. *2021 International Conference On Information Networking (ICOIN)*. 10.1109/ICOIN50884.2021.9333853

Shuchih, Chen, & Lu. (2019). Supply chain re-engineering using blockchain technology: A case of smart contract based tracking process. *Technological Forecasting and Social Change, 144*, 1–11.

Shyu, S., & Chen, Y. (2008). Threshold secret image sharing by Chinese remainder theorem. In *Proceedings of IEEE Asia-Pacific Serv Comput Conf.* (pp. 1332–1337). 10.1109/APSCC.2008.223

Sicari, S., Rizzardi, A., Cappiello, C., Miorandi, D., & Coen-Porisini, A. (2018). Toward data governance in the internet of things. In *New advances in the internet of things* (pp. 59–74). Springer.

Silva, B. N., Khan, M., & Han, K. (2018). Towards sustainable smart cities: A review of trends, architectures, components, and open challenges in smart cities. *Sustainable Cities and Society*, *38*, 697–713. doi:10.1016/j.scs.2018.01.053

Singhal, B., Dhameja, G., & Panda, P. (2018). *Beginning blockchain – A beginner's guide to building blockchain solutions*. Jenson Books Inc.

Singh, C. (2021). Blockchain and IOT integrated Smart City Architecture. *Turkish Journal of Computer and Mathematics Education*, *12*(9), 62–69.

Siyal, A. A., Junejo, A. Z., Zawish, M., Ahmed, K., Khalil, A., & Soursou, G. (2019). Applications of Blockchain Technology in Medicine and Healthcare: Challenges and Future Perspectives. *Cryptography*, *3*(1), 3. doi:10.3390/cryptography3010003

Sookhak, M., Tang, H., & Yu, F. R. (2018, June). Security and Privacy of Smart Cities: Issues and Challenge. *2018 IEEE 20th International Conference on High Performance Computing and Communications; IEEE 16th International Conference on Smart City; IEEE 4th International Conference on Data Science and Systems (HPCC/SmartCity/DSS)*, 1350-1357.

Sovbetov, Y. (2018). Factors Influencing Cryptocurrency Prices: Evidence from Bitcoin, Ethereum, Dash, Litcoin, and Monero. *Journal Of Economics And Financial Analysis*, *2*(2), 1–27.

Spielman, A. (2016). Blockchain: Digitally Rebuilding the Real Estate Industry. Vanderbilt University.

Standing Committee of the National People's Congress. (2019). *Code Law*. Retrieved from: http://www.npc.gov.cn/npc/c30834/201910/6f7be7dd5ae5459a8dc8baf36296bc74.shtml

State Council. (2016). *Notice of the State Council on the issuance of the '13th Five-Year' National Informatization Plan.* Retrieved from: http://www.gov.cn/zhengce/content/2016-12/27/content_5153411.htm

State Council. (2017). *Circular of the State Council on the issuance of the New Generation of Artificial Intelligence Development Plan.* Retrieved from: http://www.gov.cn/zhengce/content/2017-07/20/content_5211996.htm

State Council. (2019a). *Opinions on Deepening Reform and Strengthening Food Safety.* Retrieved from: http://www.gov.cn/zhengce/2019-05/20/content_5393212.htm

State Council. (2019b). *Opinions on Supporting Shenzhen in Building a Pilot Demonstration Zone for Socialism with Chinese Characteristics.* Retrieved from: http://www.gov.cn/zhengce/2019-08/18/content_5422183.htm

Suárez-Albela, M., Fernández-Caramés, T. M., Fraga-Lamas, P., & Castedo, L. (2017). A practical evaluation of a high-security energy-efficient gateway for IoT fog computing applications. *Sensors (Basel), 17*(9), 1978.

Sudha Sadasiuvam, G. (2021). A critical review on using blockchain technology in education domain. In Handbook of Deep Learning in Biomedical Engineering. Elsevier Inc. doi:10.1016/B978-0-12-823014-5.00005-3 doi:10.1016/B978-0-12-823014-5.00005-3

Sultan, K., Ruhi, U., & Lakhani, R. (2018). Conceptualizing blockchains: Characteristics & applications. Proceedings of the 11th IADIS International Conference Information Systems 2018, IS 2018, 49–57.

Sunny, J., Undralla, N., & Madhusudanan Pillai, V. (2020). Supply chain transparency through blockchain-based traceability: An overview with demonstration. *Computers & Industrial Engineering, 150*, 150. doi:10.1016/j.cie.2020.106895

Sun, X., Wang, Q., Kulicki, P., & Sopek, M. (2019). A simple voting protocol on quantum blockchain. *International Journal of Theoretical Physics, 58*(1), 275–281. doi:10.100710773-018-3929-6

Suo, H., Wan, J., Zou, C., & Liu, J. (2012, March). *Security in the internet of things: a review. In 2012 international conference on computer science and electronics engineering.* IEEE.

Surjandy, M., Hidayanto, A. N., & Prabowo, H. (2019). The latest adoption blockchain technology in supply chain management: A systematic literature review. *ICIC Express Letters, 13*, 913–920. doi:10.24507/icicel.13.10.913

Swan, M. (2015). Blockchain: Blueprint for a New Economy. In *Google Books.* O'Reilly Media, Inc. https://books.google.com/books?hl=en&lr=&id=RHJmBgAAQBAJ&oi=fnd&pg=PR3&ots=XRrEJ-WOc6&sig=Kn7hlqSVIsLoQSnvbLHAgAD6bxI

Swan, M. (2015). *Blockchain: Blueprint for a New Economy.* O'Reilly Media, Inc.

Swan, M. (2015). *Blockchain: Blueprint for a new economy.* O'Reilly Media, Inc.

Taş, R., & Tanrıöver, Ö. Ö. (2020). A Systematic Review of Challenges and Opportunities of Blockchain for E-Voting. *Symmetry, 12*(8), 1328.

Taylor, P., Dargahi, T., Dehghantanha, A., Parizi, R., & Choo, K. (2020). A systematic literature review of blockchain cyber security. *Digit. Commun. Networks, 6*(2), 147–156. doi:10.1016/j.dcan.2019.01.005

TechWeb Report. (2018). *Cloud coin network: the basic completion of the clean-up work plan to close the website on August 31.* Retrieved from: TechWeb.com.cn.

The People's Bank of China. (2011). *The Establishment meeting of Bank Card Base Payment Committee of Clearing Association of China, Network Payment Application Committee and Prepaid Card Committee of Payment.* Retrieved from: https://xueshu.baidu.com/usercenter/paper/show?paperid=7d2b16f58ef921bcfec6f8cdbbb55a60&site=xueshu_se

The People's Bank of China. (2015). *Measures for the administration of online Payment business of non-bank payment institutions*. Retrieved from: http://www.gov.cn/zhengce/2016-03/18/content_5055171.htm

Thompson, S. (2017). The preservation of digital signatures on the blockchain. The University of British Columbia ISchool Student Journal, 3(Spring).

Thurber, K. J. (2012). Do not invent buggy whips: Create! reinvent! position! disrupt! Digital Systems Press.

Tian, S., Liu, Y., Zhang, Y., & Zhao, Y. (2021) A byzantine fault-tolerant raft algorithm combined with Schnorr signature. In *Proceedings of Int. Conf. Ubiquitous Inf. Manag. Commun* (pp. 1–5). 10.1109/IMCOM51814.2021.9377376

Tourville, S. (2018). *Council Post: The Power And Promise Of Immersive*. https://www.forbes.com/sites/forbesagencycouncil/2018/12/07/the-power-and-promise-of-immersive-technology-in-brand-storytelling/

Training, J. (2021). *Cloud Computing Interview Questions & Answers*. https://www.janbasktraining.com/blog/cloud-computing-interview-questions-answers/

Treiblmaier, H., Rejeb, A., & Strebinger, A. (2020). Blockchain as a driver for smart city development: Application fields and a comprehensive research agenda. *Smart Cities*, *3*(3), 853–872. doi:10.3390martcities3030044

Turkey Annual Report General Directorate of Highways. (2020). Retrieved from https://kgm.gov.tr

TurkeyC. R. M. P. R. August 2019., Retrieved from https://ulasim.gov.tr

Tyndall, G., Gopal, C., Partsch, W., & Kamauff, J. (1998). *Supercharging supply chains. New ways to increase value through global operational excellence*. John Wiley & Sons, Incorporated. https://trid.trb.org/view/672145

University Big Data and Artificial Intelligence Promotion Alliance. (2021). *Top 10 Cases of Blockchain technology application in China in 2020*. Retrieved from: https://mp.weixin.qq.com/s/Unt3QI6a7xy-dZdvxv4q2g

Upadhyay, A., Mukhuty, S., Kumar, V., & Kazancoglu, Y. (2021). Blockchain technology and the circular economy: Implications for sustainability and social responsibility. *Journal of Cleaner Production*, *293*, 126130. doi:10.1016/j.jclepro.2021.126130

Uurtsaikh Jamsrandorj. (2017). Decentralized Access Control Using Blockchain. University of Saskatchewan. https://harvest.usask.ca/bitstream/handle/10388/8087/JAMSRANDORJ-THESIS-2017.pdf?sequence=1&isAllowed=y

Valera, A. J. J., Zamora, M. A., & Skarmeta, A. F. (2010, January). An architecture based on internet of things to support mobility and security in medical environments. In *2010 7th IEEE consumer communications and networking conference*. IEEE.

van Rooyen, J. (2017, May 3). *Blockchains for supply chains - part I*. Resolve Solution Partners. https://resolvesp.com/blockchains-supply-chains/

Varfolomeev, A., Alfarhani, L. H., & Oleiwi, Z. C. (2021). Secure-reliable smart contract applications based blockchain technology in smart cities environment. *Procedia Computer Science*, *186*, 669–676. doi:10.1016/j.procs.2021.04.188

Vertical farming. (n.d.). In *Wikipedia*. https://en.wikipedia.org/wiki/Vertical_farming

Vuppala, B., & Swarnalatha, P. (2020). Quantitative evaluation method of cloud resources based on work scheduling. *Journal of Ambient Intelligence and Humanized Computing*, *12*(7), 6969–6973. doi:10.100712652-020-02342-y

Wang, W., Hoang, D. T., Hu, P., Xiong, Z., Niyato, D., Wang, P., & Kim, D. I. (2019). A survey on consensus mechanisms and mining strategy management in blockchain networks. IEEE Access: Practical Innovations, Open Solutions, 7, 22328–22370. Advance online publication. doi:10.1109/ACCESS.2019.2896108 doi:10.1109/ACCESS.2019.2896108

Wang, X., WeiLi, J., & Chai, J. (2018, December). The research on the incentive method of consortium blockchain based on practical byzantine fault tolerant. In *2018 11th international symposium on computational intelligence and design (ISCID)* (2, 154-156). IEEE.

Wang, Y., Wang, Q., Shi, S., He, X., Tang, Z., Zhao, K., & Chu, X. (2020). Benchmarking the Performance and Energy Efficiency of AI Accelerators for AI Training. *2020 20Th IEEE/ACM International Symposium On Cluster, Cloud And Internet Computing (CCGRID)*. 10.1109/CCGrid49817.2020.00-15

Wang4185652. (2021). *Analyze the impact of the development of block chain on the financial field.* Retrieved from: https://www.doc88.com/p-17216081798929.html

Wang, D., Zhao, J., & Wang, Y. (2020). A Survey on Privacy Protection of Blockchain: The Technology and Application. *IEEE Access: Practical Innovations, Open Solutions*, *8*, 108766–108781. doi:10.1109/ACCESS.2020.2994294

Wang, S., Ouyang, L., Yuan, Y., Ni, X., Han, X., & Wang, F. Y. (2019). Blockchain-enabled smart contracts: Architecture, applications, and future trends. *IEEE Transactions on Systems, Man, and Cybernetics. Systems*, *49*(11), 2266–2277.

Wang, W., Hoang, D. T., Hu, P., Xiong, Z., Niyato, D., Wang, P., ... Kim, D. I. (2019). A survey on consensus mechanisms and mining strategy management in blockchain networks. *IEEE Access: Practical Innovations, Open Solutions*, *7*, 22328–22370.

Wang, Y., Han, J. H., & Beynon-Davies, P. (2019). Understanding blockchain technology for future supply chains: A systematic literature review and research agenda. *Supply Chain Management*, *24*(1), 62–84. doi:10.1108/SCM-03-2018-0148

Wang, Z., Jin, H., Dai, W., Choo, K. K. R., & Zou, D. (2021). Ethereum smart contract security research: Survey and future research opportunities. *Frontiers of Computer Science*, *15*(2), 1–18.

Wan, J., Li, J., Imran, M., & Li, D. (2019). A blockchain-based solution for enhancing security and privacy in smart factory. *IEEE Transactions on Industrial Informatics*, *15*(6), 3652–3660.

Wazid, M., Das, A. K., Shetty, S., & Jo, M. (2020). A tutorial and future research for building a blockchain-based secure communication scheme for internet of intelligent things. *IEEE Access: Practical Innovations, Open Solutions*, *8*, 88700–88716. doi:10.1109/ACCESS.2020.2992467

Wen, Y., & Looi, C. (2019). Review of Augmented Reality in Education: Situated Learning with Digital and Non-digital Resources. *Smart Computing and Intelligence Learning in a Digital World*, 179-193.

Wetzel. (2018). A use case of blockchain technology in the real estate industry for the rental agreement and accessory charge settlement processes. Academic Press.

What is a smart home or building. (n.d.). https://internetofthingsagenda.techtarget.com/definition/smart-home-or-building?

What is Bitcoin Blockchain and Bitcoin Blockchain Size? (n.d.). *Intellipaat Blog*. https://intellipaat.com/blog/tutorial/blockchain-tutorial/what-is-bitcoins-blockchains/

WikiB. (n.d.). *En.bitcoin.it*. https://en.bitcoin.it

Wong, M. C., Yee, K. C., & Nohr, C. (2018). Socio-technical consideration for blockchain technology in healthcare. *Studies in Health Technology and Informatics*, *247*, 636–640. doi:10.3233/978-1-61499-852-5-636 PMID:29678038

Wong, P. F., Chia, F. C., San Kiu, M., & Lou, E. C. (2020). *Potential integration of blockchain technology into smart sustainable city (SSC) developments: a systematic review*. Smart and Sustainable Built Environment.

Wu, H., Levinson, D., & Sarkar, S. (2019). How transit scaling shapes cities. *Nature Sustainability*, *2*(12), 1142–1148. doi:10.103841893-019-0427-7

Wu, T., Yang, G., Zhu, L., & Wu, Y. (2021). Privacy-preserving voluntary-tallying leader election for internet of things. *Information Sciences*, *574*, 461–472. doi:10.1016/j.ins.2021.06.028

Xi, J. P. (2019). *The 18th group study session of the Political Bureau of the CPC Central Committee*. Retrieved from: http://www.gov.cn/xinwen/2019-10/25/content_5444957.htm

Xiao, X. (2018). *Five features of the current application of blockchain in the financial field*. Internet Finance Association of China. Retrieved from: http://www.360doc.com/content/18/0623/22/44361760_764758542.shtml

Xie, J., Tang, H., Huang, T., Yu, F. R., Xie, R., Liu, J., & Liu, Y. (2019). A survey of blockchain technology applied to smart cities: Research issues and challenges. *IEEE Communications Surveys and Tutorials*, *21*(3), 2794–2830.

Xu, R., Nikouei, S., Nagothu, D., Fitwi, A., & Chen, Y. (2020). BlendSPS: A Blockchain-enabled decentralized smart public safety system. *Smart Cities*, *3*(3), 928–951. doi:10.3390martcities3030047

Yaga, D., Mell, P., Roby, N., & Scarfone, K. (2019). *Blockchain technology overview*. arXiv preprint arXiv:1906.11078.

Yang, S., Yu, X., & Zhou, Y. (2020). LSTM and GRU Neural Network Performance Comparison Study: Taking Yelp Review Dataset as an Example. *2020 International Workshop On Electronic Communication And Artificial Intelligence (IWECAI)*. 10.1109/IWECAI50956.2020.00027

Yan, N., Sun, B., Zhang, H., & Liu, C. (2016). A partial credit guarantee contract in a capital constrained supply chain: Financing equilibrium and coordinating strategy. *International Journal of Production Economics*, *173*, 122–133. doi:10.1016/j.ijpe.2015.12.005

Yao, W., Xu, K., & Li, Q. (2019). Exploring the Influence of News Articles on Bitcoin Price with Machine Learning. *2019 IEEE Symposium On Computers And Communications (ISCC)*. 10.1109/ISCC47284.2019.8969596

Yaqoob, I., Salah, K., Jayaraman, R., & Al-Hammadi, Y. (2021). Blockchain for healthcare data management: Opportunities, challenges, and future recommendations. *Neural Computing & Applications*, 1–16.

Yardim, M. S., & Akyildiz, G. (2005). Intelligent transportation systems and applications in Turkey. In *6th Transport Congress Proceedings Book*. TMMOB Chamber of Civil Engineers.

Yardım, M. S., & Erel, A. (2003). Data Requirement for Turkish Transport System. *TMMOB Transport Policies Congress Proceedings*, 171-181.

Ye, C., Li, G., Cai, H., Gu, Y., & Fukuda, A. (2018). Analysis of security in blockchain: Case study in 51%-attack detecting. 5th International Conference on Dependable Systems and Their Applications (DSA). doi:10.1109/DSA.2018.0001510.1109/DSA.2018.00015

Yi, H. (2019). Securing e-voting based on blockchain in P2P network. *EURASIP Journal on Wireless Communications and Networking*, *2019*(1), 1–9. doi:10.118613638-019-1473-6

Yong, Y., & Fei-Yue, W. (2016). *Towards blockchain-based intelligent transportation systems*. Academic Press.

Yu, G. (2020). Simple Schnorr signature with Pedersen commitment as key. IACR Cryptol ePrint Arch, 61.

Yu, P., Chen, Z., & Sun, J. (2018). *Innovative Financing: an Empirical Study on Public– Private Partnership Securitisation in China*. Australian Economic Papers. John Wiley & Sons Australia, Ltd.

Yu, B., Wright, J., Nepal, S., Zhu, L., Liu, J., & Ranjan, R. (2018). Iotchain: Establishing trust in the internet of things ecosystem using blockchain. *IEEE Cloud Computing*, *5*(4), 12–23.

Yue, X., Wang, H., Jin, D., Li, M., & Jiang, W. (2016). Healthcare data gateways: Found healthcare intelligence on blockchain with novel privacy risk control. *Journal of Medical Systems*, *40*(10), 1–8.

Yu, M. (2019). Forecasting bitcoin volatility: The role of leverage effect and uncertainty. *Physica A*, *533*, 120707. doi:10.1016/j.physa.2019.03.072

Yu, P., Gong, R., & Sampat, M. (2022). Blockchain Technology in China's Digital Economy: Balancing Regulation and Innovation. In P. Tehrani (Ed.), *Regulatory Aspects of Artificial Intelligence on Blockchain* (pp. 132–157). IGI Global. doi:10.4018/978-1-7998-7927-5.ch007

Yu, Y., Li, Y., Tian, J., & Liu, J. (2018). Blockchain-based solutions to security and privacy issues in the internet of things. *IEEE Wireless Communications*, *25*(6), 12–18.

Zadobrischi, E., Cosovanu, L. M., & Dimian, M. (2020). Traffic Flow Density Model and Dynamic Traffic Congestion Model Simulation Based on Practice Case with Vehicle Network and System Traffic Intelligent Communication. *Symmetry*, *12*(7), 1172. doi:10.3390ym12071172

Zaghloul, E., Li, T., & Ren, J. (2021). d-BAME: Distributed Blockchain-based Anonymous Mobile Electronic Voting. *IEEE Internet of Things Journal*, *8*(22), 16585–16597. doi:10.1109/JIOT.2021.3074877

Zelbst, P. J., Green, K. W., Sower, V. E., & Bond, P. L. (2019). The impact of RFID, IIoT, and Blockchain technologies on supply chain transparency. *Journal of Manufacturing Technology Management*, *31*(3), 441–457. doi:10.1108/JMTM-03-2019-0118

Zhang, J., Sethi, S. P., Choi, T.-M., & Cheng, T. C. E. (2020). Supply chains involving a mean-variance-skewness-kurtosis newsvendor: Analysis and coordination. *Production and Operations Management, 29*, 1397–1430.

Zhang, R., Xue, R., & Liu, L. (2019). Security and privacy on blockchain. ACM Computing Surveys, 52(3), 1–34. Advance online publication. doi:10.1145/3316481 doi:10.1145/3316481

Zhang, X.R. & Li, M.Y. (2020). The application of blockchain technology in the banking industry. *China Finance, 10.*

Zhang, C., Zhu, L., Xu, C., Zhang, C., Sharif, K., Wu, H., & Westermann, H. (2020). BSFP: Blockchain-enabled smart parking with fairness, reliability and privacy protection. *IEEE Transactions on Vehicular Technology*, *69*(6), 6578–6591.

Zheng, Z., Xie, S., Dai, H., Chen, X., & Wang, H. (2017, June). *An overview of blockchain technology: Architecture, consensus, and future trends. In 2017 IEEE international congress on big data (BigData congress).* IEEE.

Zhu, L., Shen, J., Xie, L., & Cheng, Z. (2017). Unsupervised Topic Hypergraph Hashing for Efficient Mobile Image Retrieval. *IEEE Transactions on Cybernetics*, *47*(11), 3941–3954. doi:10.1109/TCYB.2016.2591068 PMID:28113794

Zivic, N., Kadusic, E., & Kadusic, K. (2020). Directed Acyclic Graph as Hashgraph: An Alternative DLT to Blockchains and Tangles. 2020 19th International Symposium INFOTEH-JAHORINA, INFOTEH 2020 - Proceedings, 2019–2021. 10.1109/INFOTEH48170.2020.9066312

Zyskind, G., Nathan, O., & Pentland, A. (2015). *Enigma: Decentralized computation platform with guaranteed privacy.* arXiv preprint arXiv:1506.03471.

Zyskind, G., Nathan, O., & Pentland, A. S. (2015). Decentralizing Privacy: Using Blockchain to Protect Personal Data. *2015 IEEE Security and Privacy Workshops.*

Zyskind, G., Nathan, O., & Pentland, A. S. (2015). Decentralizing privacy: Using blockchain to protect personal data. Proceedings - 2015 IEEE Security and Privacy Workshops. SPW, 180–184. Advance online publication. 10.1109/SPW.2015.27

About the Contributors

P. Swarnalatha is an Associate Professor, in the School of Computing Science and Engineering, Vellore Institute of Technology at Vellore, India. She pursued her Ph.D degree in Image Processing and Intelligent Systems. She has published more than 120 papers in International Journals/International Conference Proceedings/National Conferences. She is having 20+ years of teaching experiences. She has filed two Patents and also awarded with Dr. APJ Abdul Kalam Award for Teaching Excellence. She is a senior member of IACSIT, CSI, ACM, IACSIT, IEEE (WIE), ACEEE.She is an Editorial board member/reviewer of rcputed International/ National Journals and Conferences. Her current research interest includes Image Processing, Remote Sensing, Artificial Intelligence and Software Engineering.

S. Prabu have completed Bachelor of Engineering in Computer Science and Engineering from Sona College of Technology (Autonomous) and Master of Technology in Remote Sensing from College of Engineering Guindy, Anna University Chennai and one more Master of Technology in Information Technology at School of Computer Science and Engineering, Bharathidasan University Trichy. Did my Doctoral studies on Integration of GIS and Artificial Neural Nctworks to Map the Landslide Susceptibility from College of Engineering Guindy, Anna University, Chennai. He was a Post-Doctoral Fellow at GISE Advanced research lab, Department of Computer Science and Engineering, Indian Institute of Technology Bombay. He is having 15+ years of teaching experiences. He is a senior member of IACSIT, CSI, ACM, IACSIT, IEEE (WIE), ACEEE. He has published more than 130 papers in International Journals/ International Conference Proceedings/National Conferences. He has filed two Patents and also awarded with Dr. APJ Abdul Kalam Award for Teaching Excellence. He served as a Division Chair for Parallel and Distributed Computing and Information Security, School of Computing Science and Engineering, Vellore Institute of Technology,Vellore.Currently, he is working as an Associate Professor, Department of Banking Technology, Pondicherry University (A Central University), Pondicherry - 605 014, India

***

A. Anitha received her PhD in Information technology from VIT University, India. At present, she is working as an Associate Professor in the School of Information Technology, Vellore Institute of Technology, Vellore, India. She has authored many international and national journal papers for Inderscience and Springer Journals. She is associated with the professional bodies CSI, Indian Science Congress. Her research interest includes data mining, fuzzy logic, neural network, Artificial Intelligence, soft hybrid techniques, rough computing, business intelligence and Deep learning.

Mohd Hammad Khan Aqeel Ahmad Khan is a computer science student at Vellore Institute of Technology, Vellore, India. With a great interest in machine learning algorithms and their applications, he plans to specialize in this field in the future. His research interests include data science and using machine learning to solve real-world problems.

S. Denis Ashok is an enthusiastic researcher who has research interests in multi-disciplinary engineering domains such as artificial intelligence and machine learning, automotive control systems; mechatronics, machine vision, and Internet of things. He has experience in handling government-sponsored research and consultancy projects in the areas of mechatronics, machine vision, mobile robotics, electric vehicle. He has published more than 50 research papers in reputed journals and conferences with H index of 14. S. Denis Ashok obtained his Bachelor's degree in Mechanical Engineering in the year 1998 from Kamaraj University. He worked as a Production Engineer in Donghee Vision Industrial Company Limited, which is an ancillary unit of Hyundai Motor India Limited during the year 1998–2000. He holds a Master's degree in Production Engineering in the year 2003 from Thiagarajar College of Engineering, affiliated to Madurai Kamaraj University, Madurai, Tamil Nadu. He received a doctoral degree from the Indian Institute of Technology, Madras in the year 2011 for the research work on the computer-aided inspection of high-speed spindles in a micromachine tool. Currently, he is working as a professor in the Department of Design and Automation, School of Mechanical Engineering at VIT University, Vellore, India. He received the research funding for the "Development of machine vision-based spindle error measurement system" from the Department of Science of Technology, Delhi under the fast track young scientist scheme in the year 2012. He received the Innovative leadership award for his contributions in the indigenous development of automated egg vending machine, from Central Poultry Development Organization, India in the year 2013. Also, he has received the sponsored research fund for the "development of a human-powered hybrid vehicle" from Science for Equity, Empowerment, and Development (SEED) Division, Department of Science of Technology, India in the year 2015. Recently, he has successfully completed the consultancy project on "Mobile controlled electrical appliances" for the TATA Power Company Limited, India. He received funding for the Industry Academy Partnership Program from The Royal Academy of Engineering, the U.K to collaborate with Loughborough University and Valeo, Chennai for the Design and development of a four-wheeled electric vehicle for research, teaching, and outreach in Tamil Nadu, India under Industry-Academia Partnership Programme during 2018-19. Recently, he received funding under teaching associateship for excellence for the project titled "Development of novel deep learning, visual servoing approaches for anti-tank mine detection using thermal vision assisted mobile robot" from the science engineering research board, New Delhi. He has guided 6 doctoral thesis works in the areas of image processing for machine tool application, automotive control systems, electro-hydraulic systems. He has filed a patent titled "Design and Development of Variable frequency drive based electro-hydraulic circuit for CNC Application".

Mannava Yesu Babu is an Associate Professor, in the Vardhaman College of Engineering Hyderabad. He completed his PhD in JNTU Anathapur.

Raghav Behal has received his Bachelor of Technology (B.Tech) degree in Computer Science and Engineering (CSE) from Vellore Institute of Technology, Vellore in 2021. His areas of interest include Blockchain, FinTech, and Data Science.

Yijun Chen is an independent researcher. Her research interests include RegTech, FinTech, Auditing and Cryptocurrencies.

Emanuela Hanes is an independent researcher. Her cooperations include the Vienna University FH BFI Campus Wien. Her research interests include China-EU business strategies, RegTech, FinTech, Cryptocurrencies, Geopolitics, Chinese Strategic Planning and Development Policies.

Dishi Jain is a final year student at Vellore Institute of Technology, Vellore pursuing a Bachelor of Technology in Computer science specialization in Bioinformatics.

Sukriti Jaitly is a computer science student at Vellore Institute of Technology, Vellore, India (310058). With a great interest in machine learning algorithms and it's applications, she plans to specialise in this field in the future. Her research interests include data science and using machine learning to solve real world problems.

Amrit Kanoi is currently pursuing a bachelor's degree in Computer science and Engineering from Vellore Institute of Technology, Vellore. His areas of interest are Application Development, Web development, UI designing, and Blockchain.

Sharmila L. is working as an HOD in the Department of Artificial Intelligence & Data Science at Dhaanish Ahmed College of Engineering, Chennai, India. She received her M.E., and Ph.D degrees in Computer Science and Engineering from Anna University and Sathyabama Institute of Science and Technology, Chennai, India. She is a reviewer for many SCI Indexed Journals like Ambient Intelligence and Humanized Computing (Springer), Journal of Supercomputing, Computer Communications (Elsevier), Microprocessor and Microsystems, ACM TOMM, Wiley and so on. She has 15 years of Teaching Experience and 6 years of research experience. Her main area of research activity is Data Mining, Medical Image processing, Wireless Networks, Network security, Big data analytics. She has published many research articles in refereed journals like Elsevier and springer. She applied for 1 patent and got published in Indian Patent Journal. She is a member of CSI, IAENG. She has been serving as an Organizing Chair and Program Chair of several International conferences and in the Program Committees of several National / International conferences.

Shengyuan Lu is an independent researcher. His research interests include RegTech, FinTech and Cryptocurrencies.

Anbarasi Masilamani received the M.Tech(CSE) from VIT University,vellore, Tamilnadu, India in 2006; she is an Assistant Professor(Senior) in the School of Computer Science and Engineering at Vellore Institute of Technology, Vellore, Tamilnadu, India. Her research interests include Data mining, Soft Computing and Bioinformatics.

Kasun Moolika Gedara is a PhD student with Auckland University of Technology, New Zealand, his research interests include blockchain, smart cities and cryptography.

Minh Nguyen received his doctoral degree from the University of Auckland, New Zealand. His research interests include Computer Vision, Deep Learning, VR/AR.

Mohak Verma Nikhilesh Verma is a pre final year computer science student at Vellore institute of technology. He has worked in the field of data engineering and in the field of statistical and well as social relevance of cryptocurrencies. He finds the field of business analytics very intriguing and wants to help the socio-economical wellbeing of the society through the computation power of machine learning and computational studies.

Kalyanaraman P. is a Professor in the School of Computer Science and Engineering at Vellore Institute of Technology, Vellore, India. He received his Masters' degree in Computer Science from PSG College of Technology, Coimbatore. He has received Doctoral Degree in Computer Science and Engineering from Vellore Institute of Technology, Vellore. He has teaching experience of more than twenty-five years. His areas of interest include Distributed Computing, E-Learning Systems, Modeling and Simulation and Knowledge Engineering. He is the life member of Computer Science and Engineering and Indian Society of Technical Education.

Swarnalatha Purushotham is an Associate Professor, in the School of Computing Science and Engineering, Vellore Institute of Technology at Vellore, India. She pursued her Ph.D degree in Image Processing and Intelligent Systems. She has published more than 120 papers in International Journals/ International Conference Proceedings/National Conferences. She is having 20+ years of teaching experiences. She is a senior member of IACSIT, CSI, ACM, IACSIT, IEEE (WIE), ACEEE.She is an Editorial board member/reviewer of reputed International/ National Journals and Conferences. Her current research interest includes Image Processing, Remote Sensing, Artificial Intelligence and Software Engineering.

Venkatesan R. received Bachelor of Engineering degree in Computer Science and Engineering from Park College of Engineering and Technology, Coimbatore, Master of Engineering degree in Computer Science and Engineering from Sona College of Technology, Salem and the Ph.D. degree in Computer Science and Engineering in Vellore Institute of Technology, Vellore. He has more than 15 years of experience in teaching. His current research includes satellite image processing and neural networks. Currently, he is an Assistant Professor in Department of Computer Science and Engineering, SRM TRP Engineering College, Trichy.

Ramani Selvanambi is an Assistant Professor (Senior) in the School of Computing Science and Engineering at Vellore Institute of Technology, Vellore, India. He received his B.E in Computer Science and Engineering from MNM Jain Engineering College, Madras University and M. Tech Computer Science and Engineering from Bharathidasan University and Ph.D. in Computer Science and Engineering from Vellore Institute of Technology, Vellore. He has 11+ years of experience in teaching and 2 years of experience in the Consultancy and Software Industry. He has published about 30 papers in International Journals on Machine Learning, Nature-Inspired Algorithms, and Health Care. His research interest includes Data Mining, Machine learning, Database Systems, Optimization Techniques, and Web Technology. He is a life member of the Computer Society of India, and IEEE.

Prabu Sevugan have completed Bachelor of Engineering in Computer Science and Engineering from Sona College of Technology (Autonomous) and Master of Technology in Remote Sensing from College of Engineering Guindy, Anna University Chennai and one more Master of Technology in Information Technology at School of Computer Science and Engineering, Bharathidasan University Trichy. Did my Doctoral studies on Integration of GIS and Artificial Neural Networks to Map the Landslide Susceptibility from College of Engineering Guindy, Anna University, Chennai. He was a Post-Doctoral Fellow at GISE Advanced research lab, Department of Computer Science and Engineering, Indian Institute of Technology Bombay. He is having 15+ years of teaching experiences. He is a senior member of IACSIT, CSI, ACM, IACSIT, IEEE (WIE), ACEEE. He has published more than 130 papers in International Journals/ International Conference Proceedings/National Conferences. He served as a Division Chair for Parallel and Distributed Computing and Information Security, School of Computing Science and Engineering, Vellore Institute of Technology,Vellore.Currently, he is working as an Associate Professor, Department of Banking Technology, Pondicherry University (A Central University), Pondicherry - 605 014, India

Devdutt Sharma is a computer science student at Vellore Institute of Technology, Vellore, India. With a great interest in machine learning algorithms and its applications, he plans to specialize in this field in the future. His research interests include data science and using machine learning to solve real-world problems.

Amedapu Srinivas working as associate professor in Sreenidhi Institute of Science and Technology, Telangana, India.

Yokesh Babu Sundaresan is currently working as an Assistant Professor (Selection Grade) in School of Computer Science and Engineering (SCOPE) at Vellore Institute of Technology, Vellore. He had around six years of industrial experience since 2005 on hardware and application domain viz., mobile application prototype development, digital set box overlay ad streaming and health monitoring, Mobile platform sensor validation etc., He has currently around eleven years of experience since 2010 in academics and his research interest includes digital circuit design, heterogeneous reconfigurable architectures, assembly programming and low cost, power efficient intelligent systems. He has authored and co-authored over ten hardware prototype papers in peer-reviewed journals.

Haritha T. worked as Assistant professor in Dept of Computer Science And Engineering at Sri Venkatesa Perumal College of Engineering & Technology (Autonomous) Puttur India. She was awarded a Graduate and Master's in Computer Science and Engineering from the University of Jawaharlal Nehru Anthapuram in 2018. She has presented International Level Journals. She was attended many conferences and workshops. She has also served on the board of the many Technical and Workshop programmes. She received best award in Academics. She is best known for her work on with the goal of bringing expertise in Computer Languages, Database Management System, Operating Systems, Internet of Things, Blockchain. So "she started working on various projects to learn her way around the profession."

Balaji Vuppala is working as an Assistant Professor in Vardhaman College of Engineering, Hyderabad. His interest areas includes Big Data Analytics, Cloud Computing, Data Base Management Systems, Internet of Things. He is working on various projects to learn his way the profession and published reasonable papers.

Abhinav Yadhav is currently pursuing a bachelor's degree in Computer science and Engineering from Vellore Institute of Technology, Vellore. His areas of interest are Stack Development, Programming, and Cyber Security.

Wei Qi Yan is the Director of Centre for Robotics & Vision (CeRV), Auckland University of Technology (AUT), his expertise is in deep learning, intelligent surveillance, computer vision, multimedia, etc. Dr. Yan has published over 250 research papers, his research distinctions include the published Springer monographs: Computational Methods for Deep Learning (2021), Introduction to Intelligent Surveillance (2019), etc. Dr. Yan is the Chair of the ACM Multimedia Chapter of New Zealand, a Member of the ACM, a Senior Member of the IEEE, TC members of the IEEE. Dr. Yan was a visiting professor with the National University of Singapore (NUS, QS Ranking #11), Singapore.

Poshan (Sam) Yu is a Lecturer in Accounting and Finance in the International Cooperative Education Program of Soochow University (China). He is also an External Professor of FinTech and Finance at SKEMA Business School (China), a Visiting Professor at Krirk University (Thailand) and a Visiting Researcher at the Australian Studies Centre of Shanghai University (China). Sam leads FasterCapital (Dubai, UAE) as a Regional Partner (China) and serves as a Startup Mentor for AIC RAISE (Coimbatore, India). His research interests include financial technology, regulatory technology, public-private partnerships, mergers and acquisitions, private equity, venture capital, start-ups, intellectual property, art finance, and China's "One Belt One Road" policy.

Index

G

I

K

L

M

P

Q

R

S

T

V

www.ingramcontent.com/pod-product-compliance
Lightning Source LLC
La Vergne TN
LVHW080525300926
849162LV00007B/1540